Neurodegenerative Diseases: Integrative PPPM Approach as the Medicine of the Future

Advances in Predictive, Preventive and Personalised Medicine

Volume 2

Series Editor:

Olga Golubnitschaja

Managing Editor:

Kristina Yeghiazaryan

For further volumes:
http://www.springer.com/series/10051

Silvia Mandel
Editor

Neurodegenerative Diseases: Integrative PPPM Approach as the Medicine of the Future

Springer

Editor
Silvia Mandel
Technion, Institute of Technology
Molecular Pharmacology
Haifa, Israel

ISSN 2211-3495 ISSN 2211-3509 (electronic)
ISBN 978-94-007-5865-0 ISBN 978-94-007-5866-7 (eBook)
DOI 10.1007/978-94-007-5866-7
Springer Dordrecht Heidelberg New York London

Library of Congress Control Number: 2012956058

Printed on acid-free paper

Springer is part of Springer Science+Business Media (www.springer.com)

What this Book Series is About . . .

Current Healthcare: What Is Behind the Issue?

For many acute and chronic disorders, the current healthcare outcomes are considered as being inadequate: global figures cry for preventive measures and personalised treatments. In fact, severe chronic pathologies such as cardiovascular disorders, diabetes and cancer are treated after onset of the disease, frequently at near end-stages. Pessimistic prognosis considers pandemic scenario for type 2 diabetes mellitus, neurodegenerative disorders and some types of cancer over the next 10–20 years followed by the economic disaster of healthcare systems in a global scale.

Advanced Healthcare Tailored to the Person: What Is Beyond the Issue?

Advanced healthcare promotes the paradigm change from delayed interventional to predictive medicine tailored to the person, from reactive to preventive medicine and from disease to wellness. The innovative Predictive, Preventive and Personalised Medicine (PPPM) is emerging as the focal point of efforts in healthcare aimed at curbing the prevalence of both communicable and non-communicable diseases such as diabetes, cardiovascular diseases, chronic respiratory diseases, cancer and dental pathologies. The cost-effective management of diseases and the critical role of PPPM in modernisation of healthcare have been acknowledged as priorities by global and regional organisations and health-related institutions such as the Organisation of United Nations, the European Union and the National Institutes of Health.

Why Integrative Medical Approach by PPPM as the Medicine of the Future?

PPPM is the new integrative concept in healthcare sector that enables to predict individual predisposition before onset of the disease, to provide targeted preventive measures and create personalised treatment algorithms tailored to the person. The expected outcomes are conducive to more effective population screening, prevention early in childhood, identification of persons at-risk, stratification of patients for the optimal therapy planning, prediction and reduction of adverse drug-drug or drug-disease interactions relying on emerging technologies, such as pharmacogenetics, pathology-specific molecular patters, sub/cellular imaging, disease modelling, individual patient profiles, etc. Integrative approach by PPPM is considered as the medicine of the future. Being at the forefront of the global efforts, the European Association for Predictive, Preventive and Personalised Medicine (EPMA, http://www.epmanet.eu/) promotes the integrative concept of PPPM among healthcare stakeholders, governmental institutions, educators, funding bodies, patient organisations and in the public domain.

Current Book Series, published by Springer in collaboration with EPMA, overview multidisciplinary aspects of advanced bio/medical approaches and innovative technologies. Integration of individual professional groups into the overall concept of PPPM is a particular advantage of this book series. Expert recommendations focus on the cost-effective management tailored to the person in health and disease. Innovative strategies are considered for standardisation of healthcare services. New guidelines are proposed for medical ethics, treatment of rare diseases, innovative approaches to early and predictive diagnostics, patient stratification and targeted prevention in healthy individuals, persons at-risk, individual patient groups, sub/populations, institutions, healthcare economy and marketing.

Prof. Dr. Olga Golubnitschaja
Book Series Editor

Dr. Golubnitschaja, Department of Radiology, Medical Faculty of the University in Bonn, Germany, has studied journalism, biotechnology and medicine and has been awarded fellowships for biomedical research in Paediatrics and Neurosciences (Medical Centres in Austria, Russia, UK, Germany, the Netherlands, and Switzerland). She is well-cited in the research fields of "gene hunting" and "subtractive hybridisation" applied to predictive prenatal and postnatal diagnostics published as *O. Labudova* in years 1990–2000. Dr. Golubnitschaja is an expert in molecular diagnostics actively publishing in the fields of perinatal diagnostics, Down syndrome, diabetes mellitus, hyperhomocysteinemia, cardiovascular disease, neurodegenerative pathologies and cancer. She is the *cofounder* of the theory of multi-pathway organ-related blood fingerprinting with specific molecular patterns at epi-genomic, transcriptional and post-translational levels, author of fundamental works in *integrative medicine*. Dr. Golubnitschaja holds appointments, at the rank of Professor, at several European Universities and in International Programmes for Personalised Medicine and is author of more than 300 international publications in the field. Awards: National and International Fellowship of the Alexander von Humboldt-Foundation; Highest Prize in Medicine and Eiselsberg-Prize in Austria. She is *Secretary-General* of the "European Association for Predictive, Preventive and Personalised Medicine" (EPMA in Brussels, www.epmanet.eu), Editor-in-Chief of *The EPMA-Journal* (BioMed Central, London); Book Editor of *Predictive Diagnostics & Personalised Treatment: Dream or Reality*, Nova Science Publishers, New York 2009; Book Co-editor *Personalisierte Medizin*, Health Academy, Dresden 2010; Book Series Editor *Advances in Predictive, Preventive and Personalised Medicine*, Springer 2012; *European Representative* in the EDR-Network at the NIH/NCI, http://edrn.nci.nih.gov/; *Advisor and Evaluator* of projects dedicated to personalised medicine at the EU-Commission in Brussels, NIH/NCI, Washington D.C., USA, and at Foundations and National Ministries of Health in several countries worldwide.

Preface

Among the diverse challenges facing clinicians during the last century is the staggering increase in its proportion of elderly members, which is directly linked to the prevalence of neurodegenerative diseases (NDDs) like Alzheimer's disease (AD) and Parkinson's disease (PD). NDDs are considered major contributors to disability and disease in the globe community. All epidemiological studies of dementia show a strong correlation between prevalence and age. In the age group of 60–64 years of age the prevalence of dementia is below 1%; it then doubles approximately every 5 years reaching a prevalence of about 50% among those aged 90 years or older. This high frequency is critical in terms of the social policy, due to the need for extensive nursing care. According to the World Alzheimer Report of 2009 and 2010, it is estimated worldwide that 36 million people are living with dementia; concomitantly expected numbers will be doubling every 20 years reaching 66 million by 2030, and 115 million by 2050. The economic burden of dementia was estimated as being US$ 604 billion in 2010. This dramatic situation can be significantly improved by the implementation of appropriate early diagnostic and preventive measures advancing healthcare and enhancing the patient life quality. In particular, the quality of primary prevention should be emphasised: preselection of predisposed individuals early in life followed by targeted preventive measures should be well regulated by innovative reimbursement programs developed by policy-makers. This is considered as the cost-effective preventive medicine of the future.

Extensive efforts of international consortia are essential in the field to consolidate the accumulated expertise and to create robust diagnostic approaches, targeted individualised treatments and effective prevention tailored to the person. This also includes a longitudinal health record for the patient that can be sustained over lifetime. Utilisation of the advanced integrative medical approach (predictive, preventive and personalised medicine, PPPM) could enable a significant portion of population to reach the 100-year age limit remaining vibrant in excellent physical and mental health as actively contributing members of the society.

The crucial role of the integrative approach by PPPM have been acknowledged as priorities by World Health Organisation, United Nations General Assembly, the

Centre for Disease Control, and the European Union among others. Standing in the forefront of PPPM related initiatives, the European Association for Predictive, Preventive and Personalised Medicine (EPMA) (http://www.epmanet.eu) has launched in 2010 *The EPMA Journal* (BioMed Central, London) for the most effective promotion of evidence-based scientific PPPM approaches in healthcare. Further, the international scientific community has highly appreciated the EPMA organisation of the EPMA World Congress dedicated to PPPM which took place in Bonn, Germany on September 15–18, 2011, hosting participants from 44 countries worldwide. This event was crowned by *The EPMA Journal* publication of the "General Report & Recommendations in Predictive, Preventive and Personalised Medicine 2012: White Paper of the European Association for Predictive, Preventive and Personalised Medicine". This was next followed by the release of the EPMA Book Series "Advances in PPPM" published by *Springer*.

This book, *Neurodegenerative Diseases: Integrative PPPM Approach as the Medicine of the Future*, is dedicated to all aspect related to the prediction, prevention and personalised treatments of NDD. This volume is intended to serve as a reference source for basic researchers and clinicians involved in the field of neurosciences with a special emphasis on healthcare promotion and innovations intended to delay or even prevent NDDs. My special thanks go to the Contributors and Publisher for the excellent performance and highest professional level in the book preparation.

Dr. Silvia Mandel
Editor

Currently vice-director of the Eve Topf Center of Excellence for Neurodegenerative Diseases Research at the Technion in Haifa, Israel, Silvia Mandel works on neurodegeneration and neuroprotection, to investigate the molecular events occurring in Parkinson's disease, Alzheimer's disease and Amyotrophic Lateral Sclerosis.

Silvia Mandel (SM) is engaged in a number of projects inside the Center and in collaboration with other centres in Israel and around the world. These include:

- Identification of peripheral biomarkers in blood for Parkinson's disease early diagnosis, differential diagnosis and therapeutic follow-up;
- Deciphering the neurorescue actions of Rasagiline (Azilect), developed in the centre as a unique anti-Parkinson drug with disease modifying activity;

- Development of novel multifunctional neuroprotective-neurorescue-iron chelators for amyothrophic lateral sclerosis;
- Development of novel bifunctional cholinesterase-monoamine oxidase inhibitor-iron chelators for the treatment of Parkinson's, Alzheimer's disease and Lewy body disease;
- Neuroprotective and neurorescue action of green tea polyphenols in models of Parkinson's disease and Alzheimer's disease.

SM has published more than 100 original contributions, book chapters and conference proceedings. SM is the acting Vice-President of the European Association for Predictive, Preventive & Personalised Medicine (EPMA in Brussels, www.epmanet.eu), responsible for Associated countries and contacts apart from Europe. She is a series/volume editor of *Handbook of Neurochemistry*, *The EPMA Journal*, *Journal of Neural Transmission and Advances in Predictive, Preventive and Personalised Medicine*. She has been serving as a grant reviewer for international organisations, is an associate editor of *Journal of Alzheimer's disease*, *Synapse* and *The EPMA Journal* and a member of the scientific advisory board of "International Conference on Neuroprotective Agents" (ICNA) and board member of the Israel Society for Neuroscience.

Contents

Contributors

Tamar Amit Eve Topf Centers of Excellence for Neurodegenerative Diseases Research and Department of Pharmacology, Rappaport Family Research Institute, Technion-Faculty of Medicine, Haifa, Israel

David Begley King's College London, London, UK

Cinzia Maria Bellettato Women's and Children's Health Department, University of Padova, Padova, Italy

Chen Benkler Felsenstein Medical Research Center, Tel Aviv University, Tel-Aviv, Israel

Adriana Ceci Fondazione Benzi Bari, Bari, Italy

Ornit Chiba-Falek Duke University, Durham, NC, USA

Donna G. Crenshaw Duke University, Durham, NC, USA

Andrea Döring NeuroCure Clinical Research Center and Clinical and Experimental Research Center for Multiple Sclerosis, Charité – Universitätsmedizin Berlin, Berlin, Germany,

Jan Dörr NeuroCure Clinical Research Center and Clinical and Experimental Research Center for Multiple Sclerosis, Charité – Universitätsmedizin Berlin, Berlin, Germany

Francesca D'Avanzo Women's and Children's Health Department, University of Padova, Padova, Italy

Vivian E. Drory Neuromuscular Service, Department of Neurology, Tel-Aviv Sourasky Medical Center, Tel-Aviv, Israel

Sackler Faculty of Medicine, Tel-Aviv University, Tel-Aviv, Israel

Josef Flammer Department of Ophthalmology, University Hospital Basel, Basel, Switzerland

Stephany Fulda Sleep and Epilepsy Center, Neurocenter of Southern Switzerland, Civic Hospital (EOC) of Lugano, Lugano, Switzerland

Daphne A. Gill Department of Biomedical Sciences, University of Prince Edward Island, Charlottetown, PEI, Canada

Olga Golubnitschaja Division of Molecular/Experimental Radiology, Department of Radiology, Rheinische Friedrich-Wilhelms-University of Bonn, Bonn, Germany

William K. Gottschalk Duke University, Durham, NC, USA

Illana Gozes Department of Human Molecular Genetics and Biochemistry, Sackler Faculty of Medicine, Tel Aviv University, Tel Aviv, Israel

The Lily and Avraham Gildor Chair for the Investigation of Growth Factors, Tel Aviv University, Tel Aviv, Israel

The Adams Super Center for Brain Studies and The Edersheim Levie-Gitter fMRI Institute, Tel Aviv University, Tel Aviv, Israel

The Dr. Diana and Zelman Elton (Elbaum) Laboratory for Molecular Neuroendocrinology, Tel Aviv University, Tel Aviv, Israel

Iris Grossman IsraGene, Ltd., Rosh HaAyin, Israel

B. Russell Huber Veterans Administration – Puget Sound Health Care System, Seattle, WA, USA

Amos D. Korczyn Tel-Aviv University Medical School, Ramat-Aviv, Israel

Peter A. LeWitt Department of Neurology, Wayne State University School of Medicine, Detroit, MI, USA

Parkinson's Disease and Movement Disorders Program, Henry Ford West Bloomfield Hospital, West Bloomfield, MI, USA

Michael W. Lutz Duke University, Durham, NC, USA

Mauro Manconi Sleep and Epilepsy Center, Neurocenter of Southern Switzerland, Civic Hospital (EOC) of Lugano, Lugano, Switzerland

Silvia Mandel Eve Topf Center for Neurodegenerative Diseases Research and Department of Molecular Pharmacology, Faculty of Medicine, Technion, Institute of Technology, Haifa, Israel

Amber L. Marriott Department of Biomedical Sciences, University of Prince Edward Island, Charlottetown, PEI, Canada

Eldad Melamed Felsenstein Medical Research Center, Tel Aviv University, Tel-Aviv, Israel

Beatrice Nefussy Neuromuscular Service, Department of Neurology, Tel-Aviv Sourasky Medical Center, Tel-Aviv, Israel

Sackler Faculty of Medicine, Tel-Aviv University, Tel-Aviv, Israel

Daniel Offen Felsenstein Medical Research Center, Tel Aviv University, Tel-Aviv, Israel

Friedemann Paul NeuroCure Clinical Research Center and Clinical and Experimental Research Center for Multiple Sclerosis, Charité – Universitätsmedizin Berlin, Berlin, Germany

Caspar F. Pfueller NeuroCure Clinical Research Center and Clinical and Experimental Research Center for Multiple Sclerosis, Charité – Universitätsmedizin Berlin, Berlin, Germany

Allen D. Roses Zinfandel Pharmaceuticals Inc., Research Triangle Park, Chapel Hill, NC, USA

Ann M. Saunders Duke University, Durham, NC, USA

Maurizio Scarpa Women's and Children's Health Department, University of Padova, Padova, Italy

Amnon Shabo (Shvo) IBM Haifa Research Lab, Haifa, Israel

Christopher A. Shaw Departments of Ophthalmology and Visual Sciences and Experimental Medicine and Graduate Program in Neurosciences, University of British Columbia, Vancouver, BC, Canada

R. Andrew Tasker Department of Biomedical Sciences, University of Prince Edward Island, Charlottetown, PEI, Canada

Rosella Tomanin Women's and Children's Health Department, University of Padova, Padova, Italy

Orly Weinreb Eve Topf Centers of Excellence for Neurodegenerative Diseases Research and Department of Pharmacology, Rappaport Family Research Institute, Technion-Faculty of Medicine, Haifa, Israel

Kristina Yeghiazaryan Division of Molecular/Experimental Radiology, Department of Radiology, Rheinische Friedrich-Wilhelms-University of Bonn, Bonn, Germany

Moussa B.H. Youdim Eve Topf Centers of Excellence for Neurodegenerative Diseases Research and Department of Pharmacology, Rappaport Family Research Institute, Technion-Faculty of Medicine, Haifa, Israel

Alessandra Zanetti Women's and Children's Health Department, University of Padova, Padova, Italy

Jing Zhang Department of Pathology, University of Washington School of Medicine, Seattle, WA, USA

Part I

Time for New Guidelines in Handling of Neurodegenerative Disorders: Optimistic Versus Pessimistic Prognosis by Application of PPPM

Silvia Mandel and Olga Golubnitschaja

Abstract **Optimistic *versus* pessimistic prognosis** depends much on diagnostic, preventive and treatment approaches which healthcare will preferably adopt in the near future. Currently there are over 350 million diabetic patients frequently affected by polyneuropathy as secondary complication, 18 million patients with Alzheimer's disease also diagnosed as Diabetes Type 3, neurodegenerative eye diseases with leading causes of blindness – diabetic retinopathy and estimated 67 million glaucoma patients worldwide, millions of patients with Parkinson's disease, Multiple Sclerosis, Epilepsy, Cerebral Palsy and Dementia in the elderly – altogether dramatically affect life quality, social and economical indexes of populations around the globe. Without innovation in healthcare, neurodegenerative disorders can reach more than 30% of global disease burden till 2020. In contrast, effective utilisation of advanced early/predictive diagnostics, preventive and personalised medical approaches could enable a significant portion of population to reach the 100-year age limit remaining vibrant in excellent physical and mental health as actively contributing members of society.

Neurodegenerative diseases are pathologies of multifactorial nature involving interplay of epigenetic and environmental risk factors. Insights into molecular pathomechanisms will enable to create the most effective targeted protective strategies and individualised treatment before manifest pathology. Multifunctional "cocktail" therapies need to be created for individual multifactorial pathologies, in

S. Mandel (✉)
Eve Topf Center for Neurodegenerative Diseases Research and Department of Molecular Pharmacology, Faculty of Medicine, Technion, Institute of Technology, Haifa, Israel
e-mail: mandel@tx.technion.ac.il

O. Golubnitschaja
Division of Molecular/Experimental Radiology, Department of Radiology, Rheinische Friedrich-Wilhelms-University of Bonn, Bonn, Germany
e-mail: Olga.Golubnitschaja@ukb.uni-bonn.de

S. Mandel (ed.), *Neurodegenerative Diseases: Integrative PPPM Approach as the Medicine of the Future*, Advances in Predictive, Preventive and Personalised Medicine 2, DOI 10.1007/978-94-007-5866-7_1,

order to advance the healthcare of the patient cohorts. Particular emphasis should be placed on primary prevention: identification of predisposed individuals early in life followed by treatments tailored to the person need regulations by innovative reimbursement programmes. This strategy creates a robust platform for the cost-effective medicine of the future.

Keywords Neurosciences • Ageing • Early/predictive diagnosis • Targeted preventive measures • Treatments tailored to the patient • Economy of personalised medicine • Healthcare industry

1 Neurodegeneration as the Common Feature of Multifactorial Physical and Cognitive Disability

Progressive neurodegeneration is the common feature of the majority of socially and economically devastating disorders and diseases with multifactorial physical and cognitive disability. Permanently increasing numbers of those patients as well as dramatic disease-related economical burden indicate that current education, research and healthcare in this field are far from optimal and should be corrected. Since neurodegeneration is a chronic process, which begins long before the onset of clinically diagnosed symptoms, currently there is an evident lack of targeted prevention in any age-category inchoate with early childhood, over adolescence till late adulthood. This is supported by current age-related dramatic increase in numbers of patients with any kind of neurodegenerative medical condition [1].

2 Time for Protection Against Neurodegeneration and Ageing in General

The task of individual prediction and well-timed targeted prevention of the life-long chronic pathologies should be given the extraordinary priority in early age, i.e. already in paediatrics [2]. Birth asphyxia is the most frequent perinatal complication [3]. Secondary to birth asphyxia, a postnatal manifestation of hypoxic-ischaemic encephalopathy is frequently observed being associated with either mild or severe organ damage in asphyxiated newborns, both leading to the development of chronic pathologies including progressive neurodegenerative processes, CNS-injury, epilepsy, schizophrenia, psychotic syndromes, diabetes mellitus, vascular pathologies, nephropathy, cardiomyopathy, cancer, senescence [4, 5]. From viewpoint of innovative perinatal/postnatal diagnostic tools to prognose pathology development and individual outcomes, there is a great potential behind new prenatal/neonatal animal models of progressive neurodegeneration [6]. Those models enable identifying targets in predictive diagnostics and pre-symptomatic treatment with relevance to epilepsy, schizophrenia, amyotrophic lateral sclerosis, Parkinson's and other neurodegenerative diseases, and open good perspectives for targeted prevention early in life.

In healthy adults, comparative subcellular imaging and "gene hunting"-investigations performed *ex vivo* with blood samples may provide pathology specific molecular patterns for individual predisposition to NDD followed by the most cost-effective form of protection namely targeted prevention [7, 8].

Recent long-term prospective studies demonstrated that 30–65% of patients with idiopathic REM sleep behaviour disorder (RBD) may develop NDD. Therefore, RBD may be one of the earliest NDD predictors, the recognition of which may enable the delay or even prevention of neurodegenerative disorders through the early application of neuroprotective or disease-modifying therapies in the future [9].

3 Alzheimer's Disease Is One of the Global Challenges of Predictive, Preventive and Personalised Medicine in the Twenty-First Century

Current prognosis for the prevalence of Alzheimer's Disease (AD) is depressing: according to the World Alzheimer Report about 110 million people around the globe or even more depending on demographic growth of the elderly, will be affected. Concomitant social and economical burden are threatening as well documented by experts elsewhere. Therefore, current philosophy behind curative medicine and healthcare of the elderly should be essentially reconsidered in favour of preventive strategies triggered early in life, preselection of predisposed individuals by predictive diagnostic approaches followed by targeted individualised treatments. Further, assistive technologies should be well developed to cover current deficits in effective supervision of already diseases people.

Due to the extensive demographic profile of ageing populations and the multi-faceted task of dementia treatment, the whole spectrum of aspects around predictive diagnostics, targeted prevention and personalised treatment of Alzheimer's patients is extensively under consideration [10]. The AD-dedicated research programmes are essential to validate currently available biomarkers for their clinical utility in supporting targeted preventive measures, paving the road for rational health management and ensuring innovative reimbursement programmes developed by policy-makers. Both neurodegenerative AD and Parkinson's Disease (PD) are considered to be progressive disorders of multifactorial origin with a long preclinical period that ensures the great impact of predictive biomarkers to diagnose pre-disposed individuals before the manifest pathology, to discriminate between neu-rodegenerative pathologies with similar clinical symptoms, to provide a measure of disease progression, and to evaluate individual drug efficacy [11]. In terms of targeted neuroprotection, current particular interest is focused on several forms of taupathies and TAU-protein related drugs to break the potentially toxic aggregates, inhibiting phosphorylation, accelerating dephosphorylation of TAU-protein and related microtubule interaction [12].

4 Moving Towards a Predictive and Personalised Clinical Approach in MS and ALS

Multiple Sclerosis (MS) is a chronic late-onset autoimmune disease with multifactorial aetiology and obvious interplay of genetic, epigenetic and environmental risk factors. MS is affecting preferably young females and being clinically characterised by manifestation of accumulating symptoms such as numbness, vertigo, depression, motor weakness and in-coordination, urinary retention, visual deficits, pain attacks, heart intolerance, cognitive dysfunction and others. Curative treatments for MS are currently unavailable. Reduction of relapses frequency is provided by the injection of immunosuppressants, however, with limited individual efficacy within the patient cohort. MS intervention therapy is associated with safety limitations and lack of a reliable measure for drug efficacy/treatment response [13]. In addition to drug-based immunomodulatory treatment, both drug-based and non-drug approaches are established as complementary strategies to alleviate existing symptoms and to prevent secondary diseases. In particular, physical therapy like exercise and physiotherapy can be customised to the individual patient‘s needs and has the potential to improve the individual outcome [14]. Extensive efforts of international consortia are essential in the field to consolidate the accumulated expertise and create robust diagnostic approaches and targeted individualised treatment of MS patients.

Amyotrophic lateral sclerosis (ALS) is the next impressive example to illustrate current absence of predictive diagnostic tools and preventive measures to identify and protect predisposed individuals against rare neurodegenerative diseases with cognitive/motor disability and fatal outcomes. An inadequate healthcare management and deficits in multidisciplinary performance to care for this patient cohort is alarmed by the experts [15]. The prevalence of over 90% of sporadic cases indicates the multifactorial character of the disease with currently unclear pathogenesis and individual risk factors. An oxidative stress seems to play the central role triggering ALS and worsening outcomes. In consensus, mitochondrial dysfunction and genetic deficits in the central detoxification pathway based on Cu-Zn superoxide-dismutase 1 activity have been reported for familial ALS cases. Therefore, several hypothetic stress factors leading to accelerated ageing and, further supported by genetic predisposition should be taken into consideration at the molecular level to select disease-specific targets, in order to develop powerful diagnostic tools for preselection and effective preventive measures for ALS-predisposed individuals [16]. New therapies consider rather a multidrug character of treatment as the most promising approach: innovative multifunctional therapeutic strategies in ALS are focused on neuroprotective approaches, CNS-related repairing gene-therapy, targeted replacements by stem-cells and others.

5 Neurodegenerative Eye-Diseases as Indicator for Chronic Medical Conditions: Can We Protectively Act in Advance?

The idea of *"**A Sick Eye in a Sick Body**"* has been formulated by F. Lagrange in 1922 [17] and still counts as being valid and supported for several pathologies by the most advanced diagnostic and follow-up technologies [18]. Worldwide the second leading cause of blindness with nearly 8 million people bilaterally blind among altogether 67 million patients is the glaucomatous optic neuropathy characterised by a chronic progressive loss of retinal ganglion cells and their neurons. To prevent blindness from glaucoma is the great challenge for innovative biomedical technologies in favour of predictive diagnosis and effective treatments tailored to the predisposed individuals [19]. Moreover, accelerated ageing triggered by an imbalanced implication of oxidative stress seems to a have a high impact for the disease progression. Consequently, if applied early enough, a protection against oxidative stress e.g. at the level of mitochondrial (dys)function effectively reduces the overall glaucoma risk and can be achieved by dietary supplements such as gingko, dark chocolate, polyphenolic flavonoids that is sufficiently present in green tea, coffee and red wine [20].

Age-related macular degeneration (AMD) results in visual deficits because of damage to the central retina. Worldwide about 30 million people mainly after 50 years of age are affected by the disease. Macular degeneration counts among multifactorial diseases; current treatment focuses on late stages of the pathology being expensive and less effective approach. To the best acknowledged risk factors of AMD belong genetic background and increased (oxidative) stress such as smoking and unfavourable/toxic environment leading to accelerated ageing of predisposed individuals. Consequently, accurately diagnosed disease's risk and progression as well as targeted personalised treatment approaches (potential targets and corresponding preventive approaches are discussed in the review-article) would ultimately delay or even prevent AMD in the elderly [21].

6 Concluding Remarks

Neurodegenerative diseases carries multifactorial character that involves interplay of epi/genetic and environmental risk factors. Insights into molecular pathomechanisms will enable to create the most effective targeted protective strategies and individualised treatment before manifest pathology. Multifunctional "cocktail" therapies need to be created for individual multifactorial pathologies, in order to advance the healthcare of the patient cohorts. Particular emphasis should be placed on primary prevention: identification of predisposed individuals early in life followed by treatments tailored to the person need regulations by innovative reimbursement programmes. This strategy creates a robust platform for the cost-effective medicine of the future. Extensive efforts of international consortia are essential in the field to consolidate the accumulated expertise. This also includes a longitudinal health record for the patient that can be sustained over lifetime. Utilisation of advanced early/predictive diagnostics, preventive and personalised medical approaches could enable a significant portion of population to reach the 100-year age limit remaining vibrant in excellent physical and mental health as actively contributing members of society.

This book volume is intended to serve as a reference source for basic researchers and clinicians involved in the field of neurosciences with a special emphasis on targeted preventive and personalised treatment approaches intended to delay or even prevent NDDs.

References

1. Golubnitschaja O (2010) Neurodegeneration: accelerated ageing or inadequate healthcare? EPMA J 1:211–215
2. Scarpa M (2011) Advanced healthcare for children and youth: application of targeted prevention, predictive diagnostics and personalised treatment approaches. EPMA J 2:137–139
3. Peeva V, Golubnitschaja O (2009) Birth asphyxia as the most frequent perinatal complication. In: Golubnitschaja O (ed) Predictive diagnostics and personalized treatment: dream or reality. Nova Science Publishers Inc, New York
4. Yeghiazaryan K, Peeva V, Morelli M, Herrera-Marshitz M, Golubnitschaja O (2009) Potential targets for early diagnosis and neuroprotection in asphyxiated newborns. In: Golubnitschaja O (ed) Predictive diagnostics and personalized treatment: dream or reality. Nova Science Publishers Inc, New York
5. Tounta G, Kolialexi A, Papantoniou N, Tsangaris G, Kanavakis E, Mavrou A (2011) Non-invasive prenatal diagnosis using cell-free fetal nucleic acids in maternal plasma: progress overview beyond predictive and personalized diagnosis. EPMA J 2:163–171
6. Tasker R, Adams-Marriott A, Shaw C (2010) New animal models of progressive neurodegeneration: tools for identifying targets in predictive diagnostics and presymptomatic treatment. EPMA J 1:217–227
7. Yeghiazaryan K, Flammer J, Orgül S, Wunderlich K, Golubnitschaja O (2009) Vasospastic individuals demonstrate significant similarity to glaucoma patients as revealed by gene expression profiling in circulating leukocytes. Mol Vis 15:2339–2348
8. Yeghiazaryan K, Flammer J, Golubnitschaja O (2010) Predictive molecular profiling in blood of healthy vasospastic individuals: clue to targeted prevention as personalised medicine to effective costs. EPMA J 1:263–272
9. Fulda S (2011) Idiopathic REM sleep behavior disorder as a long-term predictor of neurodegenerative disorders. EPMA J 2:451–458
10. Grossman I, Lutz M, Crenshaw D, Saunders A, Burns D, Roses A (2010) Alzheimer's disease: diagnostics, prognostics and the road to prevention. EPMA J 1:293–303
11. Mandel S, Morelli M, Halperin I, Korczyn A (2010) Biomarkers for prediction and targeted prevention of Alzheimer's and Parkinson's diseases: evaluation of drug clinical efficacy. EPMA J 1:273–292
12. Gozes I (2010) Tau pathology: predictive diagnostics, targeted preventive and personalized medicine and application of advanced research in medical practice. EPMA J 1:305–316

13. Grossman I, Miller A (2010) Multiple sclerosis pharmacogenetics: personalized approach towards tailored therapeutics. EPMA J 1:317–327
14. Döring A, Pfueller CF, Paul F, Dörr J (2011) Exercise in multiple sclerosis – an integral component of disease management. EPMA J 3:2
15. Nefussy B, Drory V (2010) Moving toward a predictive and personalized clinical approach in amyotrophic lateral sclerosis: novel developments and future directions in diagnosis, genetics, pathogenesis and therapies. EPMA J 1:329–341
16. Benkler C, Offen D, Melamed E, Kupershmidt L, Amit T, Mandel S, Youdim M, Weinreb O (2010) Recent advances in amyotrophic lateral sclerosis research: perspectives for personalized clinical application. EPMA J 1:343–361
17. Lagrange F (1922) Du glaucome et de l'hypotonie; leur traitement chirurgical. Doin, Paris
18. Pache M, Flammer J (2006) A sick eye in a sick body? Systemic findings in patients with primary open-angle glaucoma. Surv Ophthalmol 51:179–212
19. Golubnitschaja O, Yeghiazaryan K, Flammer J (2010) Key molecular pathways affected by glaucoma pathology: is predictive diagnosis possible? EPMA J 1:237–244
20. Mozaffarieh M, Fraenkl S, Konieczka K, Flammer J (2010) Targeted preventive measures and advanced approaches in personalised treatment of glaucoma neuropathy. EPMA J 1:229–235
21. Hasler P, Flammer J (2010) Predictive, preventive and personalised medicine for age-related macular degeneration. EPMA J 1:245–251

Part II

Individual Predispositions in Healthy Vasospastic Individuals: Patient Profiling for Targeted Prevention of "Down-Stream" Pathologies as Cost-Effective Personalised Medicine

Kristina Yeghiazaryan, Josef Flammer, and Olga Golubnitschaja

Abstract Paradigm change to cost-effective predictive and preventive medicine: A paradigm change from delayed interventional approaches to cost-effective predictive and preventive medicine presents an innovative concept for advanced healthcare. Early detection of individual predispositions in healthy populations followed by preventive treatments tailored to the person should be committed to the obligations of primary healthcare.

Vasospastic syndrome: definition, prevalence, frequent clinical picture and attributes: This chapter deals with healthy vasospastic individuals. Vasospasm is defined as an inappropriate constriction or insufficient dilatation in microcirculation. Vasospastic syndrome can be frequently observed in young populations and preferentially in females. Usually, academics are more affected by vasospastic syndrome. In addition to clinical signs of vasospastic syndrome, there is an inborn increased sensitivity to any kind of stress provocation (mechanical, cold, emotion, etc.), altered drug sensitivity, frequently cold extremities, altered sleep behaviour, reduced feeling of thirst, low blood-pressure, reduced body-mass-index, more frequent migraine compared to general population. Vasospastic individuals are considered as a healthy subpopulation predisposed, however to several pathologies including neurodegenerative diseases.

K. Yeghiazaryan
Division of Molecular/Experimental Radiology, Department of Radiology, Rheinische Friedrich-Wilhelms-University of Bonn, Bonn, Germany

J. Flammer
Department of Ophthalmology, University Hospital Basel, Basel, Switzerland

O. Golubnitschaja (✉)
Division of Molecular/Experimental Radiology, Department of Radiology, Rheinische Friedrich-Wilhelms-University of Bonn, Bonn, Germany
e-mail: Olga.Golubnitschaja@ukb.uni-bonn.de

S. Mandel (ed.), *Neurodegenerative Diseases: Integrative PPPM Approach as the Medicine of the Future*, Advances in Predictive, Preventive and Personalised Medicine 2, DOI 10.1007/978-94-007-5866-7_2,

Patient-profiling to diagnose individual predisposition before pathology manifestation: Clinical observations, subcellular imaging and "gene hunting"-investigations provide evidence for vasospasm as a predisposition to glaucoma, the second leading cause of blindness in human beings worldwide. Irrespective of glaucoma predisposition, a development of further vasospasm-related pathologies in individuals with (untreated) vasospastic syndrome cannot be excluded. Patient-profiling and predictive blood-tests detecting pathology-specific biomarker-patterns can specify individual predisposition for effective prevention at low cost.

Keywords Vascular deregulation • Blood biomarkers • Predictive diagnostics • Targeted prevention • Personalised medicine • Glaucoma

1 Vasospastic Syndrome: Definition, Prevalence, Clinical Picture, Attributes and Detection

Vasospastic syndrome or vascular deregulation (VD) is defined as an inappropriate constriction or insufficient dilatation in the microcirculation (see Fig. 1).

Generally, vasospasm is considered as both a primary and secondary one. Secondary VD is due to other diseases such as an autoimmune one. Primary VD is prevalent in the younger subpopulation and can potentially predispose to several disorders being, therefore, particularly attractive for predictive diagnostics and individualised treatment [2]. Primary VD demonstrates following particularities:

- it occurs more frequently in females and is manifested at puberty, moderating with age
- this phenomenon is even more frequent in the Japanese population compared to Caucasians [3]
- usually, academics are more affected by VD [4]
- in addition to clinical signs there is an inborn increased sensitivity to any kind of stress provocation (mechanical, cold, emotion, etc.), altered drug sensitivity, frequently cold extremities, altered sleep behaviour, reduced feeling of thirst, low

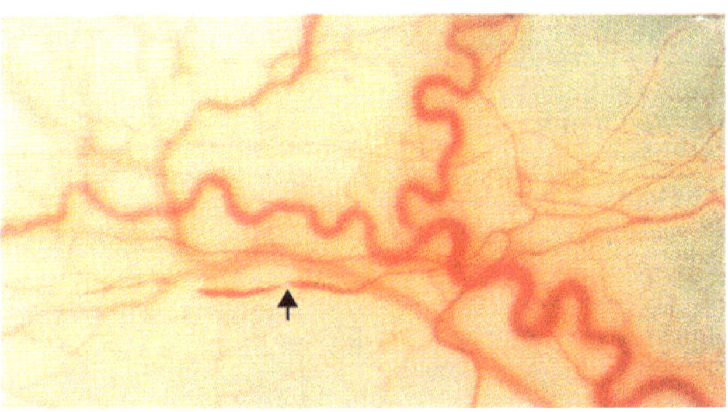

Fig. 1 Alterations in conjunctival vessels in patients with vasospasm [1]

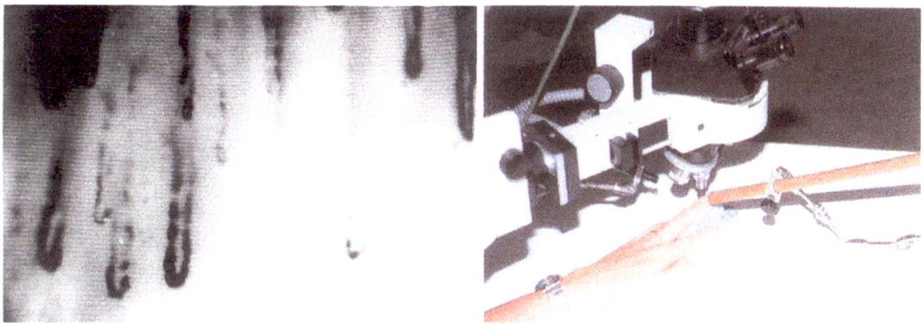

Fig. 2 Nailfold capillary microscopy (*right*) enables the monitoring of microcirculation and of cold provocation of vasospasm in fingers (*left*) [1]

blood-pressure, reduced body-mass-index, more frequent migraine compared to general population [1, 5, 6]
compared to the general population, vasospastic individuals tend to have a meticulous personality and successful professional career [7].

A valuable diagnostic tool for ascertaining vasospastic diathesis is the nailfold capillary microscopy (see Fig. 2).

The best known blood-related risk factor is an increased plasma level of endothelin-1 [1]. Since the vasospastic syndrome is of frequent occurrence in young subpopulations, this makes the task of prediction and targeted prevention of "down-stream" related pathologies particularly attractive from several points of view including economic aspects.

2 Impacts of Vascular Deregulation in Ocular Ischaemia: Individual Predispositions of Vasospastic Individuals as Revealed by Molecular Biological Examinations of Blood Samples

A wealth of literature points to the importance of haemodynamics in glaucoma pathology. Vasospasm is frequently observed in glaucoma patients [1, 8]. Ocular ischaemia resulting from blood-flow deficits may play a major role in the initiation of glaucoma: hypoxia, followed by high secretion of excitatory amino acids and elevated levels of intracellular calcium results in the process of retinal ganglion cell death [9–11]. In our previous studies, we have demonstrated stable alterations in gene expression of circulating leukocytes isolated from glaucoma patients compared to healthy controls [12–15]. Further, similar expression profiles of circulating leukocytes between vasospastic individuals and glaucoma patients have been recently published [16]. However, the same publication reports also significant dissimilarities of molecular patterns as compared to both glaucoma patients and healthy controls. Consequently, a development of both degenerative and non-degenerative pathologies different from glaucomatous optic nerve degeneration but related to primary vasospasm cannot be excluded in vasospastic individuals [17, 18].

3 Innovative Strategies for Prediction and Targeted Prevention of Glaucoma in Healthy Vasospastic Individuals: Context of Neurodegenerative Pathologies

Worldwide, 67 million patients are affected by the neurodegenerative eye disease glaucoma. Glaucomatous optic neuropathy (GON) is the second leading cause of permanent vision loss. GON is a chronic degenerative process the onset of which is not possible to monitor by currently existing diagnostic tools. Early treatment has been reported to be highly beneficial for well-timed treatment measures to slow-down the disease progression [19]. As reviewed by the authors, molecular pathomechanisms of glaucoma demonstrate both a considerable overlap and remarkable particularities to some other neurodegenerative disorders such as Alzheimer's and Parkinson's diseases [17, 20]. Hence, *versus* controls the neuronal thread protein (NTP) demonstrates enhanced expression levels in glaucoma, patients with Down Syndrome, Alzheimer's and some other neurodegenerative diseases indicating the axonal lesions. However, whereas the accumulation of *TAU*-protein is characteristic for Alzheimer's disease and other tauopathies, glaucoma patients do not demonstrate an increase in the target protein *versus* controls [19, 21]. Therefore, a potential predisposition of vasospastic individuals to related pathologies should be thoroughly examined. This examination requires innovative strategies to cover following aspects:

- identification of possible similarities as well as dissimilarities in molecular pathways between healthy vasospastic individuals and potential related pathologies developed later in life
- specificity for predictive diagnostics of glaucoma pathology in predisposed vasospastic individuals should be strictly validated against several control groups including other neuro/degenerative diseases
- selection of molecular targets should be performed for vasospastic individuals in favour of non-invasive (blood test) diagnostic approaches followed by personalised treatment towards individual predisposition to single pathologies.

A monitoring of the pathology-specific molecular patterns is particularly valuable to develop reliable diagnostic approaches before the manifest pathology. Predictive tests can specify individual predisposition for well-timed preventive measures.

4 Innovative Diagnostic Tools for Early Detection of Glaucoma Predisposition in Vasospastic Individuals: Subcellular Imaging to Quantify Damage to DNA by Ischaemic Events

Research work focused on the *ex vivo* comparative investigations of DNA damage in circulating leukocytes (CL) isolated from patients with glaucoma demonstrated significantly enhanced DNA damage compared to both healthy vasospastic and non-vasospastic individuals [22]. Comparative "Comet Assay" analysis revealed patterns of comets typical for glaucoma patients as shown in Fig. 3.

Although DNA damage in the vasospastic non-glaucomatous group is not found to be significantly increased *versus* healthy controls, DNA from vasospastic individuals showed highly group-specific comet-patterns with the degree of damage intermediate between healthy controls and glaucoma patients. These findings indicate "comet assay" profiling of DNA-damage in CL as a potentially powerful tool for the non-invasive early/predictive molecular diagnostics of glaucoma disease in vasospastic individuals [21]. Furthermore, unrepaired DNA-damage in vasospastic individuals can lead to several pathologies different from glaucomatous optic nerve degeneration. This predisposition should be thoroughly investigated and the specificity of "Comet Assay"-patterns of vasospastic individuals should be validated comparing with patterns of other degenerative and non-degenerative pathologies. Thus, "Comet Assay"-analysis as a suitable tool for biomarkers has also been suggested for another neurodegenerative disorder – Alzheimer's disease [24]. "Comet Assay"-analysis reveals enhanced DNA damage in both high- and normal-tension glaucoma [22]. Whether the level of DNA-damage correlates with disease severity, or not remains currently unclear. Further studies should also evaluate, whether a significant increase in DNA damage of leukocytes of glaucoma patients is caused by either disease specific stress factors, such as local ischaemic/reperfusion events, and/or decreased capacity of DNA-repair machinery. There is some evidence for both eventualities: simultaneous up-regulation of *p53* (stress regulated gene) and down-regulation of *XPGC* (essential member of DNA-repair machinery) have been demonstrated *ex vivo* in CL of glaucoma patients [25] and represent potential molecular blood markers for the disease.

5 Patient-Profiling: "Omics"-Technologies in Positive and Negative Prediction of Pathology Development

Whereas positive prediction for patients at high risk is important to recognise the pathology before it manifests and at early stages, the negative prediction for individuals at low risk help to avoid undesirable treatments and invasive approaches in the case of non-predisposed individuals. Individual patient profiles are utilised

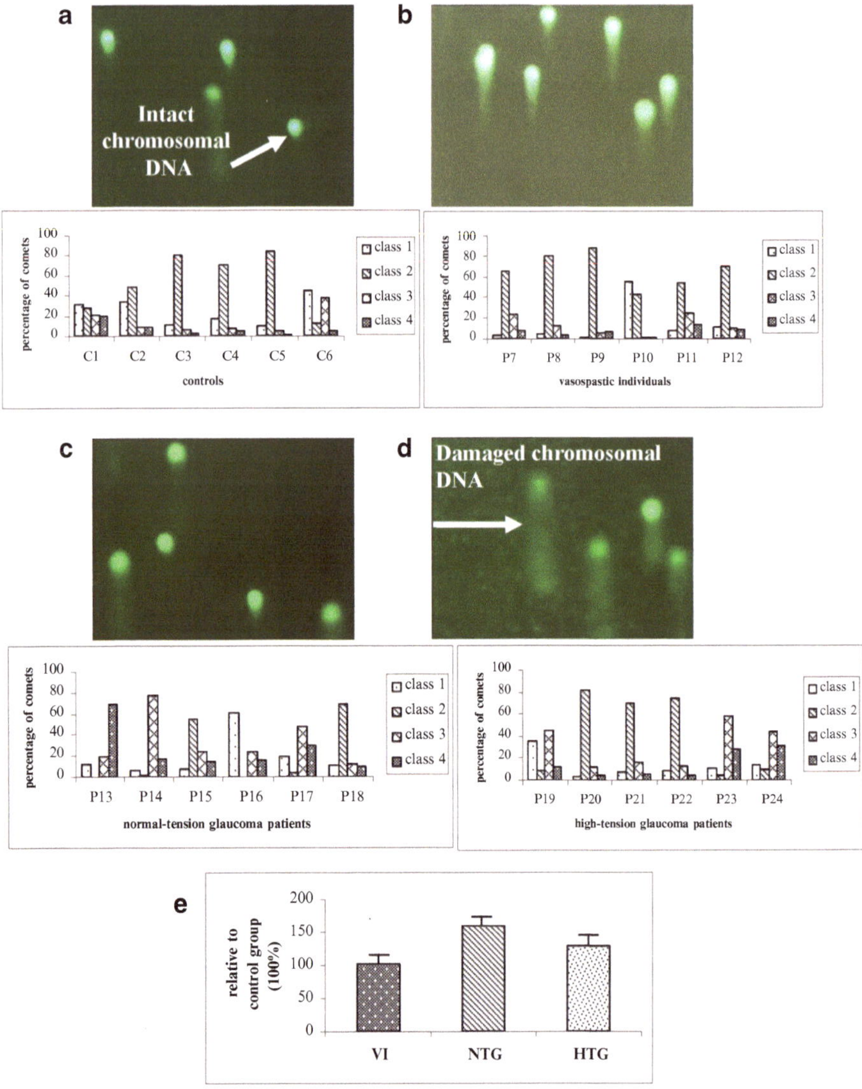

Fig. 3 *Ex vivo* "Comet Assay"-analysis of DNA damage in circulating leukocytes of glaucoma patients and non-glaucomatous vasospastic individuals *versus* healthy controls [22]. Figures (**a–d**) give examples of images typical for (**a**) control group, (**b**) group of vasospastic individuals, (**c**) group of normal-tension glaucoma patients, and (**d**) group of high-tension glaucoma patients. Comet-patterns typical for healthy controls (**a**) show that chromosomal DNA is localised mainly to heads of comets (intact DNA). In contrast, images **b**, **c**, and **d** demonstrate clearly damaged DNA (visible comet-tails and diffuse comet-heads). Evaluation of the shape of a comet allows for assessment of DNA damage. Individual migration patterns are demonstrated in groups corresponding to the images **a–d** by the evaluation of comet-classes (class number progression corresponds to increasing damage degree) as described earlier [23]. Figure (**e**) shows mean values calculated for relative DNA damage in following groups as compared to the control group (100%): VI – vasospastic individuals (102 ± 7%), NTG – normal-tension glaucoma (160 ± 13%), and HTG – high-tension glaucoma (128 ± 10%) [21]

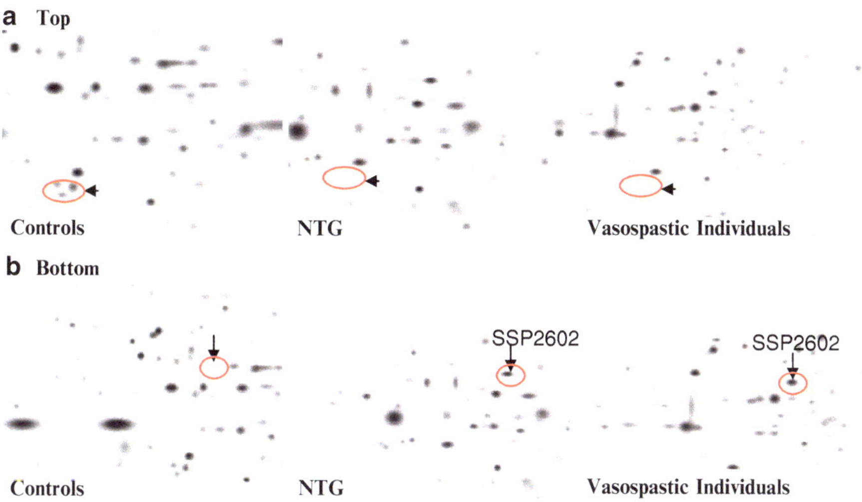

Fig. 4 Proteomics-imaging of blood-biomarkers (*ex vivo* identification in circulating leukocytes) specific for normal-tension glaucoma (NTG). (**a**) *Top*: The pathology-specific protein-cluster is completely suppressed in both NTG and vasospasm in contrast to controls. (**b**) *Bottom*: The marked protein SSP2602 is highly up-regulated in both NTG and vasospasm; this protein normally is not expressed by circulating leukocytes of controls – healthy individuals without vasospasm [25]

for patient modeling and risk assessment. "Omics"-technologies are useful for creation of accurate non-invasive positive/negative predictive approaches based on blood-test.

5.1 2D-PAGE

Protein-patterns in circulating leukocytes demonstrate clear similarities between vasospasm and normal-tension glaucoma *versus* controls. Moreover, protein-clusters can be considered for predictive imaging of healthy vasospastic individuals to glaucoma as shown in Fig. 4.

5.2 Expression Array

The image of hybridised "Atlas^{TM} Human Cardiovascular Array" revealed similarities as well as alterations in expression-patterns among NTG, HTG, and VD groups *versus* controls as summarised in Table 1.

Compared to the control group the expression of 146, 68, and 60 genes were found to be altered in NTG, HTG, and VD groups respectively; the same 53 genes

Table 1 Numbers of genes an expression of which is either differential or equal between the groups tested as shown by "Expression array"

Differential to control	VD equal to	VD Differential to
NTG → 146	Control → 528	NTG → 109
HTG → 68	NTG → 48	K → 60
VD → 60	HTG → 43	HTG → 43
NTG = HTG → 53	**NTG = HTG → 34**	NTG = HTG → 21

Thereby, 108 genes were found to be differentially expressed between NTG and HTG groups. 34 genes demonstrated similar alteration for vasospastic individuals (VD) and both glaucoma-patient groups when compared to the healthy controls (see these genes listed in Table 2)

were differentially expressed in both NTG and HTG groups *versus* controls. Among 146 genes differentially expressed specifically in NTG group we monitored 48 and 53 genes which were similarly expressed either in VD or in HTG groups, respectively. Among 68 genes differentially expressed specifically in HTG group we found 43 genes to be similarly expressed in VD group only. The highest difference – 146 genes – was found to be between the NTG and control groups. In contrast, the lowest difference – 21 genes – was demonstrated to be between VD and the overlap of NTG/HTG.

34 genes demonstrated similar expressional alterations in NTG, HTG, and VD groups *versus* controls as given in Table 2. As the differentially expressed overlap VD/NTG/HTG was compared with the control group, the following most significant differences were monitored:

- P2Y purinoreceptor 7
- Na^+/Ca^{2+} exchange protein 1 (Na^+/Ca^{2+} EP1)
- Intercellular adhesion molecule 1 (ICAM1)
- The cluster of the tissue remodelling metalloproteinases.

This group of gene-transcripts is proposed to be the reliable target for designing advanced diagnostic tools for predictive glaucoma diagnosis in healthy vasospastic individuals [16].

6 Molecular Patterns for Positive Prediction of Glaucoma Predisposition in Vasospastic Individuals

6.1 P2Y Purinoreceptor Is Upregulated in Vasospastic Individuals and Glaucoma Patients

The movement of leukocytes from blood into tissue is regulated by the local production of chemo-attractants – diverse molecules, the chemotactic signal of which is transmitted by G-protein-coupled purinoceptor family P2Y. These receptors respond

Table 2 Differentially expressed genes (altogether 34 genes as also summarised in Table 1) *versus* controls, the transcriptional levels of which were similar for VD, NTG and HTG groups [21]

Double-spot position "EA"-image / Exp. difference *vs.* control	Name of gene as given in "Atlas^{TM} Human Cardiovascular Array"	GenBank accession	SwissProt accession	Gene/protein classification
Block A				
A7d Increased	P2Y purinoceptor 7 (P2Y7); leukotriene B4 receptor; Chemoattractant receptor-like1 (CMKRL1)	U41070	Q15722 Q13305 Q92641	Other receptors (by ligands) G Protein-coupled receptors
A7e Increased	Retinoic acid (Vitamin-A1-Säure) receptor gamma 1 (RAR-gamma 1; RARG)	M24857 M38258 M57707 M32074	P13631	Transcription activator and repressors Hormone receptors Nuclear receptors
Block B				
B1n Increased	Androgen receptor coactivator 70-kDa subunit (ARA70)	L49399	Q13772	Transcription activator and repressors
B4c Increased	G protein-activated inward potassium channel 4 (GIRK4); heart K+/ATP channel (KATP1); cardiac inward rectifier (CIR); KIR3.4	U39195	P48544 Q92807	Voltage-gated ion channels
B4d Increased	Sodium/calcium exchanger 1 precursor; Na+/Ca2 + −exchange protein 1	M91368	P32418	Symporters and antiporters
B4e Increased	Multidrug resistance protein 3 (MDR3); P-Glycoprotein 3 (PGY3)	M23234	P21439	Drug-resistance proteins Xenobic transporters ABC transporters
B5f Increased	Endothelial nitric oxide synthase (EC-NOS)	M93718	P29474	Other metabolism enzymes Other intracellular transducers, effectors and modulators
B6d Increased	Intercellular adhesion molecule 1 precursor (ICAM1); major group rhinovirus receptor; CD54 antigen	J03132	P05362	Matrix adhesion receptors

(continued)

Table 2 (continued)

Double-spot position "EA"-image Exp. difference *vs.* control	Name of gene as given in "Atlas™ Human Cardiovascular Array"	GenBank accession	SwissProt accession	Gene/protein classification
7g	Calcium and integrin-binding protein (CIB)	U85611	Q99828	Calcium-binding proteins
Increased				
1g	Cadherin 7 (CDH7)	AF047826	O60574	Cell surface antigens
Increased				Cell-cell adhesion receptors
1h	Intestinal peptide-associated transporter 1 (HPT1)	U07969	Q12864	Other cell adhesion proteins
Increased				Other cell adhesion proteins
				Other facilitated diffusion proteins
2i	GAP junction alpha-5 protein	L34954	P36382	Cell-cell adhesion receptors
Increased				Other membrane channels and transporters
3m	Integrin beta 2 (ITGB2); cell surface adhesion glycoproteins LFA-1/CR3/p150, 95 beta subunit precursor; CD18 antigen;	M15395	P05107	Cell-cell adhesion receptors
Increased	Complement receptor C3 beta subunit		Q16418	
1m	Cardiac LIM domain protein; muscle LIM protein; cystein-rich protein 3 (CRP3); LIM-only protein 4	U49837	P50461	Basic transcription factors
Increased				Other transcription proteins
				DNA synthesis, recombination and repair proteins
1n	Cardiotrophin-1 (CT1)	U43030	Q16619	Growth factors, cytokines and chemokines
Increased				
2n	Matrix metalloproteinase 16 (MMP-16)	D83646	P51512	Chromatin proteins
Increased				Metalloproteinases
4a	TIMP-3	U14394	P35625	Extracellular matrix proteins
Increased				Proteinase inhibitor

4b Increased	TIMP-4	U76456	Q99727	Extracellular matrix proteins Proteinase inhibitor
4d Increased	Sterol regulatory element-binding transcription factor 1	U00968	P36956	Basic transcription factors Other apoptosis-associated proteins
4e Increased	Sterol regulatory element-binding transcription factor 2	U02031	Q12772	Basic transcription factors Other apoptosis-associated proteins
5h Increased	Rab geranylgeranyl transferase beta subunit	Y08201	P53611 Q92697	Trafficking and targeting proteins Protein modification enzymes GTP/GDP exchangers and GTPase activity modulators
6m Decreased	Muscle-specific DNase I-like precursor (DNase 1L1; DNL 1L); DNase X	X90392 L40817 U06846	P49184	DNA synthesis, recombination and repair proteins Apoptosis-associated Proteins
1b Increased	Lanosterol synthase (LSS); oxidosqualene lanosterol cyclase (OSC)	D63807	P48449	Complex lipid metabolism
3n Increased	NADPH-cytochrome p450 reductase	S90469	Q16455 P16435	Xenobic metabolism
4g Increased	Steroid 5 alpha reductase 1 (SRD5A1); 3-oxo-5-alpha steroid 4 dehydrogenase 1	M32313 M68886	P18405	Complex lipid metabolism
4h Increased	Steroid 5-alpha reductase 2 (SRD5A2); 3-oxo-5-alpha steroid 4 dehydrogenase 2	M74047	P31213	Complex lipid metabolism
2d Increased	Pregnane X receptor (PXR)	AF061056	O75469	Hormone receptors Nuclear receptors
F2e Increased	Estrogen-related receptor gamma	AF058291	O75454	Hormone receptors Nuclear receptors

(continued)

Table 2 (continued)

Double-spot position "EA"-image Exp. difference *vs.* control	Name of gene as given in "Atlas™ Human Cardiovascular Array"	GenBank accession	SwissProt accession	Gene/protein classification
F2f Increased	Nuclear receptor subfamily 4 group A member 2 (NR4A2); nuclear receptor-related protein 1 (NURR1); transcriptionally inducible nuclear receptor (TINUR); NOT	X75918	P43354	Hormone receptors Nuclear receptors Transcription activators and repressors
2i Increased	Orphan nuclear receptor TR4; nuclear receptor subfamily 2 group c member 2 (NR2C2); TAK1	U10990	P55092 P49116	Orphan receptors Nuclear receptors Transcription activators and repressors
3a Increased	RAR-related orphan receptor C	U16997	P51449	Orphan receptors Nuclear receptors Transcription activators and repressors
3e Increased	LX receptor alpha (LXR alpha)	U22662	Q13133	Orphan receptors
3i Increased	Platelet-activating factor acetylhydrolase IB alpha subunit	L13387	P43034	Other metabolism enzymes
7c Increased	Myocyte-specific enhancer factor 2A (MEF2A); serum response factor-like protein 1	X68505	Q02078 Q14223 Q14224	Basic transcription factors

to chemo-tactic signals of traumatic, infectious, post-ischaemic, autoimmune, and various toxic injuries. Extracellular nucleotides released from the activated platelets and other damaged cell types exacerbate the inflammatory response by cell specific leukotrene generation [26]. Thus, neutrophils generate leukotrienes B4 (LTB4) which are involved in the genesis of inflammation and oedema because of their effect on vascular permeability, plasma extravasation, diapedesis of white blood cells, and their important role in adaptive immune responses (as reviewed by Di Gennaro et al. [27]). Specifically, a highly enhanced concentration of leukotrienes B4 and C4 has been observed in CSF of patients with multiple sclerosis [28]. The member of leukotrienes receptors family – LTB4 receptor or P2Y purinoceptor 7 – has been isolated for the first time from the human erythroleukemia cell cDNA library [29]. The stimulation of monocytes, neutrophils and endothelial cells was suggested to be a physiological role for the LTB4 receptor [30]. There is a growing body of evidence indicating an important role of LTB4 receptors in the regulation of pathologic inflammation. Particularly using animal inflammatory models, a reduced disease severity has been shown when LTB4 receptor antagonists have been applied; the same effect has been observed in mice with target deletion of BLT1 – a high-affinity LTB4 receptor primarily expressed in leukocytes [31]. Furthermore, some studies support a potential role of P2Y receptors in controlling intraocular pressure although additional investigations of the issue are necessary [32].

6.2 ICAM-1 Is Upregulated in Vasospastic Individuals and Glaucoma Patients

Neutrophil-endothelium interactions are implicated in pathological alterations of blood vessel function, potentially leading to circulatory disturbances [33]. Interactions between blood cells and the vessel wall result in endothelial dysfunction and injury leading to increased blood–brain barrier permeability and even oedema formation [34]. Penetration of leukocytes into inflamed areas involves a complex interaction of leukocytes with the endothelium through a regulated expression of surface adhesion molecules. Found in this work to be highly expressed in VD, NTG and HTG groups ICAM-1 molecule is believed to be largely responsible for the adhesion and trans-endothelial migration of the leukocytes [35]. This is well in agreement with earlier developed strategies aimed at the inhibition of endothelial interactions with leukocytes via the use of adhesion molecule monoclonal antibodies, which successfully reduce cerebral ischaemia/reperfusion injury, infarct size, and demonstrate a neuroprotective effect generally [36–38]. In our study, highly expressed ICAM-1 was found in leukocytes of glaucoma patients; in contrast, at most, only traces of the target expression was detected in the leukocytes of healthy controls.

6.3 Sodium Calcium Exchanger (NCE)

Many studies examined the levels of cytosolic Ca^{2+} ($[Ca^{2+}]_c$) and Na^+ ($[Na^+]_c$) in human blood cells, whereby leukocytes have been the main target of studying the relationship between blood pressure and intracellular content of both ions (as reviewed by Horiguchi et al. [39].) As it has been shown by Horiguchi et al., the resting $[Ca^{2+}]_c$ correlates well with NCE expression so indicating NCE expression regulation to be an adaptive mechanism for Ca^{2+} extrusion mediation. The same study also observed a gender effect on $[Ca^{2+}]_c/[Na^+]_c$ regulation in circulating leukocytes being in relationship with blood pressure. Further, the role of the endothelial intracellular Ca^{2+} concentration in molecular mechanisms of vasoconstriction/vasodilatation has been intensively studied and the functional association between P2Y purinoceptors, endothelial NO synthesis and calcium transport in terms of vascular regulation is well documented in the literature [40, 41]. Our findings here clearly demonstrate the up-regulation of both P2Y purinoceptor and Na^+/Ca^{2+} exchanger in circulating leukocytes of glaucoma patients as well as vasospastic individuals *versus* healthy controls.

6.4 Tissue Remodelling Metalloproteinases

Significantly increased protein expression rates of both latent and active forms of metalloproteinases *MMP-9* and *MT1-MMP* in circulating leukocytes correlate well with the enhanced levels of transcription and with glaucoma diagnosis [13]. Once activated, both hydrolases necessarily contribute to remodelling or even degeneration of the tissue whereto they are secreted by circulating leukocytes. This up-regulation might be a consequence of repeated mild ischaemia/reperfusion postulated for both vasospastic individuals and glaucoma patients [21]. However, the question as to whether or not there is a correlation between an increased *MMPs* activity and glaucoma severity should be further clarified.

Furthermore, the increased synthesis of tissue-remodelling hydrolases detected in the blood of healthy vasospastic individuals can potentially lead to the development of further pathologies that have not been considered as related to vasospasm till now. To the potential list of them belong altered wound-healing, some types of organ-degeneration and increased metastases activity. It is essential that large-scale studies are performed in order to validate proposed impacts of vasospasm for the above listed pathophysiologic processes/manifested pathologies. This allows a targeted prevention at the stage of pre-lesions, upstream disease manifestation.

7 Concluding Remarks and Expert Recommendations

- The expression similarities between glaucoma and VD *versus* controls indicate, on one hand, a predisposition of VD individuals to glaucomatous damage, and, on the other hand, an important role of a vascular component in glaucoma pathology.
- Expression dissimilarities between VD and glaucoma patients might indicate some glaucoma-specific pathomechanisms that are not involved in the stage of predisposition by VD.
- Both expression similarities and dissimilarities could be useful in ascertaining the predictive diagnosis of glaucoma.
- Molecular rearrangement in leukocytes of both VD and glaucoma patients has been shown to be typical for circulating leukocytes during vascular injury as reviewed by Kunapoli and Daniel [42] and includes
  - an up-regulated adhesive protein expression via ICAM1
  - an induced chemo-taxis via P2Y purinoceptors
  - a mobilisation of intracellular Ca^{2+} response via Na^{+}/Ca^{2+} exchanger, and
  - the core of tissue remodelling metalloproteinases.
- Pathology-specific molecular and subcellular patterns may create the basis for the development of more specific and non-invasive molecular imaging technologies in early/predictive glaucoma diagnostics.
- Development of some other "down-stream" pathologies different from glaucoma but related to primary vasospastic syndrome cannot be excluded.
- Predictive molecular profiling in blood can specify individual predisposition for effective prevention to low costs [43].

References

1. Flammer J, Pache M, Resink T (2001) Vasospasm, its role in the pathogenesis of diseases with particular reference to the eye. Prog Retin Eye Res 20:319–349
2. Mozaffarieh M, Fraenkl S, Konieczka K, Flammer J (2010) Targeted preventive measures and advanced approaches in personalised treatment of glaucoma neuropathy. EPMA J 1:229–236
3. Beltrame JF, Sasayama S, Maseri A (1999) Racial heterogeneity in coronary artery vasomotor reactivity: differences between Japanese and Caucasian patients. J Am Coll Cardiol 33: 1442–1452
4. Flammer J (2006) Glaucoma. A guide for patients. An introduction for care providers. A quick reference. 3rd revised edn. Hogrefe & Huber Publishers, Bern
5. Teuchner B, Orgül S, Ulmer H, Haufschild T, Flammer J (2004) Reduced thirst in patients with a vasospastic syndrome. Acta Ophthalmol Scand 82:738–740
6. Wunderlich K, Zimmerman C, Gutmann H, Teuchner B, Flammer J, Drewe J (2003) Vasospastic persons exhibit differential expression of ABC-transport proteins. Mol Vis 9:756–761
7. Flammer J (2001) Innovative glaucoma therapy. Ophthalmologe 98:923–924
8. Finsterer J (2004) Mitochondriopathies. Eur J Neurol 11:163–186
9. Flammer J, Guthauser U, Mahler F (1987) Do ocular vasospasms help cause low-tension glaucoma? Doc Ophthalmol Proc Ser 49:397–399

10. Guthauser U, Flammer J, Mahler F (1988) The relationship between digital and ocular vasospasm. Graefes Arch Clin Exp Ophthalmol 226:224–226
11. Flammer J, Orgül S, Costa VP, Orzalesi N, Krieglstein GK, Serra LM, Renard JP, Stefánsson E (2002) The impact of ocular blood flow in glaucoma. Prog Retin Eye Res 21:359–393
12. Golubnitschaja O, Wunderlich K, Decker C, Mönkemann H, Schild HH, Flammer J (2002) Molecular imaging of perfusion disturbances in glaucoma. Amino Acids 23:293–299
13. Golubnitschaja O, Yeghiazaryan K, Liu R, Mönkemann H, Leppert D, Schild H, Haefliger IO, Flammer J (2004) Increased expression of matrix metalloproteinases in mononuclear blood cells of normal-tension glaucoma patients. J Glaucoma 13:66–72
14. Golubnitschaja-Labudova O, Liu R, Decker C, Zhu P, Haefliger IO, Flammer J (2000) Altered gene expression in lymphocytes of patients with normal-tension glaucoma. Curr Eye Res 21:867–876
15. Wunderlich K, Golubnitschaja O, Pache M, Eberle AN, Flammer J (2002) Increased plasma levels of 20S proteasome alpha-subunit in glaucoma patients: an observational pilot study. Mol Vis 8:431–435
16. Yeghiazaryan K, Flammer J, Orgül S, Wunderlich K, Golubnitschaja O (2009) Vasospastic individuals demonstrate significant similarity to glaucoma patients as revealed by gene expression profiling in circulating leukocytes. Mol Vis 15:2339–2348
17. Golubnitschaja O, Yeghiazaryan K, Flammer J (2010) Key molecular pathways affected by glaucoma pathology: is predictive diagnosis possible? EPMA J 1:237–244
18. Fraenkl SA, Mozaffarieh M, Flammer J (2010) Retinal vein occlusions: the potential impact of a dysregulation of the retinal veins. EPMA J 1:253–261
19. Golubnitschaja O, Flammer J (2007) What are the biomarkers for glaucoma? Surv Ophthalmol 52:S155–S161
20. Golubnitschaja O, Yeghiazaryan K, Flammer J (2013) Glaucomatous optic neuropathy: risk assessment and potential targets for effective prevention and treatments tailored to the patient. In: Mandel S (ed) Neurodegenerative diseases: integrative PPPM approach as the medicine of the future. Book series: Advances in PPPM. Golubnitschaja O (ed). Springer, Dordrecht, pp. 187–201
21. Golubnitschaja O, Yeghiazaryan K, Orgül S, Flammer J (2009) What are the biomarkers for glaucoma? In: Golubnitshaja O (ed) Predictive diagnostics and personalized treatment: dream or reality. Nova Science Publishers, New York
22. Moenkemann H, Flammer J, Wunderlich K, Breipohl W, Schild HH, Golubnitschaja O (2005) Increased DNA breaks and up-regulation of both G(1) and G(2) checkpoint genes p21(WAF1/CIP1) and 14-3-3 sigma in circulating leukocytes of glaucoma patients and vasospastic individuals. Amino Acid 28:199–205
23. Golubnitschaja O, Moenkemann H, Kim K, Mozaffari MS (2003) DNA damage and expression of checkpoint genes p21(WAF1/CIP1) and 14-3-3 sigma in taurine-deficient cardiomyocytes. Biochem Pharmacol 66:511–517
24. Migliore L, Fontana I, Colognato R, Coppede F, Siciliano G, Murri L (2005) Searching for the role and the most suitable biomarkers of oxidative stress in Alzheimer's disease and in other neurodegenerative diseases. Neurobiol Aging 26:587–595
25. Golubnitschaja O, Yeghiazaryan K, Wunderlich K, Schild HH, Flammer J (2007) Disease proteomics reveals altered basic gene expression regulation in leukocytes of glaucoma patients. Proteomics Clin Appl 1:1316–1323
26. Kannan S (2002) Amplification of extracellular nucleotide-induced leukocyte(s) degranulation by contingent autocrine and paracrine mode of leukotriene-mediated chemokine receptor activation. Med Hypotheses 59:261–265
27. Di Gennaro A, Carnini C, Buccellati C, Ballerio R, Zarini S, Fumagalli F, Viappiani S, Librizzi L, Hernandez A, Murphy RC, Constantin G, De Curtis M, Folco G, Sala A (2004) Cysteinylleukotrienes receptor activation in brain inflammatory reactions and cerebral edema formation: a role for transcellular biosynthesis of cysteinyl-leukotrienes. FASEB J 18:842–844
28. Rosnowska M, Cendrowski W, Sobczyk W (1997) Leukotrienes B4 and C4 in cerebrospinal of patients with multiple sclerosis. Pol Merkur Lekarski 2:254–255

29. Akbar GK, Dasari VR, Webb TE, Ayyanathan K, Pillarisetti K, Sandhu AK, Athwal RS, Daniel JL, Ashby B, Barnard EA, Kunapuli SP (1996) Molecular cloning of a novel P2 purinoceptor from human erythroleukemia cells. J Biol Chem 271:18363–18367
30. Dasari VR, Jin J, Kunapuli SP (2000) Distribution of leukotriene B4 receptors in human hematopoietic cells. Immunopharmacology 48:157–163
31. Tager AM, Luster AD (2003) BLT1 and BLT2: the leukotriene B(4) receptors. Prostaglandin Leukot Essent Fatty Acid 69:123–134
32. Pintor J, Peral A, Peláez T, Martín S, Hoyle CH (2003) Presence of diadenosine polyphosphates in the aqueous humor: their effect on intraocular pressure. J Pharmacol Exp Ther 304:342–348
33. Akopov SE, Sercombe R, Seylaz J (1995) Actions of platelet activating factor on isolated rabbit basilar artery: modulation by activated polymorphonuclear leukocytes. J Vasc Res 32:49–57
34. Akopov S, Sercombe R, Seylaz J (1996) Cerebrovascular reactivity: role of endothelium/platelet/leukocyte interactions. Cerebrovasc Brain Metab Rev 8:11–94
35. Butcher EC (1991) Leukocyte-endothelial cell recognition: three (or more) steps to specificity and diversity. Cell 67:1033–1036
36. Barone FC, Feuerstein GZ (1999) Inflammatory mediators and stroke: new opportunities for novel therapeutics. J Cereb Blood Flow Metab 19:819–834
37. Chen H, Chopp M, Zhang RL, Bodzin G, Chen Q, Rusche JR, Todd RF 3rd (1994) Anti-CD11b monoclonal antibody reduces ischemic cell damage after transient focal cerebral ischemia in rat. Ann Neurol 35:458–463
38. Connolly ES Jr, Winfree CJ, Springer TA, Naka Y, Liao H, Yan SD, Stern DM, Solomon RA, Gutierrez-Ramos JC, Pinsky DJ (1996) Cerebral protection in homozygous null ICAM-1 mice after middle cerebral artery occlusion. Role of neutrophil adhesion in the pathogenesis of stroke. J Clin Invest 97:209–216
39. Horiguchi M, Kimura M, Skurnick J, Aviv A (1998) Parameters of lymphocyte Na^{+}-Ca^{2+} regulation and blood pressure: the gender effect. Hypertension 32:869–874
40. Marrelli SP (2001) Mechanisms of endothelial P2Y(1)- and P2Y(2)-mediated vasodilatation involve differential [Ca2+]I responses. Am J Physiol Heart Circ Physiol 281:H1759–H1766
41. Moncada S, Palmer RM, Higgs EA (1991) Nitric oxide: physiology, pathophysiology, and pharmacology. Pharmacol Rev 43:109–142
42. Kunapuli SP, Daniel JL (1998) P2 receptor subtypes in the cardiovascular system. Biochem J 336:513–523
43. Yeghiazaryan K, Flammer J, Golubnitschaja O (2010) Predictive molecular profiling in blood of healthy vasospastic individuals: clue to targeted prevention as personalised medicine to effective costs. EPMA J 1:263–272

Idiopathic REM Sleep Behavior Disorder as a Long-Term Predictor of Neurodegenerative Disorders

Stephany Fulda and Mauro Manconi

Abstract REM sleep behavior disorder (RBD) is a parasomnia characterized by dream-enacting behavior and loss of muscle atonia during REM sleep. Idiopathic RBD occurs in the absence of any neurological disease or other possible cause, is male-predominant and its clinical course is generally chronic progressive. Secondary RBD may be related to neurodegenerative disorders such as multiple system atrophy, Parkinson's disease and Lewy body dementia. Recent long-term prospective studies have shown that 30–65% of patients with idiopathic RBD will eventually develop a neurodegenerative disorder with the rate of emergence depending on the length of the follow-up period. It may therefore be one of the earliest signs of and/or a long-term predictor for neurodegenerative disorders. Because RBD antecedes the development of these disorders by several years, its recognition may enable the delay or prevention of neurodegenerative disorders through the early application of neuroprotective or disease-modifying therapies in the future.

Keywords REM sleep behavior disorder • Parkinson's disease • Multiple system athrophy • Lewy body dementia

1 Introduction

Human sleep is divided into rapid eye movement (REM) sleep and non-REM (NREM) sleep. REM sleep is characterized by the prominent rapid eye movements and active paralysis of the somatic muscles. REM sleep behavior disorder (RBD)

S. Fulda (✉) • M. Manconi
Sleep and Epilepsy Center, Neurocenter of Southern Switzerland, Civic Hospital (EOC) of Lugano, Lugano, Switzerland
e-mail: stephany.fulda@gmail.com

S. Mandel (ed.), *Neurodegenerative Diseases: Integrative PPPM Approach as the Medicine of the Future*, Advances in Predictive, Preventive and Personalised Medicine 2, DOI 10.1007/978-94-007-5866-7_3,

Table 1 Diagnostic criteria for REM sleep behavior disorder (RBD)

A	REM sleep without atonia: EMG finding of excessive amounts of sustained or intermittent elevation of submental EMG tone or excessive phasic submental or limb EMG twitching
B	At least one of the following: – History of sleep related injurious, potentially injurious, or disruptive behaviors – Abnormal REM sleep behaviors documented during polysomnographic monitoring
C	Absence of EEG epileptiform activity during REM sleep unless RBD can be clearly distinguished from any concurrent REM sleep related seizure disorder
D	The sleep disorder is not better explained by another sleep disorder, medical or neurological disorder, mental disorder, medication use, or substance use disorder

Taken from [1]

is a parasomnia where the physiological atonia during REM sleep is absent or greatly diminished and is characterized by dream-enacting behavior associated with nightmares. Motor behavior ranges from vocalizations during REM sleep or simple movements to violent, complex and lasting enactment of dream content with significant injury to the patient or a bed partner. The diagnostic criteria for RBD according to the international classification of sleep disorders [1] are given in Table 1 and require a nocturnal polysomnography to document REM sleep without atonia (Fig. 1) in the presence of dream-enacting behavior.

RBD was first described in 1987 by Schenck et al. [2]. It is a male-predominant disorder and usually emerges after the age of 50 years but has also been observed in younger patients [1]. At present, its prevalence is unknown but estimates range from 0.3 to 0.8% [1]. The onset of RBD can be gradual or rapid but the course is generally chronic and progressive [2]. Complications can include injuries to the patient or the bed partner, which can be life-threatening [3]. Clonazepam, usually in doses of 0.5–1 mg/night, is efficacious in the majority of patients and is considered first-line treatment but should be used with cautions in patients with dementia, gait disorders or concomitant obstructive sleep apnea [4–6]. Melatonin has also been effective in some patients with RBD [7–9]. Modifying and maintaining a safe sleeping environment should accompany any therapeutic intervention [5].

In the idiopathic form (iRBD), RBD occurs in the absence of any other associated neurological disorder or possible cause. In contrast, RBD may be secondary to neurological diseases particularly multiple system atrophy (MSA), Parkinson's disease (PD) and dementia with Lewy bodies (DLB) (see below). In addition, medication use has been recognized as a precipitating factor, specifically antidepressants and beta-blockers [10–13].

In iRBD no evidence of neurological disease is found, however, in many iRBD patients subclinical abnormalities have been detected such as olfactory deficits [14–19], cognitive dysfunction [20–22], impaired color vision [19], and subtle cortical EEG slowing [21, 23, 24]. Other findings include dysautonomic abnormalities [25–27], reduced cardiac ^{123}I-Metaiodobenzylguanidine scintigraphy [28–31], decreased dopamine transporter imaging [32], and increased substantia nigra echogenicity [33–35]. Follow-up of patients with iRBD shows an increased risk for developing neurodegenerative diseases that strongly depends on the length

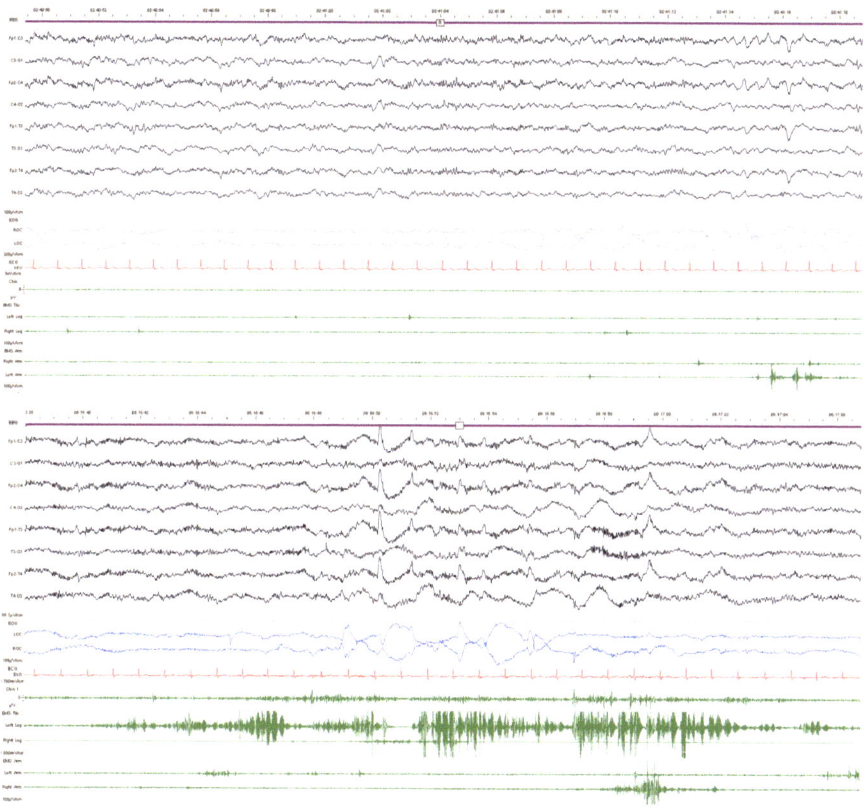

Fig. 1 Examples (30s) from nocturnal polysomnographies showing REM sleep with (*upper panel*) and without atonia (*lower panel*). Channels 1–8: electroencephalography (EEG); channels 9–10: electrooculography (EOG, eye movements); channel 11: ECG; channels 12–16: electromygraphy (EMG, muscle activation at chin, legs, and arms)

of the follow-up period. Recent evidence suggests that iRBD is an early sign of a slowly evolving neurodegenerative disorder or a long-term predictor of it. In the following this evidence will be summarized and presented focusing on the prevalence and features of RBD in neurodegenerative disorders, the emergence of neurodegenerative disorders in long-term follow-up studies of patients with iRBD and baseline differences between iRBD patients who developed a neurodegenerative disease and those who remained disease free.

2 REM Sleep Behavior Disorder in Neurodegenerative Diseases

RBD is frequently encountered in neurodegenerative disorders, particularly in MSA, PD and DLB (Table 2). The majority of patients with MSA will also have RBD with reported prevalences of 80–100% [10, 36–38]. Indeed, RBD is currently

Table 2 REM sleep behavior disorder (RBD) in neurodegenerative disorders

	Prevalence of RBD (%)	RBD preceded onset of neurodegenerative disorder		References
		In % of patients	Time interval	
Multiple system atrophy	80–100	16–54	4–7 years	[10, 36–38, 40, 41]
Parkinson's disease	30–41	18–25	3–4 years	[38, 41–46]
Dementia with Lewy bodies	40–72	71–100	6–10 years	[10, 41, 47–51]

considered a red flag for the diagnosis of MSA [39]. Interestingly, the strong male predominance encountered in the idiopathic form and in RBD associated with PD and DLB is less evident in MSA where only 33– 61% of the patients are male [36, 38, 40]. Concerning the onset of RBD in relation to the occurrence of MSA, studies with retrospective estimates have shown that in 16–54% of MSA patients RBD preceded the onset of MSA [10, 36, 38, 40, 41]. In one of the studies RBD preceded MSA by a mean of 7 years (1–38 years) in 52% of the patients [10], in another [41], RBD occurred 4 years (1–11 years) earlier in 36% of MSA patients and in a third [36], RBD preceded MSA by 1–19 years in 44% of the patients.

The prevalence of RBD in PD lies between 30 and 41% [42–45]. Between 65 and 75% of PD patients with RBD are male [38, 44]. RBD occurs both in idiopathic PD and PD secondary to genetic mutations [52] and in PD patients that are untreated or treated with dopaminergic substances [10]. RBD is more common in the rigid-akinetic subtype of PD than in the tremoric subtype [46, 53]. Non-demented patients with PD and RBD are more likely to show EEG slowing during wakefulness [54] and poorer cognitive function [55] but RBD in PD patients has also been linked to longer duration of PD [41, 44]. RBD in PD is not associated with age, disease severity, depression, or sleep architecture [10]. RBD preceded PD in 18–25% of patients [38, 41, 46], on average between 3 (1–30 years) [41] and 4 years [46]. However, in patients with *Parkin* mutations, RBD develops after PD onset [52] and RBD rarely precedes PD onset in patients with early onset PD (<50 years) [46]. Interestingly, it seems that in patients with iRBD who subsequently developed PD, the onset of PD is later (~72 years) than reported for PD patients (~62 years) [56].

The prevalence of RBD in DLB is not as well established and available studies showed prevalences between 40% [47] and 72% [48] again with a male predominance [10]. RBD preceded the onset of cognitive complaints and dementia in the vast majority of patients (71–100%) [10, 41, 49, 50] and the onset of cognitive decline followed on average 6 [49], 9 [50] or 10 years [51] later. Current consensus criteria for DLB consider RBD as a suggestive feature of this disease [57] and it is thought that the clinical combination of dementia and RBD very likely indicates underlying Lewy body disease, i.e. DLB and PD with dementia (PDD) [10].

All three disorders – MSA, PD, and DLB – are characterized by intraneural deposition of α-synuclein and therefore it has been speculated that the RBD may be linked to synuclein pathology [47]. RBD has also been observed in cases with pure autonomic failure [58, 59]. Until recently, RBD has been considered comparatively rare in tauopathies, however, newer studies have reported RBD or REM sleep without atonia (RSWA) also in progressive supranuclear palsy (27%) [60] and Alzheimer's disease (7% RBD, 29% RSWA) [61]. However, RBD has not been observed in pallindopontonigral degeneration [62]. Also, so far no case of RBD in a patient with frontotemporal dementia has been reported [63]. RBD has also been associated to autoimmune mechanisms and REM sleep without atonia is frequently found in narcolepsy [64–67], another rare sleep disorder with a possible autoimmune background. Similarly, RBD has been observed in five of six patients with non-paraneoplastic limbic encephalitis associated with antibodies to voltage-gated potassium channels [68]. Interestingly, in this study symptoms of RBD coincided with the onset of the disorder and in three of the patients immunosuppression resulted in a resolution of RBD in parallel with the remission of the limbic syndrome [68]. This fits with reports [69] in Guillain-Barrè syndrome (GBS), an acute autoimmunce polyradiculoneuritis with sensory and motor impairment, where during the acute phase in patients with mental status abnormalities REM atonia was lost and normalized when mental status abnormalities disappeared.

In summary, at present the available evidence shows that RBD is a common feature of synucleinopathies MSA, DLB, and PD. Nevertheless, among these disorders RBD is distinctly more prevalent in MSA, maybe because brainstem cell loss is widespread and severe in this disorder. Retrospective assessment indicates that RBD frequently precedes the onset of these disorders but this seems to be most pronounced in DLB probably reflecting the differences in clinical course and progression in these disorders. RBD has also been observed in tauopathies and disorders with a neuroimmunological background, however, at the moment the association with synucleinopathies outweighs that with any other.

3 RBD as a Long-Term Predictor for Neurodegenerative Disorders

There are several case studies reporting the new onset of neurodegenerative disorders years to decades after the onset of RBD. This has been documented for Parkinson's disease [70], Lewy body disorder [71–74], multiple system atrophy [73], and shy drager syndrome [75] and periods of 2–50 years [76]. Interestingly, in two of these cases Lewy body disease was only discovered at autopsy [71, 77].

In systematic long-term prospective studies the percentage of patients with iRBD who will eventually develop a neurodegenerative disorder ranges from 30 to 65% strongly depending on the length of the follow-up period (Table 3). Schenck and co-workers were the first to report that during a follow-up time of 13 years 38% of a group of 29 male iRBD patients above 50 years of age developed PD [78]. In a subsequent follow-up 7 years later, 65% had developed a neurodegenerative disorder [79]. Similar results were reported for 23 patients with iRBD of which 65% developed a neurodegenerative disorder during an average follow-up time of

Table 3 New onset of neurodegenerative disorders in prospective follow-up studies of patients with idiopathic REM sleep behavior disorder

Study	Subjects	Follow-up time	Prevalence	Time since RBD onset	Time since RBD diagnosis
Schenck et al. [78]	29 patients with iRBD ≥50 y	~13 y	38% (11|29) PD [8 definite PD, 2 probable PD, 1 possible PD]	~13 y	~4 y
Schenck et al. [79]		~20 y	65% (17|26) [13 PD, 1 PDD, 2 DLB, 1 AD]	~13 y (3–29)	
Iranzo et al. [56]	44 patients with iRBD	~5 y (2–15 y)	45% (20|44) [7 PD, 2 PDD, 6 DLB, 1 MSA, 4 MCI]	~12 y	~4 y
Iranzo et al. [81]		~7 y	64% (28|44) [6PD, 4 PDD, 8 DLB, 1 MSA, 9 MCI]		
Postuma et al. [82]	93 patients with iRBD	~5 y	30% (26|93) [14 PD, 7 DLB, 4 AD, 1 MSA]	~12 y	
Tippman-Peikert et al. [80]	23 patients with iRBD	~11 y	65% (15|23) [1 PD, 3 DLB, 1 dementia, Ten reported symptoms highly suggestive of a parkinsonian or dementing disorder]		

AD Alzheimer's disease, *DLB* dementia with Lewy bodies, *iRBD* idiopathic REM sleep behavior disorder, *MCI* minimal cognitive impairment, *MSA* multiple system atrophy, *PD* Parkinson's disease, *PDD* Parkinson's disease dementia, *y* years

11 years [80]. In an even larger case series of 44 consecutive iRBD patients, 45% developed PD, DLB, MSA or minimal cognitive impairment (MCI) after an average follow-up time of 5 years [56] and 64% after 7 years [81]. In the largest case series to date, 26 of 93 patients (30%) with iRBD developed neurodegenerative disorders after an average follow-up time of 5 years [82]. In the same study [82], the authors conducted a formal survival analysis and estimated the risk for the development of neurodegenerative disorders as 18% for a 5 year period, 41% for a 10-year period, and 52% for a 12-year period. Summarizing these four case series, across average follow-up periods of 5–20 years 86 of 189 patients (46%) with iRBD developed a neurodegenerative disorder. The average time between the onset of RBD and the onset of the neurodegenerative disorder was 12 years and for the 76 patients with a definite diagnosis the new onset disorder was Parkinson's disease in 39 (51%), Lewy body disease in 20 (26%), minimal cognitive impairment in 9 (12%), Alzheimer's disease in 5 (7%), MSA in 2 (3%) and dementia in 1 (Table 3).

Long-term studies in iRBD patients have shown that over time excessive phasic and tonic muscle activity during REM sleep increases [83], cognitive functioning declines [84], and nigrostriatal presynaptic dopaminergic function possible decreases [85] compatible with a general neurodegenerative process.

Possible predictors for the future development of neurodegenerative disorders in patients with iRBD have emerged from studies that compared baseline measurement of iRBD patients with subsequent neurodegenerative disorders to disease-free iRBD patients (Table 4). Distinguishing characteristics in these studies were an increase in REM sleep [78] or stage 1 sleep [87], increased tonic submental EMG activity [87], increased striatal dopamine dysfunction or substantia nigra hyperechogenicity [86], more severe abnormalities of olfaction and color vision [88] and a tendency towards more pronounced EEG slowing during wakefulness [24] (Table 4). Interestingly, cardiac autonomic function which differs between RBD and healthy controls [25, 26], did not distinguish between RBD patients with subsequent neurodegenerative disorders and disease free patients, suggesting that autonomic dysfunction in RBD may be independent from associated PD or LBD [27]. Currently, these promising studies await independent replication. For the future it is hoped that the timely identification of iRBD patients who will develop a neurodegenerative disorder provides a window of opportunity for prevention or early treatment of these disorders.

4 Summary and Outlook

RBD is a relatively rare sleep disorder with a male predominance that is characterized by dream-enactment and loss of REM sleep atonia. Secondary RBD is frequently encountered in neurodegenerative disorders and in particular in the synucleinopathies such as MSA, DLB, and PD. Retrospective assessment of the onset of RBD in these disorders suggests that RBD precedes the onset of these synucleinopathies in a significant proportion by several years. This is confirmed by prospective long-term studies of patients with iRBD, free of neurological disorders at the time of the diagnosis of RBD. These studies showed that up to 60% of patients with iRBD will subsequently develop a neurodegenerative disorder, most frequently PD and DLB. Again, the time span between the onset of RBD and the onset of the subsequent neurodegenerative disorder was around a decade with increasing conversion rates across longer the follow-up periods. First studies have also explored possible predictors for the development of subsequent neurodegenerative diseases in iRBD patients, but more research is needed in this area.

RBD can be seen as a long-term predictor for the development of neurodegenerative disorders. Most likely it is one of the earliest signs of these disorders. Indeed, for some of these disorders such as PD, it is now apparent that the preclinical phase can extent to 20 years or longer before the motor manifestations [89, 90] and among the symptoms that precede PD by 10–20 years are constipation, anemia, and anxiety disorders [90]. Possible predictors for the development of neurodegenerative disorders in patients with iRBD include increased REM or stage 1 sleep, increased submental EMG activity during REM sleep, striatal dopamine dysfunction and

Table 4 Possible predictors for the development of neurodegenerative disorders in RBD: baseline differences between iRBD patients who developed neurodegenerative disorders and iRBD patients who remained disease free

Study	Subjects	Follow-up time	Measured function/main findings
Schenck et al. [78]	29 male patients with iRBD ≥50 y of which 11 developed PD	~13 y	*Sleep, tonic and phasic chin EMG during REM sleep*: – Increased REM sleep and periodic leg movements at baseline – No difference in tonic or phasic EMG during REM, other sleep parameters
Iranzo et al. [24]	23 iRBD of which 10 developed MCI Ten healthy controls	2.4 ± 1.6 y	*Quantitative EEG (wake, REM sleep)*: – Both RBD groups showed slowing of EEG during wakefulness and REM sleep vs. healthy controls – No significant difference between RBD with and without MCI, but trend towards greater slowing during wakefulness
Iranzo et al. [86]	43 iRBD of which 8 developed NDD[a] 18 and 143 healthy controls	~2.5 y	*^{123}I-FP-CIT striatal binding, substantia nigra hyperechogenicity*: – 27 (63%) had reduced ^{123}I-FP-CIT striatal binding (40%) and/or substantia nigra hyperechogeneicity (36%) – All 8 with MCI had reduced ^{123}I-FP-CIT uptake or substantia nigra hyperechogenicity at baseline; 30% (8|27) of those with abnormal imaging at baseline developed NDD) – None of the 15 iRBD patients with normal neuroimaging developed NDD
Postuma et al. [87]	26 iRBD with development of NDD 26 matched iRBDD without NDD[b]	~7 y	*Sleep, tonic and phasic chin EMG during REM sleep*: – Increased percentage of tonic submental EMG activity at baseline (more severe loss of REM atonia) and increased stage 1 sleep – No difference for phasic submental EMG or other sleep parameters
Postuma et al. [88]	62 iRBD of which 21 developed NDD[c]	~5 y	*Olfaction, color vision testing*: – More severe abnormalities of olfaction and color vision at baseline
Postuma et al. [27]	42 iRBD of which 21 developed NDD[d]	~7 y	*Cardiac autonomic function*: – No difference between patient groups for any measure of cardiac autonomic function

AD Alzheimer's disease, *DLB* dementia with Lewy bodies, *iRBD* idiopathic REM sleep behavior disorder, *MCI* minimal cognitive impairment, *MSA* multiple system atrophy, *NDD* neurodegenerative disorder, *PD* Parkinson's disease, *PDD* Parkinson's disease dementia, *y* years

[a]5 PD, 2 DLB, 1 MSA

[b]12 PD, 1 MSA, 7 DLB, 6 dementia

[c]16 PDD, 4 PD, 1 dementia

[d]11 PD, 1 MSA, 5 DLB, 4 AD

substantia nigra hyperechogenicity, more severe abnormalities of olfaction and color vision and EEG slowing during wakefulness. Nocturnal video-polysomnograpy, imaging of dopamine metabolism, transcranial sonography, olfactory and color testing and daytime EEG may therefore by clinical tools to identify RBD patients at an increased risk for future neurodegenerative disorders. Because RBD precedes the development of neurodegenerative disorders by years, its timely recognition in patients may offer a temporal window of opportunity for future research programs where the early application of neuroprotective or disease modifying therapies is hoped to delay or prevent the onset of neurodegenerative disorders.

References

1. American Academy of Sleep Medicine (2005) The international classification of sleep disorders: diagnostic and coding manual, 2nd edn. American Academy of Sleep Medicine, Westchester
2. Schenck CH, Bundlie SR, Ettinger MG, Mahowald MW (1986) Chronic behavioral disorders of human REM sleep: a new category of parasomnia. Sleep 9:293–308
3. Schenck CH, Lee SA, Bornemann MAC, Mahowald MW (2009) Potentially lethal behaviors associated with rapid eye movement sleep behavior disorder: review of the literature and forensic implications. J Forensic Sci 54:1475–1484
4. Gagnon J-F, Postuma RB, Montplaisir J (2006) Update on the pharmacology of REM sleep behavior disorder. Neurology 67:742–747
5. Aurora RN, Zak RS, Maganti RK, Auerbach SH, Casey KR, Chowdhuri S, Karippot A, Ramar K, Kristo DA, Morgenthaler TI; Standards of Practice Committee; American Academy of Sleep Medicine (2010) Best practice guide for the treatment of REM sleep behavior disorder (RBD). J Clin Sleep Med 6:85–95
6. Schuld A, Kraus T, Haack M, Hinze-Selch D, Pollmächer T (1999) Obstructive sleep apnea syndrome induced by clonazepam in a narcoleptic patient with REM-sleep-behavior disorder. J Sleep Res 8:321–322
7. Boeve BF, Silber MH, Ferman TJ (2003) Melatonin for treatment of REM sleep behavior disorder in neurologic disorders: results in 14 patients. Sleep Med 4:281–284
8. Kunz D, Bes F (1999) Melatonin as a therapy in REM sleep behavior disorder patients: an open-labeled pilot study on the possible influence of melatonin on REM-sleep regulation. Mov Disord 14:507–511
9. Kunz D, Mahlberg R (2010) A two-part, double-blind, placebo-controlled trial of exogenous melatonin in REM sleep behaviour disorder. J Sleep Res 19:591–596
10. Iranzo A, Santamaria J, Tolosa E (2009) The clinical and pathophysiological relevance of REM sleep behavior disorder in neurodegenerative diseases. Sleep Med Rev 13:385–401
11. Hoque R, Chesson AL (2010) Pharmacologically induced/exacerbated restless legs syndrome, periodic limb movements of sleep, and REM behavior disorder/REM sleep without atonia: literature review, qualitative scoring, and comparative analysis. J Clin Sleep Med 6:79–83
12. Morrison I, Frangulyan R, Riha RL (2011) Beta-blockers as a cause of violent rapid eye movement sleep behavior disorder: a poorly recognized but common cause of violent parasomnias. Am J Med 124:e11
13. Teman PT, Tippmann-Peikert M, Silber MH, Slocumb NL, Auger RR (2009) Idiopathic rapid-eye-movement sleep disorder: associations with antidepressants, psychiatric diagnoses, and other factors, in relation to age of onset. Sleep Med 10:60–65
14. Fantini ML, Postuma RB, Montplaisir J, Ferini-Strambi L (2006) Olfactory deficit in idiopathic rapid eye movements sleep behavior disorder. Brain Res Bull 70:386–390

15. Miyamoto T, Miyamoto M, Iwanami M, Hirata K (2010) Olfactory dysfunction in Japanese patients with idiopathic REM sleep behavior disorder: comparison of data using the university of Pennsylvania smell identification test and odor stick identification test for Japanese. Mov Disord 25:1524–1526
16. Miyamoto T, Miyamoto M, Iwanami M, Hirata K, Kobayashi M, Nakamura M, Inoue Y (2010) Olfactory dysfunction in idiopathic REM sleep behavior disorder. Sleep Med 11:458–461
17. Miyamoto T, Miyamoto M, Iwanami M, Suzuki K, Inoue Y, Hirata K (2009) Odor identification test as an indicator of idiopathic REM sleep behavior disorder. Mov Disord 24:268–273
18. Stiasny-Kolster K, Clever S-C, Möller JC, Oertel WH, Mayer G (2007) Olfactory dysfunction in patients with narcolepsy with and without REM sleep behaviour disorder. Brain 130:442–449
19. Postuma RB, Lang AE, Massicotte-Marquez J, Montplaisir J (2006) Potential early markers of Parkinson disease in idiopathic REM sleep behavior disorder. Neurology 66:845–851
20. Ferini-Strambi L, Di Gioia MR, Castronovo V, Oldani A, Zucconi M, Cappa SF (2004) Neuropsychological assessment in idiopathic REM sleep behavior disorder (RBD): does the idiopathic form of RBD really exist? Neurology 62:41–45
21. Massicotte-Marquez J, Décary A, Gagnon J-F, Vendette M, Mathieu A, Postuma RB, Carrier J, Montplaisir J (2008) Executive dysfunction and memory impairment in idiopathic REM sleep behavior disorder. Neurology 70:1250–1257
22. Terzaghi M, Sinforiani E, Zucchella C, Zambrelli E, Pasotti C, Rustioni V, Manni R (2008) Cognitive performance in REM sleep behaviour disorder: a possible early marker of neurodegenerative disease? Sleep Med 9:343–351
23. Fantini ML, Gagnon J-F, Petit D, Rompré S, Décary A, Carrier J, Montplaisir J (2003) Slowing of electroencephalogram in rapid eye movement sleep behavior disorder. Ann Neurol 53:774–780
24. Iranzo A, Isetta V, Molinuevo JL, Serradell M, Navajas D, Farre R, Santamaria J (2011) Electroencephalographic slowing heralds mild cognitive impairment in idiopathic REM sleep behavior disorder. Sleep Med 11:534–539
25. Ferini-Strambi L, Oldani A, Zucconi M, Smirne S (1996) Cardiac autonomic activity during wakefulness and sleep in REM sleep behavior disorder. Sleep 19:367–369
26. Lanfranchi PA, Fradette L, Gagnon J-F, Colombo R, Montplaisir J (2007) Cardiac autonomic regulation during sleep in idiopathic REM sleep behavior disorder. Sleep 30:1019–1025
27. Postuma RB, Lanfranchi PA, Blais H, Gagnon J-F, Montplaisir JY (2010) Cardiac autonomic dysfunction in idiopathic REM sleep behavior disorder. Mov Disord 25:2304–2310
28. Kashihara K, Imamura T, Shinya T (2010) Cardiac 123I-MIBG uptake is reduced more markedly in patients with REM sleep behavior disorder than in those with early stage Parkinson's disease. Parkinsonism Relat Disord 16:252–255
29. Miyamoto T, Miyamoto M, Inoue Y, Usui Y, Suzuki K, Hirata K (2006) Reduced cardiac 123I-MIBG scintigraphy in idiopathic REM sleep behavior disorder. Neurology 67:2236–2238
30. Miyamoto T, Miyamoto M, Iwanami M, Hirata K (2011) Cardiac (123)I-MIBG accumulation in Parkinson's disease differs in association with REM sleep behavior disorder. Parkinsonism Relat Disord 17:219–220
31. Miyamoto T, Miyamoto M, Suzuki K, Nishibayashi M, Iwanami M, Hirata K (2008) 123I-MIBG cardiac scintigraphy provides clues to the underlying neurodegenerative disorder in idiopathic REM sleep behavior disorder. Sleep 31:717–723
32. Eisensehr I, Linke R, Noachtar S, Schwarz J, Gildehaus FJ, Tatsch K (2000) Reduced striatal dopamine transporters in idiopathic rapid eye movement sleep behaviour disorder. Comparison with Parkinson's disease and controls. Brain 123:1155–1160
33. Iwanami M, Miyamoto T, Miyamoto M, Hirata K, Takada E (2010) Relevance of substantia nigra hyperechogenicity and reduced odor identification in idiopathic REM sleep behavior disorder. Sleep Med 11:361–365

34. Stockner H, Iranzo A, Seppi K, Serradell M, Gschliesser V, Sojer M, Valldeoriola F, Molinuevo JL, Frauscher B, Schmidauer C, Santamaria J, Högl B, Tolosa E, Poewe W; SINBAR (Sleep Innsbruck Barcelona) Group (2009) Midbrain hyperechogenicity in idiopathic REM sleep behavior disorder. Mov Disord 24:1906–1909
35. Unger MM, Möller JC, Stiasny-Kolster K, Mankel K, Berg D, Walter U, Hoeffken H, Mayer G, Oertel WH (2008) Assessment of idiopathic rapid-eye-movement sleep behavior disorder by transcranial sonography, olfactory function test, and FP-CIT-SPECT. Mov Disord 23:596–599
36. Plazzi G, Corsini R, Provini F, Pierangeli G, Martinelli P, Montagna P, Lugaresi E, Cortelli P (1997) REM sleep behavior disorders in multiple system atrophy. Neurology 48:1094–1097
37. Tachibana N, Kimura K, Kitajima K, Shinde A, Kimura J, Shibasaki H (1997) REM sleep motor dysfunction in multiple system atrophy: with special emphasis on sleep talk as its early clinical manifestation. J Neurol Neurosurg Psychiatr 63:678–681
38. Iranzo A, Santamaría J, Rye DB, Valldeoriola F, Martí MJ, Muñoz E, Vilaseca I, Tolosa E (2005) Characteristics of idiopathic REM sleep behavior disorder and that associated with MSA and PD. Neurology 65:247–252
39. Köllensperger M, Geser F, Seppi K, Stampfer-Kountchev M, Sawires M, Scherfler C, Boesch S, Mueller J, Koukouni V, Quinn N, Pellecchia MT, Barone P, Schimke N, Dodel R, Oertel W, Dupont E, Østergaard K, Daniels C, Deuschl G, Gurevich T, Giladi N, Coelho M, Sampaio C, Nilsson C, Widner H, Sorbo FD, Albanese A, Cardozo A, Tolosa E, Abele M, Klockgether T, Kamm C, Gasser T, Djaldetti R, Colosimo C, Meco G, Schrag A, Poewe W, Wenning GK; European MSA Study Group (2008) Red flags for multiple system atrophy. Mov Disord 23:1093–1099
40. Vetrugno R, Provini F, Cortelli P, Plazzi G, Lotti EM, Pierangeli G, Canali C, Montagna P (2004) Sleep disorders in multiple system atrophy: a correlative video-polysomnographic study. Sleep Med 5:21–30
41. Olson EJ, Boeve BF, Silber MH (2000) Rapid eye movement sleep behaviour disorder: demographic, clinical and laboratory findings in 93 cases. Brain 123:331–339
42. Gagnon JF, Bédard MA, Fantini ML, Petit D, Panisset M, Rompré S, Carrier J, Montplaisir J (2002) REM sleep behavior disorder and REM sleep without atonia in Parkinson's disease. Neurology 59:585–589
43. De Cock VC, Vidailhet M, Leu S, Texeira A, Apartis E, Elbaz A, Roze E, Willer JC, Derenne JP, Agid Y, Arnulf I (2007) Restoration of normal motor control in Parkinson's disease during REM sleep. Brain 130:450–456
44. Wetter TC, Trenkwalder C, Gershanik O, Högl B (2001) Polysomnographic measures in Parkinson's disease: a comparison between patients with and without REM sleep disturbances. Wien Klin Wochenschr 113:249–253
45. Diederich NJ, Vaillant M, Mancuso G, Lyen P, Tiete J (2005) Progressive sleep "destructuring" in Parkinson's disease. A polysomnographic study in 46 patients. Sleep Med 6:313–318
46. Kumru H, Santamaria J, Tolosa E, Iranzo A (2007) Relation between subtype of Parkinson's disease and REM sleep behavior disorder. Sleep Med 8:779–783
47. Boeve BF, Silber MH, Ferman TJ, Lucas JA, Parisi JE (2001) Association of REM sleep behavior disorder and neurodegenerative disease may reflect an underlying synucleinopathy. Mov Disord 16(4):622–630
48. Ferman TJ, Smith GE, Boeve BF, Ivnik RJ, Petersen RC, Knopman D, Graff-Radford N, Parisi J, Dickson DW (2004) DLB fluctuations: specific features that reliably differentiate DLB from AD and normal aging. Neurology 62:181–187
49. Boeve BF, Silber MH, Ferman TJ, Kokmen E, Smith GE, Ivnik RJ, Parisi JE, Olson EJ, Petersen RC (1998) REM sleep behavior disorder and degenerative dementia: an association likely reflecting lewy body disease. Neurology 51:363–370
50. Ferman TJ, Boeve BF, Smith GE, Silber MH, Kokmen E, Petersen RC, Ivnik RJ (1999) REM sleep behavior disorder and dementia: cognitive differences when compared with AD. Neurology 52:951–957

51. Boeve BF, Silber MH, Parisi JE, Dickson DW, Ferman TJ, Benarroch EE, Schmeichel AM, Smith GE, Petersen RC, Ahlskog JE, Matsumoto JY, Knopman DS, Schenck CH, Mahowald MW (2003) Synucleinopathy pathology and REM sleep behavior disorder plus dementia or parkinsonism. Neurology 61:40–45
52. Kumru H, Santamaria J, Tolosa E, Valldeoriola F, Muñoz E, Marti MJ, Iranzo A (2004) Rapid eye movement sleep behavior disorder in parkinsonism with parkin mutations. Ann Neurol 56:599–603
53. Postuma RB, Gagnon JF, Vendette M, Charland K, Montplaisir J (2008) REM sleep behaviour disorder in Parkinson's disease is associated with specific motor features. J Neurol Neurosurg Psychiatr 79:1117–1121
54. Gagnon J-F, Fantini ML, Bédard M-A, Petit D, Carrier J, Rompré S, Décary A, Panisset M, Montplaisir J (2004) Association between waking EEG slowing and REM sleep behavior disorder in PD without dementia. Neurology 62:401–406
55. Vendette M, Gagnon J-F, Décary A, Massicotte-Marquez J, Postuma RB, Doyon J, Panisset M, Montplaisir J (2007) REM sleep behavior disorder predicts cognitive impairment in Parkinson disease without dementia. Neurology 69:1843–1849
56. Iranzo A, Molinuevo JL, Santamaría J, Serradell M, Martí MJ, Valldeoriola F, Tolosa E (2006) Rapid-eye-movement sleep behaviour disorder as an early marker for a neurodegenerative disorder: a descriptive study. Lancet Neurol 5:572–577
57. McKeith IG, Dickson DW, Lowe J, Emre M, O'Brien JT, Feldman H, Cummings J, Duda JE, Lippa C, Perry EK, Aarsland D, Arai H, Ballard CG, Boeve B, Burn DJ, Costa D, Del Ser T, Dubois B, Galasko D, Gauthier S, Goetz CG, Gomez-Tortosa E, Halliday G, Hansen LA, Hardy J, Iwatsubo T, Kalaria RN, Kaufer D, Kenny RA, Korczyn A, Kosaka K, Lee VM, Lees A, Litvan I, Londos E, Lopez OL, Minoshima S, Mizuno Y, Molina JA, Mukaetova-Ladinska EB, Pasquier F, Perry RH, Schulz JB, Trojanowski JQ, Yamada M; Consortium on DLB (2005) Diagnosis and management of dementia with Lewy bodies: third report of the DLB consortium. Neurology 65:1863–1872
58. Sforza E, Krieger J, Petiau C (1997) REM sleep behavior disorder: clinical and physiopathological findings. Sleep Med Rev 1:57–69
59. Weyer A, Minnerop M, Abele M, Klockgether T (2006) REM sleep behavioral disorder in pure autonomic failure (PAF). Neurology 66:608–609
60. Arnulf I, Merino-Andreu M, Bloch F, Konofal E, Vidailhet M, Cochen V, Derenne JP, Agid Y (2005) REM sleep behavior disorder and REM sleep without atonia in patients with progressive supranuclear palsy. Sleep 28:349–354
61. Gagnon J-F, Petit D, Fantini ML, Rompré S, Gauthier S, Panisset M, Robillard A, Montplaisir J (2006) REM sleep behavior disorder and REM sleep without atonia in probable Alzheimer disease. Sleep 29:1321–1325
62. Boeve BF, Lin S-C, Strongosky A, Dickson DW, Wszolek ZK (2006) Absence of rapid eye movement sleep behavior disorder in 11 members of the pallidopontonigral degeneration kindred. Arch Neurol 63:268–272
63. Gagnon J-F, Postuma RB, Mazza S, Doyon J, Montplaisir J (2006) Rapid-eye-movement sleep behaviour disorder and neurodegenerative diseases. Lancet Neurol 5:424–432
64. Dauvilliers Y, Rompré S, Gagnon J-F, Vendette M, Petit D, Montplaisir J (2007) REM sleep characteristics in narcolepsy and REM sleep behavior disorder. Sleep 30:844–849
65. Mayer M-E (1993) Motor dyscontrol in sleep of narcoleptic patients (a lifelong development?). J Sleep Res 2:143–148
66. Nightingale S, Orgill JC, Ebrahim IO, de Lacy SF, Agrawal S, Williams AJ (2005) The association between narcolepsy and REM behavior disorder (RBD). Sleep Med 6:253–258
67. Schenck CH, Mahowald MW (1992) Motor dyscontrol in narcolepsy: rapid-eye-movement (REM) sleep without atonia and REM sleep behavior disorder. Ann Neurol 32:3–10
68. Iranzo A, Graus F, Clover L, Morera J, Bruna J, Vilar C, Martínez-Rodriguez JE, Vincent A, Santamaría J (2006) Rapid eye movement sleep behavior disorder and potassium channel antibody-associated limbic encephalitis. Ann Neurol 59:178–181

69. Cochen V, Arnulf I, Demeret S, Neulat ML, Gourlet V, Drouot X, Moutereau S, Derenne JP, Similowski T, Willer JC, Pierrot-Deseiligny C, Bolgert F (2005) Vivid dreams, hallucinations, psychosis and REM sleep in Guillain-Barré syndrome. Brain 128:2535–2545
70. Tan A, Salgado M, Fahn S (1996) Rapid eye movement sleep behavior disorder preceding Parkinson's disease with therapeutic response to levodopa. Mov Disord 11:214–216
71. Uchiyama M, Isse K, Tanaka K, Yokota N, Hamamoto M, Aida S, Ito Y, Yoshimura M, Okawa M (1995) Incidental Lewy body disease in a patient with REM sleep behavior disorder. Neurology 45:709–712
72. Turner RS, D'Amato CJ, Chervin RD, Blaivas M (2000) The pathology of REM sleep behavior disorder with comorbid Lewy body dementia. Neurology 55:1730–1732
73. Tison F, Wenning GK, Quinn NP, Smith SJ (1995) REM sleep behaviour disorder as the presenting symptom of multiple system atrophy. J Neurol Neurosurg Psychiatry 58:379–380
74. Schenck CH, Mahowald MW, Anderson ML, Silber MH, Boeve BF, Parisi JE (1997) Lewy body variant of Alzheimer's disease (AD) identified by postmortem ubiquitin staining in a previously reported case of AD associated with REM sleep behavior disorder. Biol Psychiatry 42:527–528
75. Wright BA, Rosen JR, Buysse DJ, Reynolds CF, Zubenko GS (1990) Shy-Drager syndrome presenting as a REM behavioral disorder. J Geriatr Psychiatry Neurol 3:110–113
76. Claassen DO, Josephs KA, Ahlskog JE, Silber MH, Tippmann-Peikert M, Boeve BF (2010) REM sleep behavior disorder preceding other aspects of synucleinopathies by up to half a century. Neurology 75:494–499
77. Boeve BF, Dickson DW, Olson EJ, Shepard JW, Silber MH, Ferman TJ, Ahlskog JE, Benarroch EE (2007) Insights into REM sleep behavior disorder pathophysiology in brainstem-predominant Lewy body disease. Sleep Med 8(1):60–64
78. Schenck CH, Bundlie SR, Mahowald MW (1996) Delayed emergence of a parkinsonian disorder in 38% of 29 older men initially diagnosed with idiopathic rapid eye movement sleep behavior disorder. Neurology 46:388–393
79. Schenck CH, Bundlie SR, Mahowald MW (2003) REM sleep behaviour disorder (RBD) delayed emergence of parkinsonism and/or dementia in 65% of older men initially diagnosed with idiopathic RBD, and an analysis of the maximum and minimum tonic and/or phasic electromyographic abnormalities found during REM sleep. Sleep 26(Abstract Suppl):A316
80. Tippman-Peikert M, Olson EJ, Boeve B, Silber MH (2006) Idiopathic REM sleep behavior disorder: a follow-up of 39 patients. Sleep 29:A272
81. Iranzo A, Molinuevo J, Santamaria J, Serradell M, Marti M, Valledeoriola F, Tolosa E (2008) Sixty-four percent of patients with idiopathic REM sleep behavior disorder developed a neurological disorder after a mean clinical follow-up of seven years. Sleep 31:A280
82. Postuma RB, Gagnon JF, Vendette M, Fantini ML, Massicotte-Marquez J, Montplaisir J (2009) Quantifying the risk of neurodegenerative disease in idiopathic REM sleep behavior disorder. Neurology 72:1296–1300
83. Iranzo A, Ratti PL, Casanova-Molla J, Serradell M, Vilaseca I, Santamaria J (2009) Excessive muscle activity increases over time in idiopathic REM sleep behavior disorder. Sleep 32:1149–1153
84. Fantini ML, Farini E, Ortelli P, Zucconi M, Manconi M, Cappa S, Ferini-Strambi L (2011) Longitudinal study of cognitive function in idiopathic REM sleep behavior disorder. Sleep 34:619–625
85. Miyamoto T, Orimo S, Miyamoto M, Hirata K, Adachi T, Hattori R, Suzuki M, Ishii K (2010) Follow-up PET studies in case of idiopathic REM sleep behavior disorder. Sleep Med 11:100–101
86. Iranzo A, Lomeña F, Stockner H, Valldeoriola F, Vilaseca I, Salamero M, Molinuevo JL, Serradell M, Duch J, Pavía J, Gallego J, Seppi K, Högl B, Tolosa E, Poewe W, Santamaria J; Sleep Innsbruck Barcelona (SINBAR) group (2010) Decreased striatal dopamine transporters uptake and substantia nigra hyperechogenicity as risk markers of synucleinopathy in patients with idiopathic rapid-eye-movement sleep behaviour disorder: a prospective study. Lancet Neurol 9:1070–1077

87. Postuma RB, Gagnon JF, Rompré S, Montplaisir JY (2010) Severity of REM atonia loss in idiopathic REM sleep behavior disorder predicts Parkinson disease. Neurology 74:239–244
88. Postuma RB, Gagnon J-F, Vendette M, Desjardins C, Montplaisir JY (2011) Olfaction and color vision identify impending neurodegeneration in rapid eye movement sleep behavior disorder. Ann Neurol 69:811–818
89. Ahlskog JE (2007) Beating a dead horse: dopamine and Parkinson disease. Neurology 69:1701–1711
90. Savica R, Rocca WA, Ahlskog JE (2010) When does Parkinson disease start? Arch Neurol 67:798–801

New Animal Models of Progressive Neurodegeneration: Tools for Developing Predictive Diagnostics and Identifying Presymptomatic Therapeutic Targets

Amber L. Marriott, Daphne A. Gill, Christopher A. Shaw, and R. Andrew Tasker

Abstract Mental and neurological disorders are increasingly prevalent and constitute a major societal and economic burden worldwide. Many of these diseases and disorders are characterized by progressive deterioration over time that ultimately results in identifiable symptoms that in turn dictate therapy. Disease-specific symptoms, however, often occur late in the degenerative process. A better understanding of the presymptomatic events could allow for the development of new diagnostics and earlier interventions that could slow or stop the disease process. Such studies of progressive neurodegeneration require the use of animal models that are characterized by delayed or slowly developing disease phenotype(s). This brief review describes several examples of such animal models that have recently been developed with relevance to epilepsy, schizophrenia, autism spectrum disorders, amyotrophic lateral sclerosis and Parkinson's disease.

1 Introduction

Mental and neurological disorders are increasingly prevalent and constitute a major societal and economic burden worldwide. The impact of these disorders has traditionally been underestimated by epidemiological data that have relied on mortality rather than morbidity. According to the World Health Organization mental and neurological disorders are responsible for only about 1% of deaths but account

A.L. Marriott • D.A. Gill • R.A. Tasker (✉)
Department of Biomedical Sciences, University of Prince Edward Island,
550 University Avenue, Charlottetown C1A4P3, PEI, Canada
e-mail: tasker@upei.ca

C.A. Shaw
Departments of Ophthalmology and Visual Sciences and Experimental Medicine and Graduate Program in Neurosciences, University of British Columbia, Vancouver, BC, Canada

S. Mandel (ed.), *Neurodegenerative Diseases: Integrative PPPM Approach as the Medicine of the Future*, Advances in Predictive, Preventive and Personalised Medicine 2, DOI 10.1007/978-94-007-5866-7_4,

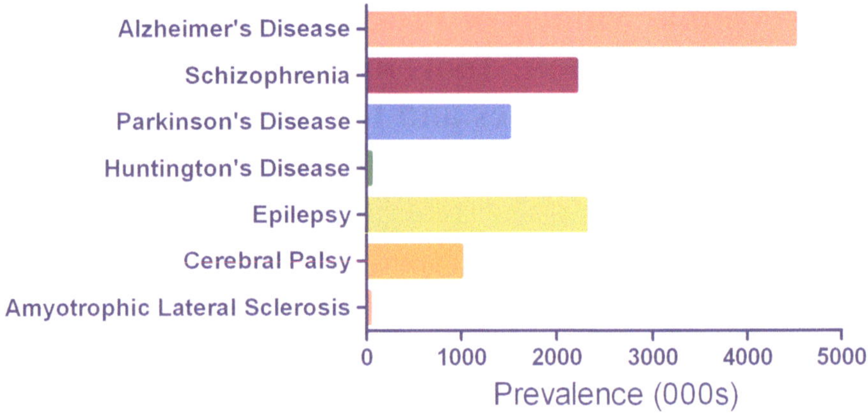

Fig. 1 Prevalence in the USA of several important neurological diseases characterized by progressive neurodegeneration. Numbers are expressed in thousands (000s) of persons affected (data derived from [2])

for almost 11% of the global disease burden [1]. Further, due to increased life expectancy and the aging of general populations in both developed and developing countries this number is expected to rise to 14.7% by 2020 [1]. Many of the most socially and economically devastating neurological diseases and disorders are characterized by progressive neurodegeneration. The prevalence of some of the most common of these diseases in the United States is depicted in Fig. 1. By extrapolation the prevalence worldwide is probably about 20× that of the USA.

Symptoms of many forms of progressive disease often appear in late adolescence or early adulthood and then become increasingly severe with increasing age. The neurodegenerative process however, usually begins long before the onset of clinically diagnosed symptoms. For example, it is estimated that up to 60% of the dopaminergic neurons in the substantia nigra need to be lost before the first clinical signs of Parkinson's disease appear [3]. Thus, in many cases a precipitating event occurring much earlier in life, often around the time of birth or in early childhood, appears to initiate a degenerative cascade that proceeds undetected for some time prior to the onset of clinical signs. Presymptomatic detection of disease represents, therefore, a largely unexplored opportunity for therapeutic intervention. By detecting the disease process earlier, and initiating appropriate therapy to arrest the neurodegenerative process prior to the onset of symptoms, the disease process could be slowed or even stopped long before the patient becomes debilitated by both the primary disease process and secondary complications. This general concept is depicted in Fig. 2. While conceptually simple, such an approach to preventive therapy is by no means easy to implement. Even if presymptomatic biomarkers relevant to individual diseases could be identified, health budgets cannot possibly support sustainable screening of all members of the population, and the incidence of these diseases is too large to permit accurate presymptomatic risk analysis (e.g., Framingham). Further, most neurological diseases are presumed

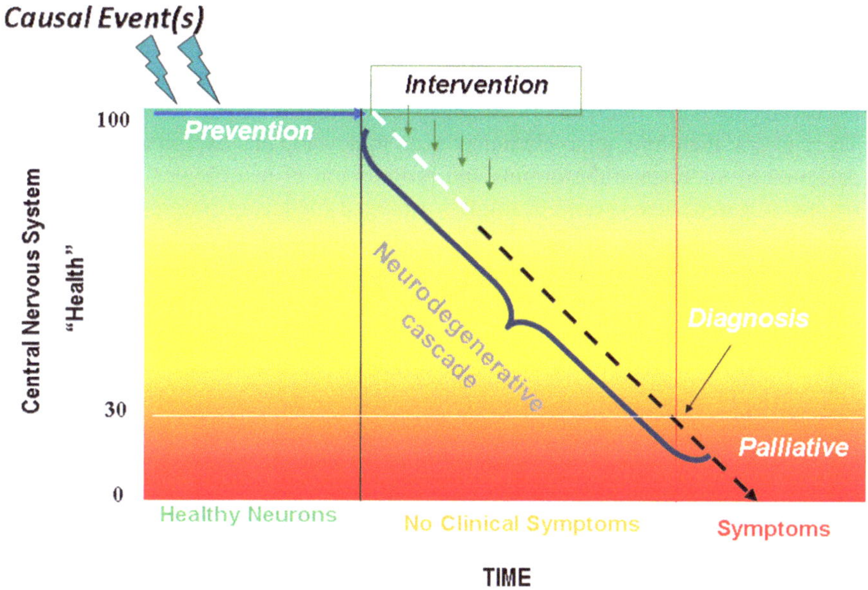

Fig. 2 Representation of the concept of progressive neurodegeneration and presymptomatic therapeutic intervention. Events, often unidentified, initiate a progressive deterioration of brain health that continues for some time prior to the appearance of symptoms associated with a particular neurological disease. Traditionally therapy is initiated at the time of symptom onset and diagnosis, but earlier detection of deteriorating brain health would allow for presymptomatic intervention that could slow or prevent disease progression

to not result from a single precipitating event, but rather arise from a complex interaction between initiating stimuli, genetic predisposition, environmental factors and lifestyle [4]. Nonetheless, the overall goal of identifying disease earlier is both socially and economically responsible and should be vigorously pursued. Genetic predispositions and gene-environment interactions can be identified for individual diseases through careful study, and it is quite conceivable that the early indicators of neurodegeneration may be common or overlap between different diseases. This latter point gives rise to hope, and is supported by the knowledge that many neurodegenerative diseases demonstrate comorbidity. In other words patients with a diagnosis of one disorder often show signs of others. For example schizophrenia patients often suffer from epilepsy and/or autism spectrum disorder (ASD), Parkinson's disease patients often develop dementia, and epilepsy and/or ASD patients often present with cognitive impairment. Thus, while the ultimate diagnosis reflects the predominant symptomatology (which in turn dictates treatment), it is reasonable to assume that disease phenotypes overlap to some extent and may, therefore, have some common origin(s).

Understanding the etiology and initiation of disease often relies on animal models, as does the development of new therapeutic strategies. But while there are a multitude of animal models that mimic specific features of end-stage disease for almost all neurological conditions, most of these models have been created to identify new therapeutants to alleviate symptoms, and hence do not allow study of the development of disease. Understanding disease development is the key to identifying early intervention strategies consistent with the concept in Fig. 2. There exists, therefore, a largely unmet need to develop and exploit animal models designed to understand the origins and progression of neurological disease. This review will introduce several examples of animal models of progressive neurodegenerative disease. Because many diseases are neurodevelopmental in origin, those models that have been developed largely, but not exclusively, involve altering brain development in experimental animals. The descriptions below reflect that bias, and are not intended to be all-encompassing or even to be an exhaustive review of all such models or models of all important neurological diseases. They are intended, rather, simply as examples to illustrate the concept of modeling disease progression as a means to identify predictive and/or preventive therapeutic strategies. Table 1 summarizes the models that are described in greater detail below.

2 Progressive Models of Neurological Disease

2.1 Brain Development

Brain development in mammals is a prolonged process beginning early in gestation and continuing into young adulthood. Abnormal development occurs as a result of changes in cellular processes that normally work in a delicate functional and temporal sequence. Whether these changes arise from physical damage or from neurotoxic insult, the impact of the injury depends upon the organism's stage of growth [5, 6]. Generally, periods of rapid development are more vulnerable to damage and the extent of a given effect depends in large part on exposure timing. These critical periods, or "windows" of development have been divided by Vorhees [5] into four stages (although the final three have some overlap); namely, (1) implantation and all events preceding fertilization, (2) organogenesis, (3) differentiation and specialization of cells, and (4) functional organization, a period that encompasses both late fetal and early postnatal life and includes gene and receptor expression and post-translational modification of proteins that may be either programmed or be activity-dependent. It is this fourth developmental phase that manifests the greatest response to behavioral teratogens and is a prolonged period of vulnerability [5]. In the rat, the period from postnatal day (PND) 0 to PND 21 displays a particularly high rate of differentiation and acceleration of brain growth, with half of the total cell population forming during this time. In addition, processes such as synaptogenesis, developmental cell death (apoptosis), myelination, selective removal of excess synapses, and maturation of all major neurotransmitter systems are on-going at this time [6, 7]. This so-called "brain growth spurt" is, therefore, a period of particular vulnerability to neurological insult, and disruptions during

Table 1 Summary of non-genetic animal models of progressive neurodegeneration

Disease/ Disorder	Species	Age at induction	Treatment/ Manipulation	Primary outcomes	References
Epilepsy	Rat	Neonatal	Inflammation (LPS or Poly I:C)	Reduced seizure threshold	[23–25]
	Rat	Neonatal	Subconvulsive domoic acid	Low grade seizures as adults Neurochemical/ histochemical changes Reduced seizure threshold	[32–35]
Schizophrenia	Rat	Neonatal	Ventral Hippocampal Lesion	Delayed hyperactivity Changes in PPI, LI, memory, social interaction Neurochemical changes	[54–62]
	Rat	Prenatal	Inflammation (Poly I:C)	Hyperactivity Changes in PPI, LI, novel object recognition, social interaction Neurochemical changes	[66–74]
	Rat	Neonatal	NMDA antagonist (MK-801)	Hyperactivity Changes in PPI, memory Histopathological changes	[75, 76]
	Rat	Neonatal	Kainic acid (i.c.v.)	Delayed histopathological changes	[77–80]
	Rat	Prenatal	Subconvulsive domoic acid	Changes in learning and memory	[82, 83]
	Rat	Neonatal	Subconvulsive domoic acid	Changes in response to novelty, PPI, memory, cognition, social interaction	[84–89]
Autism	Rat	Prenatal	Inflammation (Poly I:C or LPS)		[110, 112]
	Rat	Prenatal	Valproic acid	Changes in social interaction, activity, stereotypies, anxiety and PPI Malformations Histopathological changes	[114–121]
ALS-PDC	Mouse	Prenatal/ Neonatal	Steryl glucoside toxins	Motor impairments Histopathological changes	[125]
	Mouse	Adult	Cycad seed flour/steryl glucosides	Progressive motor neuron loss Motor impairments Histopathological changes	[128]

See text for additional details

Abbreviations: *LI* Latent Inhibition, *LPS* lipopolysaccharide, *Poly I:C* polyriboinosinic: polyribocytidilic acid, *PPI* prepulse inhibition this time may be manifested through a variety of behavioral and pathological abnormalities. In humans, the equivalent period occurs roughly between the third trimester of gestational development and the second year of life [8].

2.2 *Neonatal Models of Epileptogenesis*

Epilepsy is a complex, symptomatically-defined cluster of disorders that affects 1–1.5% of the population worldwide [9]. Characterized by recurrent seizures manifesting either as motor convulsions or abnormalities in electroencephalographic (EEG) recordings, epilepsy can be either generalized (affecting the whole brain) or partial (affecting discrete brain regions such as the temporal lobe). Partial epilepsy may or may not progress to generalized seizures [10]. Epilepsy is also frequently neurodevelopmental in origin, whereby a strong correlation exists between early-life CNS trauma (e.g. hypoxia, closed head trauma, febrile seizures) and the subsequent development of seizure disorders such as temporal lobe epilepsy (TLE) later on [11]. While childhood epilepsy is not uncommon, early trauma is often followed by a latent, or "silent", period ranging from years to decades before the onset of epileptic seizures. This delayed onset is a time when the brain is undergoing progressive changes in structure and function that ultimately culminate in seizures, and is referred to as "epileptogenesis" (i.e., the process of becoming epileptic) [10].

The process of epileptogenesis is poorly understood, but because it is presymptomatic, the epileptogenic period represents a window in which therapeutic intervention could slow or even stop the development of epilepsy. Understanding the process of epileptogenesis and identification of potential biomarkers for presymptomatic intervention, however, requires the development of appropriate animal models, and while there are a number of epilepsy models that involve chemical or electrical induction of spontaneous recurrent seizures in adult animals (for review see [11, 12]), modeling epilepsy, and particularly epileptogenesis, in neonatal animals has proven more difficult. This is due, in part, to what superficially appears to be a paradox. It is well established that the immature brain is more prone than the adult brain to seizures arising from chemical or environmental intervention [13–16], probably because of a difference in the developmental maturation of excitatory and inhibitory systems [17]. Despite this increased susceptibility to seizure genesis, however, the neonatal brain has generally been regarded as largely resistant to the development of long-term consequences consistent with epilepsy (e.g., spontaneous recurrent seizures, mossy fiber sprouting, hippocampal cell loss) [18–20], although others have reported long-lasting consequences of neonatal seizures particularly when multiple administrations of convulsants are administered during early development [21, 22].

Of particular relevance to understanding the neurodevelopmental origins of epilepsy and epileptogenesis with an eye on early intervention, however, are two recently described rodent models of epilepsy that produce slowly-developing or delayed onset changes in epilepsy-relevant morphology, neurochemistry and behavior. These are the neonatal inflammation model and the neonatal domoic acid model; each of which is described in more detail below.

2.3 Neonatal Inflammation

Recently an interesting model linking neonatal inflammation and seizure susceptibility has been described [23–25]. Sprague–Dawley rat pups injected intraperitoneally with the bacterial endotoxin lipopolysaccharide on PND14 demonstrate enhanced susceptibility to seizures induced by lithium-pilocarpine, kainic acid (KA) or the GABA antagonist pentylenetetrazol when challenged at 6–8 weeks of age. The effect is antagonized by concurrent administration of an antibody to tumor necrosis factor α (TNFα). The authors also investigated whether the same result was obtained following LPS injection at other stages of development, and reported that injections on PND 7 also reduced seizure threshold, but that LPS injections on PND 1 or PND 20 did not; strongly implicating the second postnatal week of life as a "critical window" for the lasting effects. A follow-up paper by the same group described how direct intracerebroventricular (i.c.v.) injections of the viral mimetic polyinosinic:polycytidylic acid into PND 14 rat pups produces central inflammation and also results in a reduced seizure threshold when tested at 7–8 weeks of age [24]. Thus both peripheral and central inflammation during the second postnatal week of development in the rat appears to initiate a chronic neurodegenerative condition that culminates in reduced seizure threshold.

2.3.1 Neonatal Domoic Acid

The other neurodevelopmental model of epileptogenesis that has been described is the neonatal domoic acid model. Domoic acid is a naturally occurring excitotoxin that is structurally similar to KA and an analogue of glutamate, and is recognized as a selective, but not specific, agonist at the kainate subclass of non-NMDA glutamate receptors [26, 27]. In the late 1980s domoic acid was identified as the causative agent in an outbreak of human toxicity in which patients experienced dose-dependent neurotoxicity culminating in seizures and death [28, 29]. One of the patients who survived subsequently went on to develop temporal lobe epilepsy [30]. Building on this background of clinical data, Doucette and co-workers investigated the response of neonatal rats to very low doses of domoic acid [16, 31] and in 2004 they described a unique low-grade seizure response that occurred in adult rats exposed to novel and/or stressful environments even though the only drug treatment had been as neonates [32]. In addition to manifesting seizures in adulthood, post-mortem analysis of these animals at 15 months of age revealed many hallmark features of clinical TLE including hippocampal cell loss, mossy fiber sprouting (MFS) in the dentate gyrus and area CA3, and regionally-selective

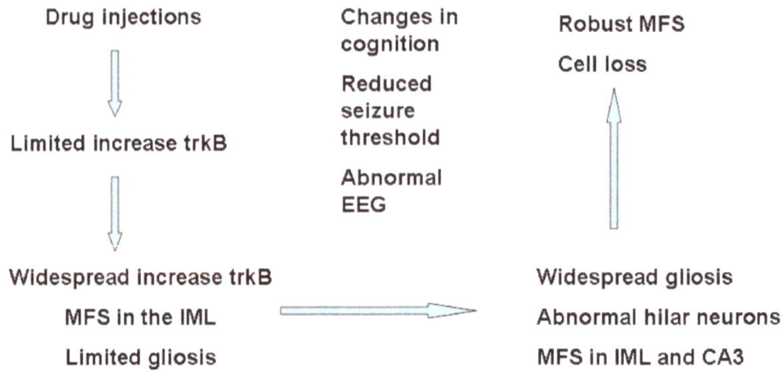

Fig. 3 Preliminary schematic depiction of identified changes over time in a neurodevelopmental rat model of temporal lobe epilepsy. Low doses of domoic acid during the second postnatal week lead to a progressive series of changes in hippocampal neurotrophin signaling, astrogliosis, sprouting of dentate granule cell axons and cell loss, that are accompanied by changes in cognition, seizure threshold and EEG although the precise time of onset of each is currently unknown. Changes depicted are correlative only and not necessarily causally related (data derived from [32–36])

elevations of the neurotrophin BDNF [32]. Of particular interest in the context of the current review, however, is that these changes did not occur at the time of drug administration (unpublished) but were progressive in nature whereby younger mature animals (PND75) displayed less severe MFS and no loss of hippocampal cells [33] and adolescent rats (PND 29) showed no hippocampal pathology with the exception of mild astrogliosis [34]. Further, confirming the relevance of this model to understanding epilepsy and epileptogenesis are two reports by Gill et al. [35, 36] demonstrating alterations in electroencephalographically (EEG) recorded sleep patterns and reductions in seizure threshold that accompany selective loss of inhibitory neurons in the hippocampus. A general overview of the current understanding of hippocampal epileptogenesis in this model is shown in Fig. 3.

2.4 *Neonatal Models of Schizophrenia and Related Disorders*

Schizophrenia is a complex and devastating mental disorder affecting approximately 1% of the general population. While the term schizophrenia was first introduced by Eugen Bleuler in 1911, it is believed that the disorder has been present in all cultures for centuries [37]. Characterized by impairments in the perception and expression of reality, the symptoms of schizophrenia are most often categorized into positive, negative and more recently, cognitive symptoms [38]. Schizophrenia is debilitating and results in great emotional cost to the person afflicted, as well as to friends and relatives. It also results in great financial burden to the worldwide economy with the estimated cost to the US alone being $62.7 billion in 2002 [39]. While the outlook for those affected has improved somewhat, a diagnosis of schizophrenia is still likely to result in substantial disability and may reduce life expectancy due to suicide [40–42].

Although schizophrenia has been present for centuries and progress has been made through research, much about the disorder is still unknown. This is probably due to the complexity of the proximal cause(s) of the disorder and the apparent delay of the onset of definitive symptoms. Although the cause(s) of schizophrenia is/are not yet known, a prominent hypothesis is that it is neurodevelopmental in origin and arises due to a combination of environmental factors and genetic susceptibility. According to this theory, factors which occur long before symptoms are recognized and diagnosed (i.e. during gestational or early developmental stages), disrupt normal brain development, resulting in severe and long-lasting changes in the functional integrity of the CNS [43]. Early life events that have been implicated in conferring a higher-than-average risk of developing schizophrenia include maternal illness during gestation [44], obstetric complications [45, 46] and toxin exposure [47]. Such events can lead to subtle alterations in CNS function, which may result in an increased vulnerability to environmental triggers later in life [48]. However, while environmental factors likely cause an individual to be more susceptible to the illness or play a role in triggering the onset of symptoms, specific genetic vulnerabilities also appear to contribute to the manifestation of schizophrenia. Although the prevalence of schizophrenia may only be up to 1% in the general population, the incidence of the disorder is much higher in those who have a schizophrenic relative. Additionally, though the disorder can occur in people who have no family history of its presence, the risk increases dramatically with the degree of genetic closeness to a person who is afflicted [37].

A variety of genes have been associated with schizophrenia including neuregulin (NRG1) [49], dysbindin (DTNBP1) [50], Catechol-O-methyl transferase (COMT) [51] and DISC1 [52], and consequently a number of genetically-based animal models exist for studying schizophrenia. A thorough discussion of these models and their use in schizophrenia research is beyond the scope of this review, but has been discussed in detail elsewhere (for review see [53]).

Of more relevance to the current discussion, however, are several well characterized and/or emerging rodent models of schizophrenia that involve experimental manipulation in the perinatal period. Initially asymptomatic, these changes result in schizophrenia-like symptomatology that develops over time. Such slowly progressing animal models provide good platforms for both understanding disease development and identifying potential early markers of disease progression.

2.4.1 Lesion Models

First proposed by Lipska and co-workers in 1993 [54], the neonatal ventral hippocampal (NVH) lesion model is one of the most widely studied animal models of schizophrenia. By using ibotenic acid (a glutamate agonist) to lesion the ventral hippocampus of neonatal rats on PND 7 of development, many aspects of the disorder have been demonstrated. Compared to controls, rats with lesions demonstrated a higher degree of activity in response to a novel environment and an amphetamine challenge in a novel environment when tested in early adulthood (PND 56), but not at earlier time points (PND 35) [54]. Deficits have also been observed in other behavioral measures including prepulse inhibition (PPI) [55], latent inhibition (LI) [56], measures of working memory [57], set shifting [58] and social interaction [59]. Lesioned animals have also shown increased alcohol seeking and consumption behavior [60].

Neurobiological factors such as an increase in hippocampal dopamine (DA) have also been found in this model [61] and a number of studies have indicated that antipsychotic drugs normalize some of the lesion-induced behaviors [54, 62].

In addition to the NVH model, other lesion based models of schizophrenia have been described, including the neonatal amygdalar lesion model [63] where ibotenic acid is used to lesion the amygdala on PND 7, and models in which ibotenic acid is used to lesion the prefrontal cortex in neonatal or adult animals [64].

2.4.2 Immunological Models

Prenatal exposure to infection has also been implicated in an increased risk of developing schizophrenia in adulthood [65]. Presently it is believed that the maternal immune response, particularly the pro-inflammatory cytokines which are released in such situations, can interfere with the normal development of the CNS. As such, immuno-precipitated models seek to imitate this potential risk factor [66]. Seeking to elicit the maternal immune response by mimicking viral exposure, Zuckerman and Weiner in 2003 created such a model by injecting pregnant rats with polyriboinosinic-polyribocytidilic acid (polyI:C). This procedure resulted in adult animals displaying a variety of schizophrenia-like anomalies including deficits in PPI [67, 68] and LI [69], as well as increased sensitivity to the locomotor effects of amphetamine [70], impairments in novel object recognition [68] and reduced social interaction [71]. Neuroanatomical and neurochemical abnormalities have also been observed including alterations to hippocampal morphology, increases in striatal DA release [70] and reduced dopamine D1 and D2 receptors in the prefrontal cortex and increased tyrosine hydroxylase in the striatum [72–74].

2.4.3 Pharmacological Models

Another animal model of schizophrenia proposed by Harris et al. [75], used the administration of a glutamate receptor antagonist to neonatal rats, to produce a schizophrenia-like phenotype in adulthood. The procedure involved the subcutaneous (s.c.) injection of the NMDA receptor antagonist MK-801 (2×0.5 mg/kg) to Sprague–Dawley rat pups on PND 7. As adults, the animals displayed reduced volume and neuronal numbers in the hippocampus, and female rats displayed PPI deficits and increased locomotor activity in the open field [75].

In another study, Kawabe et al. [76] administered lower doses of MK-801 twice daily to neonatal rats from PND 7–20. It was found that when they reached adulthood, these animals displayed significant deficits in radial arm maze (RAM) learning. Interestingly, when adult animals who had not received the neonatal MK-801 treatment neonatally were given the same treatment in adulthood, no changes to RAM performance were noted [76].

In addition to the aforementioned work on NMDA receptors, neonatal manipulation of non-NMDA receptors has also been shown to result in schizophrenia-like symptoms that develop as animals age. In a series of papers published over the last 10 years, the Csernansky research group at Washington University in St. Louis has described how i.c.v. injection of a sub-convulsive dose of kainic acid on PND 7 induces delayed histopathology consistent with schizophrenia. Looking at time points ranging from 8 to 75 days of age, these authors have described progressive loss of neurons in the CA1 and CA3 hippocampal subfields [77, 78]. Interestingly, at a low dose of KA this cell loss did not begin until around PND40 (i.e. mimicking the delayed onset of schizophrenia) and appeared to be due to apoptosis rather than the necrosis typically associated with KA-induced excitotoxicity [78]. Subsequent papers by the same group have also reported changes in hippocampal interneuron populations [79] and hippocampal neurogenesis [80] in this model.

Closely related to these findings are the recent data emerging from two independent groups that have administered the AMPA/kainate agonist domoic acid (DOM) during perinatal development. DOM is a naturally occurring environmental toxin produced by various marine organisms such as red algae [81]. A 2005 study by Levin and co-workers [82] revealed that a single prenatal exposure to DOM during midgestational development (with doses ranging from 0.3 to 1.2 mg/kg) produced dose related changes which included an increase in response latency in a spontaneous alternation task in the T-maze, as well as an increase in initial activity and a larger habituation score in the Figure-8 maze. Additionally, a later study by the same group [83] considered the effect of administering DOM (0.025–0.1 mg/kg) to rat pups twice per day, on PND 1 and 2. While the higher doses proved to be lethal, the authors did report some effects at lower doses, but these were less frequently observed than those seen following DOM exposure during mid-gestation.

In a related series of experimentation Doucette et al. [31] demonstrated that when tested during the neonatal period, administration of very low doses of DOM to Sprague–Dawley rat pups did not alter a number of developmental milestones but did alter conditioned place preference in an olfactory conditioning paradigm. The authors interpreted these findings as confirmation that this low dose of DOM (20 μg/kg) was centrally active but not overtly toxic. The same group then went on to administer low-dose DOM throughout the second postnatal week (PND 8–14) and reported a number of schizophrenia-like changes in behavior when the animals were tested as adults. Treated animals were shown to have abnormal responses to novel spatial environments [32] as well as impairments in prepulse inhibition [84], spatial cognition [85] and working memory [86]. Alterations in emotionality (elevated plus maze) as well as changes in response to novelty and increased sensitivity to nicotine have also been found in this paradigm [86–88]; as have increased sensitivity to reward (in a choice task) and psychomotor agitation (in a novelty conditioning paradigm) and social withdrawal [89] .

Thus, there are a number of different rodent models that utilize changes in perinatal brain development resulting in slowly progressing deterioration of central function relevant to schizophrenia.

2.5 Prenatal/Neonatal Models of Autism Spectrum Disorder

With symptoms present prior to age 3, ASD is an early-onset collection of heterogeneous neurodevelopmental disorders characterized by stereotypic behaviors and impairments in social and communicative skills. Affecting 1 in 200 people, autism is also very gender biased, occurring four times more often in males than in females [90]. It is also highly heritable, although it is now known that environmental factors, particularly during critical developmental periods, play a key role as well [91]. Due to its early onset, its persistence throughout the lifespan, and its impact on cognitive function, ASD exerts a major financial impact on society, with an estimated total annual cost of approximately $35 billion dollars in the United States alone [92].

Although historically linked to schizophrenia (the term "autism" was first used by Bleuler in 1912 to describe socially withdrawn schizophrenia patients), these two disorders have occupied distinct diagnostic categories in all versions of the Diagnostic and Statistical Manual of Mental Disorders (DSM) since 1974. This decision is primarily based on distinct differences in age of onset, with ASD symptoms reliably occurring before 3 years, while schizophrenia is typically diagnosed between the late teens and mid 30s, and is very rarely noted prior to adolescence (never prior to age 3). Further, motor stereotypies are a prevalent feature in ASD, while delusions and other signs of disordered thinking (positive symptoms) are the hallmark of schizophrenia [93]. However, while positive symptoms are absent in ASD, there are still many symptomatic overlaps found in adulthood, most notably the negative symptoms characteristic of disorganized type schizophrenia [94]. Additionally, individuals with ASD have been shown to experience difficulties in sensorimotor gating as measured through PPI testing [95]. From the perspective of animal modeling, similarities such as these hold the most promise for the elucidation of a common underlying mechanism(s), and thus enhance the potential for developing early intervention strategies.

In this respect, ASD may be a particularly valuable developmental disorder to model, as it is highly comorbid with many other neurological/neuropsychiatric disorders, including motor impairments [96, 97], mood (affective) disorders (e.g. depression, mania [98–100]), anxiety disorders (e.g. generalized anxiety disorder, obsessive compulsive disorder [98, 99, 101]), attention-deficit/hyperactivity disorder (ADHD) [98, 99, 102], and a variety of conduct disorders (e.g. impulsivity, aggression, oppositional defiant disorder [98, 103]). Indeed, in some studies, as many as 70% of children with ASD have been diagnosed with at least one comorbid disorder, and 41% with two or more [98]. Interestingly, comorbidity between ASD and epilepsy appears to be particularly prevalent, with more than 30% of children with epilepsy also diagnosed with ASD [104] and 30% of individuals with ASD subsequently diagnosed with epilepsy [105]. Children with both ASD and epilepsy are also far more likely to experience developmental delays and cognitive impairments [106, 107]. This consistently high comorbidity rate has prompted a flurry of studies aimed at identifying potentially shared cellular, molecular, and genetic mechanisms in these two disorders (for review, see [108]).

2.5.1 Animal Models

In spite of the number of comorbid disorders associated with ASD, or perhaps because of them, animal models of autism are scarce, and those that do exist often model only one of the three core behavioral symptoms. Development of a definitive ASD model is further hampered by difficulties in assessing social interactions and communicative abilities in animals that provide any kind of valid representation of the human condition. Currently, there are no specific pharmacological therapies available for direct treatment of ASD, although there are a variety of interventions that target its comorbid symptoms (for review, see [109, 110]).

Animal models of ASD fall into two broad categories: those produced through genetic manipulations, and those which represent environmental influences. Genes that have been associated with ASD include SH3 and multiple ankyrin (ANK) repeat domain *(SHANK3),* fragile X mental retardation 1 (*FMR1*), and contractin-associated protein-like 2 (*CNTNAP2*). However, most genes implicated in ASD also confer a risk for many other disorders as well (for review, see [111]), and to date, no specific ASD gene has been identified.

Environmental models include those involving maternal infection with either polyI:C (with intervention at a different gestational age than that used to model schizophrenia) [112] or lipopolysaccharide (for review, see [110]). In addition, exposing rats to teratogens such as valproic acid, ethanol, thalidomide, or misoprostol during gestation results in offspring that display various ASD-like features (for review, see [113]). Of these, the VPA model is the one most commonly utilized, and dependent upon dose and time of administration, has been shown to produce behavioral alterations that include impairments in social interactions [114–117], hyperactivity [114, 116], increased repetitive/stereotypic activity [115, 116, 118], heightened anxiety levels [115, 118, 119], and decreased PPI [116]. Malformations, particularly those involving the face, are common with this treatment, as well as a number of neuronal alterations [120, 121]. However, models like the VPA rat represent only a small proportion of the diversity seen in the human condition, and much more work is necessary. Nevertheless, these models have definitively proved that investigation of susceptibility genes alone is not enough. As an extraordinarily complex and multifaceted disorder, ASD research will require the production of carefully designed developmental animal models that can reflect its extensive diversity.

2.6 *Models of Movement Disorders*

2.6.1 Prenatal/Neonatal Models

Most *in vivo* studies of ALS and Parkinson's disease have examined associated movement disorders primarily in adult animals. There are good theoretical reasons and practical reasons for this emphasis. First, most cases of both diseases in humans are of the late onset form and are usually sporadic in nature. Gene-based models in fact have not been able to duplicate early onset human neurological diseases. For example, the most common ALS models of the familial (genetic) forms of the disease, mutant superoxide dismutase (mSOD) and the alsin protein deletion associated with juvenile ALS 2, do not show clear behavioral or pathological outcomes until well into postnatal life, if at all. For mSOD, the rapidly progressive G93A mutation only begins so show pathological features at about 50 days postnatal life [122]. The G37R variant takes up to a year to demonstrate behavioral or pathological deficits [123]. ALS2/alsin –/– mice show almost no disease phenotype at all at any age [124]. Similarly, Parkinson's genetic models show few early striking behavioral outcomes or neuronal pathologies [4].

Studies of environmental factors typically use adult animals (that is post sexual maturity and older) for a variety of factors such as ease of handling and testing and toxin administration. These factors may be one reason why no good age-dependence animal studies exist for either disease. In spite of these difficulties, for both ALS and Parkinson's disease, there have been some attempts to probe the impact of some of the toxins studied in adulthood in prenatal mice. A recent series of experiments by Shaw and Banjo [125] began to examine the neonatal and postnatal outcomes of steryl glucoside neurotoxins administered to pregnant dams during the period of greatest differentiation of the nigro-striatal system. This study also sought to determine if the impact of a second toxin exposure in adulthood was influence by the prenatal one. Finally, the experiments separated male from female mice to determine the role that sex might play.

Prenatal exposure to steryl glucoside toxins (related to those previously implicated in ALS-PDC [126, 127] and known to cause neurodegeneration following adult exposure [128]) showed that mice exposed *in utero* had greater body weights that persisted throughout life and manifested behavioral deficits associated with damage in the nigro-striatal system. Male mice were more profoundly impacted than females. A second adult toxin exposure had more negative impacts on the nigro-striatal system than in the spinal cord.

These data mirror the earlier findings of Barlow et al. [129] which revealed that exposure of pregnant dams to the pesticides maneb and paraquat cause damage to the nigro-striatal system, that the effects were sex specific as above, and that an adult toxin exposure was more likely to trigger a Parkinson's phenotype.

The results of these experiments reaffirm the notion that prenatal and likely neonatal life are potentially crucial periods that might set the conditions for later neurodegenerative disorders.

2.6.2 Adult Models

There are a variety of adult models of ALS and Parkinson's disease, based generally on genetic or environmental models.

For ALS, the primary genetic models are based on various mutations of SOD that produce, with different time courses, many of the features of ALS, including progressive motor deficits and the loss of spinal cord motor neurons. Alsin mutations have not produced significant ALS phenotypes to date. The mSOD model has been used with great success in isolating the putative stages of neurodegeneration which is now believed to have a partial origin in the loss of neuromuscular junction endplates in mutant animals [130].

Environmental links to ALS have arisen from a variety of sources, with the most well studied cases related to the cluster of ALS found on Guam and elsewhere in the Western Pacific during the 1950s and 1960s. This form of ALS, linked to the sometimes co-expression of a form of parkinsonism (parkinsonism dementia complex), forms a spectrum of neurological disorders termed ALS-PDC [131]. The strongest epidemiological correlate for ALS-PDC was the consumption of a local food product, the seeds of the cycad palm [132, 133]. Cycad seed has been shown to contain a variety of neurotoxic compounds, some of them water insoluble. Mice fed washed cycad seed flour exhibit a progressive form of motor neuron loss accompanied by motor behavioral deficits, later followed by gait disorders associated with pathological outcomes in the nigro-striatal system, the latter including the loss of tyrosine hydroxylase and apoptosis in both the striatum and substantia nigra pars compacta. Increases in reactive astrocytes and microglia were components of both areas. In marked contrast, rats fed cycad seed flour showed profound parkinsonism outcomes of gait disturbances that initially presented unilaterally, later freezing, and highly significant losses of neurons in substantia nigra. Treated rats did not show any losses of motor neurons in the spinal cord [134]. Table 2 summarizes the human and animal data.

Mice fed the putative neurotoxic factor in cycad, variant steryl glucosides, showed many of the same features as those fed cycad [128]. Preliminary studies using steryl glucosides in rats suggests that they exhibit only parkinsonism phenotypes (unpublished).

Attempts have been made to duplicate ALS-PDC in its various forms in monkeys [135] using BMAA and earlier by Dastur and colleagues using washed and unwashed cycad seed flour [136]. The Spencer results have not been repeated and the original data could not be duplicated in mice fed the free plant amino acid, beta-methyl amino alanine (BMAA) [137, 138].

In regard to Parkinson's disease, a great variety of studies have focused on either genetic links to the disease including mutations in α-synuclein and others (for review see [139]) or to various toxins linked to some forms of parkinsonism. These models rarely produce full blown Parkinson's disease phenotypes. The latter have included toxins such as MPTP, linked originally to rapid parkinsonism onset following the use of a synthetic opioid, 6 hydroxydopamine, and various pesticides as well as cycad and steryl glucosides as cited above.

Table 2 Comparison of features of ALS-PDC of Guam with animal models of the disorder

Disease outcomes	ALS-parkinsonism dementia complex	Rodent models of ALS-PDC (cycad or steryl glucoside)
Behavioral changes motor function	Expresses as ALS, PD, or both for motor function losses	Mice: ALS-phenotype is dominant; Rats: parkinsonism only
Cognitive function	Dementia accompanies parkinsonism	Cognitive decline on various memory tests both models
Olfactory function	ALS-PDC shows early olfactory disturbance	Mice fail to discriminate odours
Motor neuron loss	Yes, all levels of motor system	Mice, yes; Rats, no
Nigro-striatal neuron loss	PDC	Mice: significant but not dominant feature; Rats: significant losses of cells in SNpc and striatum
Loss of neurons in hippocampus/cortex	Features of PDC	Mice at late stages
Expression of hyperphosphorylated tau	Both ALS and PDC	Mice at late stages in lumbar spinal cord motor neurons with steryl glucoside feeding
A-synuclein expression	Approx. 20% PDC cases	Rats with cycad
TDP-43 translocation	Noted	Mice at late stages in lumbar cord motor neurons with steryl glucoside feeding
Dopamine loss	PDC	Mice and rats
D2 receptor upregulation; DAT down regulation	unk	Mice; Rats unk.
Increases in reactive astrocytes	ALS-PDC	Both mice and rats in various affected regions of CNS
Increases in microglial proliferation	ALS-PDC	Both mice and rats in various affected regions of the CNS

The various features of the Guamanian spectrum of disorders are compared to two rodent models fed flour made from seeds of the local cycad palm, *Cycas micronesica.* Cycad consumption is still considered to be the strongest epidemiological link to ALS-PDC. Both model systems have also examined dietary exposure to isolated cycad neurotoxins, notably two variant steryl glucoside, B-sitosterol B-D glucoside (BSSG) and stigmasterol glucoside (SG). Combined, the two models faithfully reproduce the salient behavioral and pathological features of the ALS-PDC spectrum. Data derived from [126–136]

One of the more interesting animal models of Parkinson's disease is based on the cluster on the island of Guadeloupe linked to annonacin neurotoxins contained in the fruit of the soursop (summarized in [4]). Here, as in ALS-PDC, an environmental neurotoxin epidemiologically linked to a human neurodegenerative disease condition can duplicate many of the disease phenotypes in *in vivo* models.

3 Conclusions

Animal models that allow examination of the developmental processes underlying neurodegenerative disease represent an important opportunity to develop therapies directed at prevention or amelioration of the disease phenotype. As stated at the beginning of this review, the task of identifying presymptomatic markers of disease in human populations and then implementing appropriate screening and treatment options is daunting, but potentially of enormous value. Further development and analysis of slowly progressing animal models will allow for better estimation of the feasibility of this approach; and if factors common to multiple diseases can be discovered, the value and practicality of presymptomatic intervention will be greatly enhanced. While seeming fanciful to some, some of the examples described above give credence to this possibility. Both neonatal inflammation and neonatal glutamatergic stimulation have been used to model both epilepsy and schizophrenia; two conditions known to exhibit comorbidity but currently treated with widely differing classes of drugs. Administration of steryl glucosides results in symptoms common to both ALS and parkinsonism with the relative proportion of phenotype influenced by the species of choice (mice versus rats). Continued study and development through the use of animal models of disease progression appears to be a viable option for advancing the agenda of predicting and preventing, rather than simply treating, some of society's most devastating neurological diseases.

Acknowledgements Amber Marriott is supported by a Canada Graduate Scholarship awarded through the Natural Sciences and Engineering Research Council of Canada and Daphne Gill is supported by a Post-doctoral fellowship from Innovation PEI. Research funding to the investigators was provided by an NINDS RO1 grant to CAS and an Atlantic Innovation Fund grant to RAT.

References

1. World Health Organization (2006) Neurology and public health. http://www.who.int/mental_health/neurology/en. Accessed 15 Mar 2012
2. OHSU Brain Institute (2010) Disease statistics. Oregon Health & Science University. http://www.ohsu.edu/xd/health/services/brain/in-community/brain-awareness/brain-health/disease-statistics.cfm?WT_rank=1. Accessed 15 Mar 2012
3. Schulz JB, Falkenburger BH (2004) Neuronal pathology in Parkinson's disease. Cell Tissue Res 318:135–137
4. Shaw CA, Hoglinger GU (2008) Neurodegenerative diseases: neurotoxins as sufficient etiological agents? Neuromol Med 10:1–9
5. Vorhees CV (1986) Principles of behavioural teratology. In: Riley EP, Vorhees CV (eds) Handbook of behavioural teratology. Plenum Press, New York
6. Kaufmann W (2000) Developmental neurotoxicity. In: Krinkle GJ (ed) The laboratory rat. Academic, London
7. Ben-Ari Y (2001) Developing networks play a similar melody. Trends Neurosci 24:353–360
8. Dobbing J, Smart JL (1974) Vulnerability of developing brain and behaviour. Br Med Bull 30:164–168

9. Murray CJL, Lopez AD (eds) (1994) Global comparative assessment in the health sector: disease burden, expenditures, and intervention packages. World Health Organization, Geneva
10. Engel J Jr, Schwartzkroin PA (2006) What should be modeled? In: Pitkanen A, Schwartzkroin PA, Moshe SL (eds) Models of seizures and epilepsy. Academic, London
11. Jefferys JGR (2003) Models and mechanisms of experimental epilepsies. Epilepsia 44:44–50
12. Pitkanen A, Schwartzkroin PA, Moshe SL (eds) (2006) Models of seizures and epilepsy. Academic Press, London
13. Cavalheiro EA, Silva DF, Turski WA, Calderazzo-Filho LS, Bortolotto ZA, Turski L (1987) The susceptibility of rats to pilocarpine-induced seizures is age-dependent. Dev Brain Res 37:43–58
14. Stafstrom CE, Thompson JL, Holmes GL (1992) Kainic acid seizures in the developing brain: status epilepticus and spontaneous recurrent seizures. Brain Res Dev Brain Res 65:227–236
15. Stafstrom CE, Chronopoulos A, Thurber S, Thompson JL, Holmes GL (1993) Age-dependent cognitive and behavioral deficits after kainic acid-induced seizures. Epilepsia 34:420–432
16. Doucette TA, Strain SM, Allen GV, Ryan CL, Tasker RA (2000) Comparative behavioural toxicity of domoic acid and kainic acid in neonatal rats. Neurotox Teratol 22:863–869
17. Ben-Ari Y, Khazipov R, Leinekugel X, Caillard O, Gaiarsa JL (1997) GABA-A, NMDA and AMPA receptors: a developmentally regulated 'ménage a trois'. Trends Neurosci 20:523–529
18. Nitecka L, Tremblay E, Charton G, Bouillot JP, Berger ML, Ben-Ari Y (1984) Maturation of kainic acid seizure-brain damage syndrome in the rat. II. Histopathological sequelae. Neuroscience 13:1073–1094
19. Sperber EF, Haas KZ, Stanton PK, Moshé SL (1991) Resistance of the immature hippocampus to seizure-induced synaptic reorganization. Brain Res Dev Brain Res 60:88–93
20. Haas KZ, Sperber EF, Opanashuk LA, Stanton PK, Moshé SL (2001) Resistance of immature hippocampus to morphologic and physiologic alterations following status epilepticus or kindling. Hippocampus 11:615–625
21. Holmes GL, Sarkisian M, Ben-Ari Y, Chevassus-Au-Louis N (1999) Mossy fiber sprouting after recurrent seizures during early development in rats. J Comp Neurol 22:537–553
22. Liu Z, Yang Y, Silveira DC, Sarkisian MR, Tandon P, Huang LT, Stafstrom CE, Holmes GL (1999) Consequences of recurrent seizures during early brain development. Neuroscience 92:1443–1454
23. Galic MA, Riazi K, Heida JG, Mouihate A, Fournier NM, Spencer SJ, Kalynchuk LE, Teskey GC, Pittman QJ (2008) Postnatal inflammation increases seizure susceptibility in adult rats. J Neurosci 28:6904–6913
24. Galic MA, Riazi K, Henderson AK, Tsutsui S, Pittman QJ (2009) Viral-like brain inflammation during development causes increased seizure susceptibility in adult rats. Neurobiol Dis 36:343–351
25. Riazi K, Galic MA, Pittman QJ (2010) Contributions of peripheral inflammation to seizure susceptibility: cytokines and brain excitability. Epilepsy Res 89:34–42
26. Verdoorn TA, Johansen TH, Drejer J, Nielsen EO (1994) Selective block of recombinant GluR6 receptors by NS-102; a novel non-NMDA receptor antagonist. Eur J Pharmacol 269:43–49
27. Tasker RA, Strain SM, Drejer J (1996) Selective reduction in domoic acid toxicity in vivo by a novel non-N-methyl-D-aspartate receptor antagonist. Can J Physiol Pharmacol 74:1047–1054
28. Perl TM, Bedard L, Kosatsky T, Hockin JC, Todd EC, Remis RS (1990) An outbreak of toxic encephalopathy caused by eating mussels contaminated with domoic acid. N Engl J Med 322:1775–1780
29. Teitelbaum J, Zatorre RJ, Carpenter S, Gendron D, Cashman NR (1990) Neurologic sequelae of domoic acid intoxication due to ingestion of contaminated mussels. N Engl J Med 322:1781–1787
30. Cendes F, Andermann F, Carpenter S, Zatorre RJ, Cashman NR (1995) Temporal lobe epilepsy caused by domoic acid intoxication: evidence for glutamate receptor-mediated excitotoxicity in humans. Ann Neurol 37:123–126

31. Doucette TA, Bernard PB, Yuill PC, Tasker RA, Ryan CL (2003) Low doses of non-NMDA glutamate receptor agonists alter neurobehavioural development in the rat. Neurotox Teratol 25:473–479
32. Doucette TA, Bernard PB, Husum H, Perry MA, Ryan CL, Tasker RA (2004) Low doses of domoic acid during postnatal development produce permanent changes in rat behaviour and hippocampal morphology. Neurotox Res 6:555–563
33. Bernard PB, Macdonald DS, Gill DA, Ryan CL, Tasker RA (2007) Hippocampal mossy fiber sprouting and elevated trkB receptor expression following systemic administration of low dose domoic acid during neonatal development. Hippocampus 17:1121–1133
34. MacDonald DS, Bernard PB, Ramsay Gill DA, Tasker RA (2008) Progressive changes in hippocampal cytoarchitecture in a non-convulsive developmental model of TLE. Program No. 449.8 Neuroscience Meeting Planner. Washington, DC: Society for Neuroscience
35. Gill DA, Bastlund JF, Anderson NJ, Tasker RA (2009) Reductions in paradoxical sleep time in adult rats treated neonatally with low dose domoic acid. Behav Brain Res 205:564–569
36. Gill DA, Ramsay SL, Tasker RA (2010) Selective reductions in subpopulations of GABAergic neurons in a developmental rat model of epilepsy. Brain Res 1331:114–123
37. Tamminga CA, Holcomb HH (2005) Phenotype of schizophrenia: a review and formulation. Mol Psychiatry 10:27–39
38. Tandon R, Nasrallah HA, Keshavan MS (2009) Schizophrenia "just the facts" 4: clinical features and conceptualization. Schizophr Res 110:1–23
39. Wu EQ, Birnbaum HG, Shi L, Ball DE, Kessler RC, Moulis M, Aggarwal J (2005) The economic burden of schizophrenia in the United States in 2002. J Clin Psychiatry 66:1122–1129
40. Caldwell CB, Gottesman II (1990) Schizophrenics kill themselves too: a review of risk factors for suicide. Schizophr Bull 16:571–589
41. Radomsky ED, Haas GL, Mann JJ, Sweeney JA (1999) Suicidal behavior in patients with schizophrenia and other psychotic disorders. Am J Psychiatry 156:1590–1595
42. Valevski A, Zalsman G, Tsafrir S, Lipschitz-Elhawi R, Weizman A, Shohat T (2012) Rate of readmission and mortality risks of schizophrenia patients who were discharged against medical advice. Eur Psychiatry 27:496–499
43. Rapoport JL, Addington AM, Frangou S (2005) The neurodevelopmental model of schizophrenia: update 2005. Mol Psychiatry 10:434–449
44. Mednick SA, Machon RA, Huttunen MO (1998) Adult schizophrenia following prenatal exposure to an influenza epidemic. Arch Gen Psychiatry 45:189–192
45. Cannon M, Jones PB, Murray RM (2002) Obstetric complications and schizophrenia: historical and meta-analytic review. Am J Psychiatry 159:1080–1092
46. Geddes JR, Verdoux H, Takei N, Lawrie SM, Bovet P, Eagles JM, Heun R, McCreadie RG, McNeil TF, O'Callaghan E, Stöber G, Willinger U, Murray RM (1999) Schizophrenia and complications of pregnancy and labor: an individual patient data meta-analysis. Schizophr Bull 25:413–423
47. Fiore M, Grace AA, Korf J, Stampachiacchiere B, Aloe L (2004) Impaired brain development in the rat following prenatal exposure to methylazoxymethanol acetate at gestational day 17 and neurotrophin distribution. Neuroreport 15:1791–1795
48. Lieberman JA, Perkins D, Belger A, Chakos M, Jarskog F, Boteva K, Gilmore J (2001) The early stages of schizophrenia: speculations on pathogenesis, pathophysiology, and therapeutic approaches. Biol Psychiatry 50:884–897
49. Stefansson H, Sigurdsson E, Steinthorsdottir V, Bjornsdottir S, Sigmundsson T, Ghosh S, Brynjolfsson J, Gunnarsdottir S, Ivarsson O, Chou TT, Hjaltason O, Birgisdottir B, Jonsson H, Gudnadottir VG, Gudmundsdottir E, Bjornsson A, Ingvarsson B, Ingason A, Sigfusson S, Hardardottir H, Harvey RP, Lai D, Zhou M, Brunner D, Mutel V, Gonzalo A, Lemke G, Sainz J, Johannesson G, Andresson T, Gudbjartsson D, Manolescu A, Frigge ML, Gurney ME, Kong A, Gulcher JR, Petursson H, Stefansson K (2002) Neuregulin 1 and susceptibility to schizophrenia. Am J Hum Genet 71:877–892

50. Straub RE, Jiang Y, MacLean CJ, Ma Y, Webb BT, Myakishev MV, Harris-Kerr C, Wormley B, Sadek H, Kadambi B, Cesare AJ, Gibberman A, Wang X, O'Neill FA, Walsh D, Kendler KS (2002) Genetic variation in the 6p22.3 gene DTNBP1, the human ortholog of the mouse dysbindin gene, is associated with schizophrenia. Am J Hum Genet 71:337–348
51. Egan MF, Goldberg TE, Kolachana BS, Callicott JH, Mazzanti CM, Straub RE, Goldman D, Weinberger DR (2001) Effect of COMT Val108/158 Met genotype on frontal lobe function and risk for schizophrenia. Proc Natl Acad Sci U S A 98:6917–6922
52. Millar JK, Wilson-Annan JC, Anderson S, Christie S, Taylor MS, Semple CA, Devon RS, St Clair DM, Muir WJ, Blackwood DH, Porteous DJ (2000) Disruption of two novel genes by a translocation co-segregating with schizophrenia. Hum Mol Genet 9:1415–1423
53. O'Tuathaigh CM, Babovic D, O'Meara G, Clifford JJ, Croke DT, Waddington JL (2007) Susceptibility genes for schizophrenia: characterisation of mutant mouse models at the level of phenotypic behaviour. Neurosci Behav Rev 31:60–78
54. Lipska BK, Jaskiw GE, Weinberger DR (1993) Postpubertal emergence of hyperresponsiveness to stress and to amphetamine after neonatal excitotoxic hippocampal damage: a potential animal model of schizophrenia. Neuropsychopharmacology 9:67–75
55. Lipska BK, Swerdlow NR, Geyer MA, Jaskiw GE, Braff DL, Weinberger DR (1995) Neonatal excitotoxic hippocampal damage in rats causes post-pubertal changes in prepulse inhibition of startle and its disruption by apomorphine. Psychopharmacology (Berl) 122:35–43
56. Grecksch G, Bernstein HG, Becker A, Höllt V, Bogerts B (1999) Disruption of latent inhibition in rats with postnatal hippocampal lesions. Neuropsychopharmacology 20:525–532
57. Lipska BK, Aultman JM, Verma A, Weinberger DR, Moghaddam B (2002) Neonatal damage of the ventral hippocampus impairs working memory in the rat. Neuropsychopharmacology 27:47–54
58. Brady AM (2009) Neonatal ventral hippocampal lesions disrupt set-shifting ability in adult rats. Behav Brain Res 205:294–298
59. Becker A, Grecksch G, Bernstein HG, Höllt V, Bogerts B (1999) Social behaviour in rats lesioned with ibotenic acid in the hippocampus: quantitative and qualitative analysis. Psychopharmacology (Berl) 144:333–338
60. Brady AM, Saul RD, Wiest MK (2010) Selective deficits in spatial working memory in the neonatal ventral hippocampal lesion rat model of schizophrenia. Neuropharmacology 59: 605–611
61. Alquicer G, Silva-Gómez AB, Peralta F, Flores G (2004) Neonatal ventral hippocampus lesion alters the dopamine content in the limbic regions in postpubertal rats. Int J Dev Neurosci 22:103–111
62. Becker A, Grecksch G (2003) Haloperidol and clozapine affect social behaviour in rats postnatally lesioned in the ventral hippocampus. Pharmacol Biochem Behav 76:1–8
63. Daenen EW, Wolterink G, Van Ree JM (2003) Hyperresponsiveness to phencyclidine in animals lesioned in the amygdala on day 7 of life. Implications for an animal model of schizophrenia. Eur Neuropsychopharmacol 13:273–279
64. Miner LA, Ostrander M, Sarter M (1997) Effects of ibotenic acid-induced loss of neurons in the medial prefrontal cortex of rats on behavioral vigilance: evidence for executive dysfunction. J Psychopharmacol 11:169–178
65. Brown AS, Derkits EJ (2010) Prenatal infection and schizophrenia: a review of epidemiologic and translational studies. Am J Psychiatry 167:261–280
66. Zuckerman L, Weiner I (2005) Maternal immune activation leads to behavioral and pharmacological changes in the adult offspring. J Psychiatr Res 39:311–323
67. Wolff AR, Bilkey DK (2008) Immune activation during mid-gestation disrupts sensorimotor gating in rat offspring. Behav Brain Res 190:156–159
68. Ozawa K, Hashimoto K, Kishimoto T, Shimizu E, Ishikura H, Iyo M (2006) Immune activation during pregnancy in mice leads to dopaminergic hyperfunction and cognitive impairment in the offspring: a neurodevelopmental animal model of schizophrenia. Biol Psychiatry 59:546–554

69. Zuckerman L, Weiner I (2003) Post-pubertal emergence of disrupted latent inhibition following prenatal immune activation. Psychopharmacology (Berl) 169:308–313
70. Zuckerman L, Rehavi M, Nachman R, Weiner I (2003) Immune activation during pregnancy in rats leads to a postpubertal emergence of disrupted latent inhibition, dopaminergic hyperfunction, and altered limbic morphology in the offspring: a novel neurodevelopmental model of schizophrenia. Neuropsychopharmacology 28:1778–1789
71. Shi L, Fatemi SH, Sidwell RW, Patterson PH (2003) Maternal influenza infection causes marked behavioral and pharmacological changes in the offspring. J Neurosci 23:297–302
72. Meyer U, Feldon J, Schedlowski M, Yee BK (2005) Towards an immuno-precipitated neurodevelopmental animal model of schizophrenia. Neurosci Biobehav Rev 29:913–94773
73. Meyer U, Nyffeler M, Engler A, Urwyler A, Schedlowski M, Knuesel I, Yee BK, Feldon J (2006) The time of prenatal immune challenge determines the specificity of inflammation-mediated brain and behavioural pathology. J Neurosci 26:4752–4762
74. Meyer U, Engler A, Weber L, Schedlowski M, Feldon J (2008) Preliminary evidence for a modulation of fetal dopaminergic development by maternal immune activation during pregnancy. Neuroscience 154:701–709
75. Harris LW, Sharp T, Gartlon J, Jones DN, Harrison PJ (2003) Long-term behavioural, molecular and morphological effects of neonatal NMDA receptor antagonism. Eur J Neurosci 18:1706–1710
76. Kawabe K, Iwasaki T, Ichitani Y (2007) Repeated treatment with N-methyl-d-aspartate antagonists in neonatal, but not adult, rats causes long-term deficits of radial-arm maze learning. Brain Res 1169:77–86
77. Montgomery EM, Bardgett ME, Lall B, Csernansky CA, Csernansky JG (1999) Delayed neuronal loss after administration of intracerebroventricular kainic acid to preweanling rats. Dev Brain Res 112:107–116
78. Humphrey WM, Dong H, Csernansky CA, Csernansky JG (2002) Immediate and delayed hippocampal neuronal loss induced by kainic acid during early postnatal development in the rat. Dev Brain Res 137:1–12
79. Dong H, Csernansky CA, Chu Y, Csernansky JG (2003) Intracerebroventricular kainic acid administration to neonatal rats alters interneuron development in the hippocampus. Dev Brain Res 145:81–92
80. Dong H, Csernansky CA, Goico B, Csernansky JG (2003) Hippocampal neurogenesis follows kainic acid-induced apoptosis in neonatal rats. J Neurosci 23:1742–1749
81. Pulido OM (2008) Domoic acid toxicologic pathology: a review. Mar Drugs 6:180–219
82. Levin ED, Pizarro K, Pang WG, Harrison J, Ramsdell JS (2005) Persisting behavioral consequences of prenatal domoic acid exposure in rats. Neurotox Teratol 27:719–725
83. Levin ED, Pang WG, Harrison J, Williams P, Petro A, Ramsdell JS (2006) Persistent neurobehavioral effects of early postnatal domoic acid exposure in rats. Neurotoxi Teratol 28:673–680
84. Adams AL, Doucette TA, Ryan CL (2008) Altered pre-pulse inhibition in adult rats treated neonatally with domoic acid. Amino Acid 35:157–160
85. Adams AL, Doucette TA, James R, Ryan CL (2009) Persistent changes in learning and memory in rats following neonatal treatment with domoic acid. Physiol Behav 96:505–512
86. Doucette TA, Ryan CL, Tasker RA (2007) Gender-based changes in cognition and emotionality in a new rat model of epilepsy. Amino Acid 32:317–322
87. Burt MA, Ryan CL, Doucette TA (2008) Altered responses to novelty and drug reinforcement in adult rats treated neonatally with domoic acid. Physiol Behav 93:327–336
88. Burt MA, Ryan CL, Doucette TA (2008) Low dose domoic acid in neonatal rats abolishes nicotine induced place preference during late adolescence. Amino Acid 35:247–24989
89. Ryan CL, Robbins MA, Smith MT, Gallant IC, Adams-Marriott AL, Doucette TA (2011) Altered social interaction in adult rats following neonatal treatment with domoic acid. Physiol Behav 102:291–295
90. Autism Society Canada (2012) http://www.autismsocietycanada.ca/ Accessed 24 Feb 2012

91. Iwata K, Matsuzaki H, Takei N, Manabe T, Mori N (2010) Animal models of autism: an epigenetic and environmental viewpoint. J Cent Nerv Syst Dis 2:37–44
92. Ganz ML (2007) The lifetime distribution of the incremental societal costs of autism. Arch Pediatric Adolescent Med 161:343–349
93. American Psychiatric Association (2000) Diagnostic and statistical manual of mental disorders. In: Section A 09 - autism spectrum disorder, 4th edn. American Psychiatric Association, Washington, DC
94. Konstantareas MM, Hewitt T (2001) Autistic disorder and Schizophrenia: diagnostic overlaps. J Autism Dev Disord 31:19–28
95. Perry W, Minassian A, Lopez B, Maron L, Lincoln A (2007) Sensorimotor gating deficits in adults with autism. Biol Psychiatry 61:482–486
96. Macneil LK, Mostofsky SH (2012) Specificity of dyspraxia in children with autism. Neuropsychology 26:165–171
97. Maski KP, Jeste SS, Spence SJ (2011) Common neurological co-morbidities in autism spectrum disorders. Curr Opin Pediatr 23:609–615
98. Simonoff E, Pickles A, Charman T, Chandler S, Loucas T, Baird G (2008) Psychiatric disorders in children with autism spectrum disorders: prevalence, comorbidity, and associated factors in a population-derived sample. J Am Acad Child Adolesc Psychiatr 47:921–929
99. Leyfer OT, Folstein SE, Bacalman S, Davis NO, Dinh E, Morgan J, Tager-Flusberg H, Lainhart JE (2006) Comorbid psychiatric disorders in children with autism: interview development and rates of disorders. J Autism Dev Dis 36:849–861
100. Mukaddes NM, Hergűner S, Tanidir C (2010) Psychiatric disorders in individuals with high-functioning autism and Asperger's disorder: similarities and differences. World J Biol Psychiatr 11:964–971
101. Skokauskas N, Gallagher L (2010) Psychosis, affective disorders and anxiety in autistic spectrum disorder: prevalence and nosological considerations. Psychopathology 43:8–16
102. Charnsil C, Sriapai P (2011) Attention deficit hyperactivity symptoms in children with autistic disorder: a cross-sectional descriptive study. J Med Assoc Thail 94:231–234
103. Sipes M, Matson JL, Horovitz M, Shoemaker M (2011) The relationship between autism spectrum disorders and symptoms of conduct problems: the moderating effect of communication. Dev Neurorehab 14:54–59
104. Clarke DF, Roberts W, Daraksan M, Dupuis A, McCabe J, Wood H, Snead OC 3rd, Weiss SK (2005) The prevalence of autistic spectrum disorder in children surveyed in a tertiary care epilepsy clinic. Epilepsia 46:1970–1977
105. Tuchman R, Rapin I (2002) Epilepsy in autism. Lancet Neurol 1:352–358
106. Tuchman R, Moshé SL, Rapin I (2009) Convulsing toward the pathophysiology of autism. Brain Dev 31:95–103
107. Amiet C, Gourfinkel-An I, Bouzamondo A, Tordjman S, Baulac M, Lechat P, Mottron L, Cohen D (2008) Epilepsy in autism is associated with intellectual disability and gender: evidence from a meta-analysis. Biol Psychiatry 64:577–582
108. Brooks-Kayal A (2010) Epilepsy and autism spectrum disorders: are there common developmental mechanisms? Brain Dev 32:731–738
109. Nazeer A (2011) Psychopharmacology of autistic spectrum disorders in children and adolescents. Pediatr Clin North Am 58:85–97
110. Patterson PH (2011) Modeling autistic features in animals. Pediatr Res 69:34R–40R
111. Geschwind DH (2011) Genetics of autism spectrum disorders. Trends in Cognit Sci 15: 409–416
112. Malkova NV, Yu CZ, Hsiao EY, Moore MJ, Patterson PH (2012) Maternal immune activation yields offspring displaying mouse versions of the three core symptoms of autism. Brain Behav Immun 26:607–616
113. Dufour-Rainfray D, Vourc'h P, Tourlet S, Guilloteau D, Chalon S, Andres CR (2011) Fetal exposure to teratogens: evidence of genes involved in autism. Neurosci Biobehav Rev 35:1254–1265

114. Dufour-Rainfray D, Vourc'h P, Le Guisquet AM, Garreau L, Ternant D, Bodard S, Jaumain E, Gulhan Z, Belzung C, Andres CR, Chalon S, Guilloteau D (2010) Behavior and serotonergic disorders in rats exposed prenatally to valproate: a model for autism. Neurosci Lett 470:55–59
115. Schneider T, Roman A, Basta-Kaim A, Kubera M, Budziszewska B, Schneider K, Przewłocki R (2008) Gender-specific behavioral and immunological alterations in an animal model of autism induced by prenatal exposure to valproic acid. Psychoneuroendocrinology 33:728–740
116. Schneider T, Przewlocki R (2005) Behavioral alterations in rats prenatally exposed to valproic acid: animal model of autism. Neuropsychopharmacology 30:80–89
117. Kim KC, Kim P, Go HS, Choi CS, Yang SI, Cheong JH, Shin CY, Ko KH (2011) The critical period of valproate exposure to induce autistic symptoms in Sprague–Dawley rats. Toxicol Lett 201:137–142
118. Markram K, Rinaldi T, La Mendola D, Sandi C, Markram H (2008) Abnormal fear conditioning and amygdala processing in an animal model of autism. Neuropsychopharmacology 33:901–912
119. Schneider T, Ziolkowska B, Gieryk A, Tyminska A, Przewłocki R (2007) Prenatal exposure to valproic acid disturbs the enkephalinergic system functioning, basal hedonic tone, and emotional responses in an animal model of autism. Psychopharmacology 193:547–555
120. Tashiro Y, Oyabu A, Imura Y, Uchida A, Narita N, Narita M (2011) Morphological abnormalities of embryonic cranial nerves after in utero exposure to valproic acid: implications for the pathogenesis of autism with multiple developmental anomalies. Int J Dev Neurosci 29:359–364
121. Kuwagata M, Ogawa T, Shioda S, Nagata T (2009) Observation of fetal brain in a Rat valproate-induced autism model: a developmental neurotoxicity study. Int J Dev Neurosci 27:399–405
122. Gurney ME, Pu H, Chiu AY, Dal Canto MC, Polchow CY, Alexander DD, Caliendo J, Hentati A, Kwon YW, Deng HX, Chen W, Zhai P, Sufit RL, Siddique T (1994) Motor neuron degeneration in mice that express a human Cu, Zn superoxide dismutase mutation. Science 264:1772–1775
123. Wong PC, Pardo CA, Borchelt DR, Lee MK, Copeland NG, Jenkins NA, Sisodia SS, Cleveland DW, Price DL (1995) An adverse property of familial ALS-linked SOD1 mutation causes motor neuron disease characterized by vacuolar degeneration of mitochondria. Neuron 14:1105–1116
124. Devon RS, Orban PC, Gerrow K, Barbieri MA, Schwab C, Cao LP, Helm JR, Bissada N, Cruz-Aguado R, Davidson TL, Witmer J, Metzler M, Lam CK, Tetzlaff W, Simpson EM, McCaffery JM, El-Husseini AE, Leavitt BR, Hayden MR (2006) Als2-deficient mice exhibit disturbances in endosome trafficking associated with motor behavioral abnormalities. PNAS 103:9595–9600
125. Shaw CA, Banjo OC (2008) Fetal exposure of mice to cycad neurotoxins impacts postnatal behaviour. Program No. 445.25. Neuroscience Meeting Planner, Washington, DC: Society for Neuroscience. (Online)
126. Wilson JM, Khabazian I, Wong MC, Seyedalikhani A, Bains JS, Pasqualotto BA, Williams DE, Andersen RJ, Simpson RJ, Smith R, Craig UK, Kurland LT, Shaw CA (2002) Behavioral and neurological correlates of ALS-parkinsonism dementia complex in adult mice fed washed cycad flour. J Neuromol Med 1:207–222
127. Khabazian I, Bains JS, Williams DE, Cheung J, Wilson JM, Pasqualotto BA, Pelech SL, Andersen RJ, Wang YT, Liu L, Nagai A, Kim SU, Craig UK, Shaw CA (2002) Isolation of various forms of sterol β-d-glucoside from the seed of cycas circinalis: neurotoxicity and implications for ALS-PDC. J Neurochem 83:1–13
128. Tabata RC, Wilson JM, Ly P, Zwiegers P, Kwok D, Van Kampen JM, Cashman N, Shaw CA (2008) Chronic exposure to dietary sterol glucosides is neurotoxic to motor neurons and induces an ASL-PDC phenotype. Neuromol Med 10:24–39
129. Barlow BK, Richfield EK, Cory-Slechta DA, Thiruchelvam M (2004) A fetal risk factor for Parkinson's disease. Dev Neurosci 26:11–23

130. Fisher LR, Culver DG, Tennant P, Davis AA, Wang M, Castellano-Sanchez A, Khan J, Polak MA, Glass JD (2004) Amyotrophic lateral sclerosis is a distal axonopathy: evidence in mice and man. Exp Neurol 185:232–240
131. Kurland LT (1988) Amyotrophic lateral sclerosis and Parkinson's disease complex on Guam linked to an environmental neurotoxin. Trends Neurol Sci 11:51–54
132. Whiting MG (1964) Food practices in ALS foci in Japan, the Marianas, and New Guinea. Federation Proc. Third Conference on Toxicity of Cycads 23: 1343–1345
133. Borenstein AR, Mortimer JA, Schofield E, Wu Y, Salmon DP, Gamst A, Olichney J, Thal LJ, Silbert L, Kaye J, Craig UL, Schellenberg GD, Galasko DR (2007) Cycad exposure and risk of dementia, MCI, and PDC in the Chamorro population of Guam. Neurology 68:1764–1767
134. Shen W-B, McDowell KA, Siebert AA, Clark SM, Dugger NV, Valentino KM, Jinnah HA, Sztalryd C, Fishman PS, Shaw CA, Jafri MS, Yarowsky PJ (2010) Environmental neurotoxin-induced progressive model of parkinsonism in rats. Ann Neurol 68(6):70–80
135. Spencer PS, Hugon J, Ludolph A, Nunn PB, Ross SM, Roy DN, Schaumburg HH (1987) Discovery and partial characterization of primate motor-system toxins. CIBA Found Symp 126:221–238
136. Dastur DK (1964) Cycad toxicity in monkeys: clinical, pathological, and biochemical aspects. Fed Proc 23:1368–1369
137. Perry TL, Bergeron C, Biro AJ, Hansen S (1989) Beta-N-methylamino-l-alanine. Chronic oral administration is not neurotoxic to mice. J Neurol Sci 94:173–180
138. Cruz-Aguado R, Winkler D, Shaw CA (2006) Lack of behavioral and neuropathological effects of dietary b-methylaminoalanine (BMAA) in mice. Pharmacol Biochem Behav 84:294–299
139. Martin I, Dawson VL, Dawson TM (2011) Recent advances in the genetics of Parkinson's disease. Ann Rev Genomic Hum Genet 12:301–325

Tau Pathology: A Selected View on the Current Status

Illana Gozes

Abstract Tau (tubulin associated unit) is a microtubule regulatory protein forming pathological aggregates in multiple neurodegenerative disease termed tauopathies. Microtubules are key cytoskeletal elements found in all eukaryotic cells, comprising the mitotic spindle in dividing cells and the skeleton and transport system of the nerve cell. The polymerized microtubule shaft (a hollow cylinder of ~250 nm) is composed of the heterodimer protein, tubulin. Microtubules are decorated with multiple microtubule associated regulatory proteins (MAPs). Tau (MAPT), among the first microtubule associated proteins to be identified, was implicated in microtubule initiation as well as assembly, with increased expression in neurons and specific association with axonal microtubules. Alzheimer's disease (AD) is the most prevalent tauopathy, exhibiting tau-neurofibrillary tangles associated with cognitive dysfunction. AD is also characterized by β-amyloid plaques. An abundance of tau inclusions, in the absence of β-amyloid deposits, can be found in Pick's disease, progressive supranuclear palsy (PSP), corticobasal degeneration (CBD) and other diseases, collectively described as tauopathies. The increase in tau pathology in AD

I. Gozes (✉)
Department of Human Molecular Genetics and Biochemistry, Sackler Faculty of Medicine, Tel Aviv University, Tel Aviv 69978, Israel
e-mail: igozes@post.tau.ac.il

The Lily and Avraham Gildor Chair for the Investigation of Growth Factors, Tel Aviv University, Tel Aviv 69978, Israel

The Adams Super Center for Brain Studies and The Edersheim Levie-Gitter fMRI Institute, Tel Aviv University, Tel Aviv 69978, Israel

The Dr. Diana and Zelman Elton (Elbaum) Laboratory for Molecular Neuroendocrinology, Tel Aviv University, Tel Aviv 69978, Israel

S. Mandel (ed.), *Neurodegenerative Diseases: Integrative PPPM Approach as the Medicine of the Future*, Advances in Predictive, Preventive and Personalised Medicine 2, DOI 10.1007/978-94-007-5866-7_5,

correlates with the associated cognitive decline. This book chapter is an update on several recent reviews. It is not a comprehensive review of the literature, but rather an updating point of view.

Keywords Tau • MAPT • Tubulin • Tauoparhy • Diagnostic tools • Drug candidates • Davunetide

1 Tau Background a Selected Overview

As previously described, the understanding of the multiple triggers of neurodegeneration is bound to lead to better diagnostics and improved therapeutics. In this respect and looking for unifying pathways, the accumulation of intracellular neurofibrillary lesions composed of abnormally phosphorylated and aggregated tau protein characterizes many neurodegenerative disorders [1–6].

The microheterogeneity of tubulin [7, 8], the building block of the microtubule shaft, expressing multiple isotypes with brain specificity [9], coupled to the intricate complexity of the microtubule interacting proteins leads to enhanced control of neuronal function. Mis-metabolism and mutations in tubulin or in the interacting proteins leads to a variety of disease conditions, e.g. [10]. Here, the emphasis is on tau.

Tau (tubulin associated unit) or MAPT (microtubule associated protein tau) [11] (Fig. 1) is required for microtubule initiation as well as assembly [12], with increased expression in neurons and specific association with axonal microtubules [13]. MAPs binding to tubulin/microtubules enables them to play a fundamental role in promoting microtubule assembly and stability [11, 14, 15]. Functionally, microtubules are essential for cell division, neuronal development, maintenance of neuronal shape and neuronal plasticity (e.g. plasticity of dendritic spines and axonal transport), with tau participating in neurite extension and axonal transport [16–19].

Tau is primarily a neuronal protein, though not exclusively. In an adult human brain, there are six major isoforms of tau derived from a single gene through alternative splicing [20, 21]. These isoforms differ by the presence or absence of one or two short inserts in the amino terminal half and in whether they contain three or four tubulin binding domain repeats in the carboxy terminal half. The alternative splicing of tau has been suggested to impact axonal functional homeostasis [22], and three repeat tau was suggested to have a stronger effect on axon transport dynamics. Additionally, there are two alternatively spliced exons at the N-terminal side of the tau mRNA.

Using cultured Aplysia neurons and online confocal imaging of human tau, it was shown that over-expression of tau generates the hallmarks of human tau pathogenesis. It was further demonstrated that the tau-induced impairment of organelle transport is because of polar reorientation of the microtubules along the axon or their displacement to sub-membrane domains establishing that tau over-expression leads to impaired retrograde and anterograde organelle transport [23].

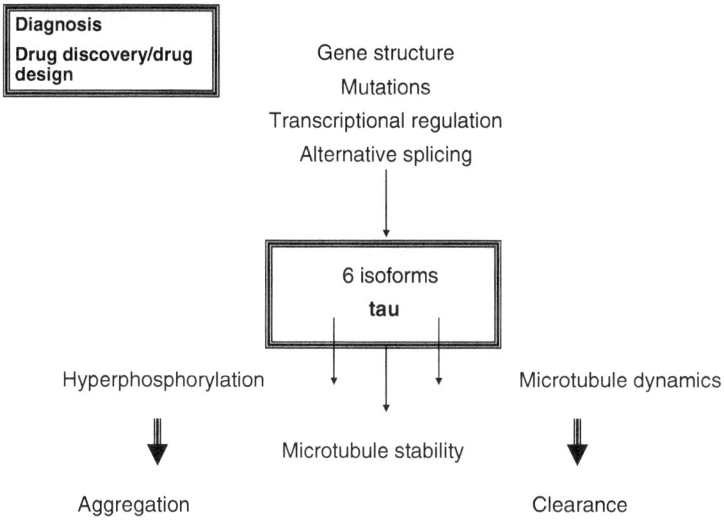

Fig. 1 Tau regulation and potential points of diagnosis and intervention. Tau is subject to multiple points of regulation including the intrinsic regulatory elements of the gene (gene structure), tau mutations and transcriptional controls. Following transcription, the tau mRNA is regulated by alternative splicing, yielding six unique isoforms and post-translational modifications which is highly pronounced at the multiple phosphorylation sites. The tau protein is contributing to microtubule dynamics and stability and is a subject to clearance mechanisms. Under pathological conditions, tau may undergo hyperphosphorylation as well as self-aggregation resulting in cell death. Tau pathology may be propagated from cell to cell (spreading pathology)

Relatively increased microtubule stability along the shaft in a single neurite specifies axonal formation and neuronal polarization [24], thus, implicating microtubule dynamics and stabilization at center stage regarding neuronal differentiation and plasticity. This microtubule stabilization/destabilization is associated with tau and tau phosphorylation at the level of the tau phosphorylating enzyme activated glycogen synthase kinase (GSK) [25]. Recent literature showed that hyperphosphorylated tau can be mis-localized to dendritic spines, where it accumulates and disrupts synaptic function by impairing glutamate receptor trafficking or synaptic anchoring. This may cause early synaptic dysfunction residing in dendritic spines [26]. When tau is overexpressed – hyperphosphorylated, vesicle-associated tau is localizing to dendrites, which is saturating dendritic microtubules, followed by tau secretion, loss of microtubules and perpetuation of the degeneration by localized microtubule loss [27]. In transgenic animals over expressing tau, this is translated into age-dependent changes in dendritic spine morphology with 12-month-old mice exhibiting more complex arborization patterns, coupled with a shift in spine morphology with fewer mushroom and more thin spines in both apical and basal dendrites as a function of human tau accumulation in layer III pyramidal neurons from the prefrontal cortex [28]. Together with transsynaptic spread of tau, these results place mis-regulated tau and microtubule loss at center stage for disease propagation [29, 30].

Tubulin microheterogeneity [7] and tau heterogeneity [31] have been correlated with brain development. Recent studies found that the immature neurons, identified by doublecortin (DCX) and neurogenic differentiation factor (neuroD), were only immunoreactive to phospho-tau (p-tau) but not to the non-p-tau in adult rat brain and human patients with AD, and the p-tau was co-expressed temporally and spatially with DCX and neuroD in the hippocampal dentate gyrus (DG) of the rat brain during postnatal development. A correlative increase of immature neuron markers and tau phosphorylation was induced in rat hippocampal DG by upregulating GSK-3 (wortmanin), and the increased neurogenesis was due to an enhanced proliferation but not survival or differentiation of the newborn neurons. The hippocampal neurogenesis was severely impaired in tau knockout mice and activation of GSK-3 in these mice did not rescue the deficits [32]. Together, an essential role for tau phosphorylation in neurogenesis and adult hippocampal neurogenesis is suggested. Furthermore, spatial/temporal manipulation of tau phosphorylation may be compensatory for the neuronal loss in neurodegenerative diseases, including AD.

Tau protein interaction is not limited to interaction with tubulin. Tau bridges tubulin (through the tau-tubulin binding domains) and actin (through the tau proline-rich domain) – associating major cellular organelles, the microtubules and the microfilaments [33] Interaction of tau, especially in its hyperphosphorylated form, was found with the brain abundant FKBP52, a member of the FKBP (FK506-binding protein) family that comprises intracellular protein effectors of immunosuppressive drugs (such as FK506 and rapamycin). The relevance of this observation was confirmed by the co-localization of both proteins in the distal part of the axons of cortical neurons and by the antagonistic effect of FKBP52 on the ability of tau to promote microtubule assembly. Overexpression of FKBP52 in differentiated PC12 cells prevented the accumulation of tau and resulted in reduced neurite length [34]. Interestingly, in undifferentiated hippocampal neurons, FKBP52, Hsp90 and p23 are located in the cell nucleus, forming a ring-like structure that disassembles when the differentiation process is triggered by FK506 (also called tacrolimus, molecular formula, C44H69NO12). This annular structure of chaperones is reassembled after neuronal damage, whereas FK506 prompts their rapid regeneration, a process linked to the subcellular redistribution of the heterocomplex [35].

Another tau – interacting molecule is the tyrosine kinase Fyn [36], which plays a critical role in mediating synaptic toxicity and neuronal loss in response to beta amyloid in models of Alzheimer's disease [37]. The Pro216 within the PXXP(213–216) motif in tau mediates in part the interaction of tau with Fyn, furthermore, the Fyn-tau interaction is also influenced by phosphorylation of tau at Tyr18, a reported target for Fyn [38].

In this respect, beta amyloid causes tau missorting, loss of neuronal cell polarity and loss of dendritic microtubules. This in turn leads to impaired organelle/mitochondria transport, whereby synapses cannot be maintained properly and eventually decay [39].

The complexity of tau alternative splicing, phosphorylation and disease association is further dealt with below.

2 Tauopathies

As reviewed before, disruption of the microtubule network is a hallmark of neurodegeneration, including the most prevalent tauopathy, Alzheimer's disease (AD). AD is characterized by the deposition of intracellular fibrillar structures forming paired helical filaments (PHFs) followed by larger aggregates termed neurofibrillary tangles (NFTs) [40] as well as the β-amyloid deposits.

Tau was originally identified as the primary component of PHF [41–47]. NFTs are composed mainly of PHFs with a minority of straight filament (SF) [48–50]. PHFs appear to consist of two filaments that are wound helically around one another, with a longitudinal spacing between crossovers of about 80 nm and a width of 30 nm at the widest point and 15 nm at the narrowest [51]. PHFs morphology distinguishes them from microtubules (250 nm diameter), neurofilaments (~100 nm diameter) and microfilaments (~60 nm diameter) which are the major cytoskeletal elements in healthy neurons [48–50, 52]. The biochemistry of tau is a subject of multiple reviews [53].

Phosphorylation occurs on a number of different parts of tau and hyperphosphorylation generally reduces tau-tubulin binding. A link to a slide delineating different tau domains and tau mutations can be accessed via: http://www.alzforum.org/res/com/mut/tau/table1.asp. This site outlines the various mutations that have been found in tau that are associated with multiple tauopathies including CBD = corticobasal degeneration. DDPAC = disinhibition-dementia-parkinsonism-amyotrophy complex; FTD = frontotemporal dementia. FTDP-17 = FTD with Parkinsonism linked to chromosome 17; HFDT = hereditary frontotemporal dementia; MSTD = multiple system tauopathy dementia; PPND = pallidopontonigral dementia; and PSP = progressive supranuclear palsy. The different tau mutations are associated with different morphology of the tau pathology (http://www.alzforum.org/res/com/mut/tau/table1.asp). The original identification of mutations in tau in familial forms of FTD was described in 1998 [54–57].

While phosphorylation is part of the normal regulation of tau (as alluded to above), hyperphosphorylation in the aging brain is associated with tangle formation and the majority of tau in PHFs and NFTs is hyperphosphorylated [58, 59]. This hyperphosphorylation may be related to either an increase in kinase activity or a decrease in phosphatase activity [60]. Additionally, tau undergoes a specific type of serine–threonine O-glycosylation, and these modifications can reduce the extent of tau phosphorylation [61]. As indicated above, tau can also be tyrosine phosphorylated [62], as well as sumoylated and nitrated [63], although the effects of these modifications on tau require further investigations. Tau hyperphosphorylation appears to precede its accumulation in the affected neurons in AD [64, 65]. Hyperphosphorylated tau shows impaired axonal transport [17, 66], defective microtubule binding [67–70], failure to promote microtubule assembly [71, 72], and self-assembly into NFTs [73, 74]. Additionally, hyperphosphorylation of tau might make it more resistant to caspase proteolysis as well as degradation and thereby more likely to accumulate in neurons, forming PHFs and NFTs [75].

As suggested before, original studies indicated that a down-regulation of protein phosphatase 2A (PP2A), the major tau phosphatase in human brain, contributes to tau hyperphosphorylation in AD. PP2A dephosphorylated tau at several phosphorylation sites with different efficiencies. Among the sites studied, Thr205, Thr212, Ser214, and Ser262 were the most favorable sites, and Ser199 and Ser404 were the least favorable sites for PP2A in vitro. In addition to its direct effect on tau, inhibition of PP2A with okadaic acid in metabolically active rat brain slices caused inhibition of GSK-3beta via an increase in its phosphorylation at Ser9. GSK-3beta phosphorylates tau at many sites, with Ser199, Thr205, and Ser396 being the most favorable sites. The overall alterations in tau phosphorylation induced by PP2A inhibition were the result of the combined effects of both reduced tau dephosphorylation due to PP2A inhibition directly and reduced phosphorylation by GSK-3beta due to its inhibition [76]. Tau phosphorylation impacts its biological activity and neurofibrillary degeneration in association with the specific phosphorylation site (for a comprehensive web site on tau hyperphosphorylation, including the kinases involved see http://www.alzheimer-adna.com/Gb/Tau/TauPhosphoSeq.htm). Thus, the interplay between phosphorylating enzymes and phosphatases plays a role in tauopathy progression. It should be noted that other kinases have been implicated in tau phosphorylation including the protein kinases of the microtubule affinity regulating kinase (MARK)/Par-1 family playing important roles in the establishment of cellular polarity, cell cycle control, and intracellular signal transduction [77, 78] as well as cdk-5 [79].

Most of the tau phosphorylation sites that have been characterized were Ser and Thr residues. More recent reports showed that tau can be phosphorylated at Tyr residues by kinases including Fyn, Syk, and c-abl (Abl), (as alluded above). Proteomic analyses show that tau phosphorylated at Tyr394 (Y394) exists within AD PHF samples. It was further shown that Abl phosphorylated this particular site on tau. A most recent report showed that Arg, the other member of the Abl family of tyrosine kinases, also phosphorylates tau at Y394 in a manner independent of Abl activity. Given the reported role of Arg in oxidative stress response and neural development, the ability to phosphorylate tau at Y394 implicated Arg as a potential player in the pathogenesis of AD and other tauopathies [80].

The MAPT gene on human chromosome 17 appears in two main extended haplotypes. In a total of approximately 200 unrelated Caucasian individuals, there was complete disequilibrium between polymorphisms which span the gene (which covers approximately 100 kb of DNA). This showed an establishment of the two haplotypes (H1 and H2) [81]. It was further shown that the more common haplotype (H1) is significantly over-represented in PSP patients [82].

Single nucleotide polymorphisms mapped linkage disequilibrium in the regions flanking MAPT and have established the maximum extent (~2 Mb) of the haplotype block on chromosome 17q21.31 [83] and a 900 kb inversion which suppresses recombination [84]. This ~2 Mb gene-rich region extends centromerically slightly beyond (~400 kb) the corticotrophin releasing hormone receptor 1 gene. The telomeric end was defined by a ~150 kb region beyond the WNT3 gene. This study of Pittman et al., showed that the entire H1 haplotype is associated with PSP, which may implicate several other genes in addition to MAPT, as candidate pathogenic loci [83]. Additional studies by Pittman et al., resolved multiple variants of the H1 haplotype, reflecting a greater diversity of MAPT than can be explained by the H1 and H2 clades alone [85].

Mild cognitive impairment (MCI) is often considered a transitional condition prodromal to AD. Therefore, the genotypes of 7 polymorphisms tagging the major tau haplotypes were assayed on 186 patients with amnestic MCI and 191 unrelated controls. Association study was conducted by logistic regression including apoliprotein E genotype and age as covariates (with the aplipoprotein E4 and age representing major risk factors for AD). The common H1 haplotype was found to be significantly overrepresented in amnestic MCI patients. This finding was confirmed when the aplipoprotein E4 allele was taken into account. These results suggest (1) that the risk of MCI is influenced by tau protein gene variations and (2) that MCI shares a common genetic background with AD. The results may help elucidating the genetic risk to cognitive decline and designing effective clinical trials, future diagnostics and future therapies [86].

Interestingly, in trying to evaluate the rate of progression and the predictors of worsening in frontotemporal lobar degeneration (FTLD) patients, 127 FTLD patients entered a study and were re-evaluated at 1-year follow-up. A statistical driven approach on wide neuropsychological, behavioral, and functional data was applied to identify homogeneous groups both at baseline and at follow-up within FTLD, taking into account: (i) the demographic characteristics, (ii) the genetic background, i.e. apolipoprotein E genotype, tau haplotype, and functional polymorphisms affecting serotonin and dopamine pathways, and (iii) the clinical phenotype. On the basis of the overall assessment (disease severity), the results recognized two groups of patients, "good performers" and "bad performers". At 1-year follow-up, almost 30% of FTLD patients progressed from "good" to "bad" performances, whilst 70% maintained stable "good" performances. Apolipoprotein E4 allele, Tau H2 haplotype and behavioral variant FTD phenotype were associated with worse prognosis over time, suggesting specific genetic and clinical predictors in FTLD progression [87] that converge and differ from those of other tauopthies (e.g. amnestic MCI/AD and PSP).

Assorted reviews describe the different tauopathies, e.g. [53, 88]. Thus, tau comprises the Pick bodies found in Pick's disease (PiD) with aggregates of three microtubule binding repeat tau (3R tau) and frontal atrophy associated with cognitive clinical dysfunction of frontal dysexcutive syndrome, progressive nonfluent aphasia and semantic dementia [53, 88].

Like other tauopathies, progressive supranuclear palsy (PSP) is characterized by accumulation of abnormally phosphorylated tau [89, 90]. Predominantly, the four microtubule binding domain (4R) form of tau accumulates in straight NFTs in neurons and glial cells in affected brain regions, [90], [82, 91]. PSP is characterized by NFTs in basal ganglia, diencephalon, and brain stem with frontal executive cognitive impairment and movement disorders characterized by supranuclear gaze palsy, falls and Parkinsonism-like behavior [88, 92–94]. It is a rare progressive disease [95–97]. Like in AD and PiD, brain atrophy occurs in patients with PSP.

While some degree of brain atrophy occurs with normal aging, the atrophy in PSP is more pronounced [98, 99], correlating with clinical disease progression [100] and assessed by structural brain imaging [98, 101].

Cortico-basal ganglionic degeneration (CBD) is characterized by parietofrontal or frontotemporal atrophy and pallor in substantia nigra and 4R tau-aggregates (also found in glial lesions) that is manifested by cognitive dysfunction – cortical sensory loss, apraxia (loss of the ability to execute or carry out learned purposeful movements) and asymmetric akinetic rigid syndrome [53, 88].

PiD, PSP and CBD are subtypes within the broader category of called frontotemporal dementias (FTD). Another FTD is frontotemporal dementia with Parkinsonism linked to chromosome 17 (FTDP-17) that is characterized by frontal atrophy often seen with tau-positive neuronal and glial inclusions with behavioral changes, cognitive decline and Parkinsonism [102]. It is important to stress that the FTD group of diseases include additional different protein pathologies, other than tau, which were recently reviewed [103].

Many of the tau alterations that are characteristic of AD have also been identified in PSP and CBD. Tau truncation at Asp421 is an alteration that is unique to neuronal lesions, occurring in Pick bodies as well as in NFTs, but not in lesions associated with glia. Conversely, phosphorylation at Ser422 is not only present in all these lesions, but identifies additional glial and neuronal pathology in disease-susceptible cortical regions. The results suggest that the molecular alterations of tau that occur during the initial process of tangle formation in AD are similar in non-AD tauopathies, but the middle and later changes are not common to all diseases [104].

Tauopathies include also the early-onset dementia observed in Down syndrome (DS; trisomy 21). Splicing misregulation of adult-specific exon 10 results in expression of abnormal ratios of tau isoforms, leading to FTDP. Positions +3 to +19 of the intron downstream of exon 10 define a hotspot: Point mutations in it result in tauopathies. All these mutations increase the inclusion of exon 10 except for mutation +19, which almost entirely excludes exon 10. To investigate the tau connection between DS and AD, studies examined splicing factors located on chromosome 21 for their effect on tau exon 10. In these experiments, the splicing factor located on chromosome 21, heterogeneous nuclear ribonucleoprotein E3 – hnRNPE3 (PCBP3), was found to modestly activate the splicing of exon 10 by interacting with its proximal downstream intron around position +19 [105]. Other studies found decreased levels of Tra2beta, an RNA splicing factor responsible for tau exon 10 inclusion, in both in cortical cell cultures exposed to MG132 (proteasome inhibitor) and in cerebral cortex after ischemic injury, suggesting that transient focal cerebral ischemia reduces tau exon 10 splicing through a mechanism involving proteasome-ubiquitin dysfunction and down-regulation of Tra2beta [106].

Interestingly, single cell gene expression profiling coupled with custom-designed cDNA array analysis to evaluate tau expression and other cytoskeletal elements within individual neuronal populations in patients with no cognitive impairment (NCI), MCI, and AD, revealed a shift in the ratio of three-repeat tau to four-repeat tau mRNAs within individual human cholinergic basal forebrain neurons within nucleus basalis and CA1 hippocampal neurons during the progression of AD, but not during normal aging. A shift in tau alternative splicing may precipitate a cascade of events in the selective vulnerability of neurons, ultimately leading to tauopathies including AD [107] .

3 Diagnostic Tools

Cerebrospinal fluid (CSF) total tau levels vary widely in neurodegenerative disorders, thus not being useful in their discrimination over AD. It has been suggested that total tau alongside with amyloid beta can predict conversion from MCI to AD [108–111]. AD is characterized by a signature of phosphorylated tau and amyloid beta in the CSF [110], coupled with brain imaging technologies this can follow AD disease progression [112–114] (http://www.labtestsonline.org/understanding/analytes/tau/test.html). Commercial companies like Applied NeuroSolutions, Inc. (http://www.appliedneurosolutions.com/) and Innogenetics (http://www.innogenetics.com/neurodegeneration.html) offer measurements of phosphorylated tau in the CSF, specifically P-tau 231 and P-tau 181 determination, respectively. Furthermore, characterization of tau versions present in CSF with respect to their molecular mass and isoelectric point also facilitates advanced diagnosis [115].

A recent study characterized and measured tau forms in order to verify the differential patterns among neurodegenerative disorders. A quantitative immunoprecipitation showed an extended (55 kDa), and a truncated (33 kDa) forms of tau in the CSF with differential expression and was suggested as a potential marker for PSP [116] however, this finding is disputed by another group of researchers [117].

Another study suggested that CSF concentrations of the 42 amino acid fragment of amyloid-beta (Abeta42), neurofilament light chain (NFL), neurofilament heavy chain (pNFH), tau protein, glial fibrillary acidic protein (GFAP), neuron specific enolase (NSE), S-100B protein, and myelin basic protein (MBP) that are released into the CSF after brain tissue damage caused by a variety of neurological diseases could be of value in the differential diagnosis of neurodegenerative disorders [118].

CSF t-tau/Abeta(1–4) ratio, MRI volumetric biomarkers and neuropsychological tests are used to associate with disease progression and conversion for MCI to AD [119, 120].

In an interesting proteomics study, 18 signaling proteins in blood plasma were suggested to be used to classify blinded samples from AD and control subjects with close to 90% accuracy and to identify patients who had MCI that progressed to AD 2–6 years later. Biological analysis of the 18 proteins points to systemic dysregulation of hematopoiesis, immune responses, apoptosis and neuronal support in presymptomatic AD. Further studies should evaluate whether this test could be implemented for other tauopathies and whether distinction can be made based on this highly desirable plasma evaluation tool [121].

Further blood-based biomarkers of microvascular pathology in Alzheimer's disease were reviewed [122]. This is coupled to recent developments of gene expression arrays in peripheral blood [123, 124].

While amyloid beta detection by specific intercalating compounds correlating with CSF amyloid beta load are being developed and used [125] (http://eon.businesswire.com/news/eon/20120611006243/en/Navidea-Biopharmaceuticals/Inc.-%28NYSE-Amex%3A-NAVB%29/AZD4694), (http://www.amyvid.com/Pages/index.aspx) NFT in vivo labeling is still lagging. However, some studies also used a molecular imaging probe for plaques and tangles, 2-(1-{6-[(2-[F-18]fluoroethyl)(methyl)amino]-2-naphthyl}ethylidene)malonon itrile ([(18)F] FDDNP) relating cognitive ability to positron emission tomography (PET) and suggesting this an a potential early diagnostic tool [126]. (18)F-THK523 was recently described as a novel tau imaging radiotracer and would be interesting to follow-up on [127].

It has been long suggested that neurofibrillary tangles but not senile plaques parallel duration and severity of AD [128, 129]. Furthermore, in cognitively normal older individuals, Abeta-associated clinical decline during a mean of 3 years may occur only in the presence of ongoing downstream neurodegeneration and requires the presence of phosphorylated tau [130].

Tau as a drug target is a relatively young field and future clinical studies should aim at evaluating tau distribution in biological fluids as well as develop imaging technologies to identify potential changes in tau following candidate drug application.

4 Drug Candidates Aimed at Tauopathy

Several authors [131–133] have previously reviewed the status of Alzheimer's drug development. A major effort has been put into research and development of amyloid-targeting vaccine(s) [134, 135], future potential tau directed therapies have been also reviewed [5, 131, 136–142] and are updated in short below. Several aspects of tauopathy are currently targeted. Those include, targeting tangles to break the potentially toxic aggregates, inhibiting tau phosphorylation, accelerating tau dephosphorylation, and accelerating – tau microtubule interactions for an additional review please see [143].

Advanced models of tauopathy have been developed for translational research, including a model for rapid analysis of tau-related degeneration in zebrafish [144] and a model of tauopathy in drosophila that dissociates tau toxicity and phosphorylation at the level of GSK-3beta, MARK and Cdk5. The drosophila model suggests that, in addition to tau phosphorylation, microtubule binding plays a crucial role in the regulation of tau toxicity when misexpressed. These data have important implications for the understanding and interpretation of animal models of tauopathy [145]. Models of drosophila are also developed to enhance validation and show the importance of GSK3 in axonal transport [146]. Advanced mouse models are many as recently reviewed [147]. Preferred models encompass the ability to inactivate the transgene at will (e.g. [148]). Interestingly, the propagation of protein misfolding, such as tau, may occur through mechanisms similar to those that underlie prion pathogenesis [149–151].

Non-related studies have shown tau enrichment in metastatic tumors and an ability of tau to promote tumor cell reattachment through tubulin microtentacles formation, supporting a model in which tau-induced microtubule stabilization provides a selective advantage during tumor metastasis [152] and suggesting a potential cross talk between cancer therapeutics/diagnosis and tauopathies.

4.1 Drug Candidates

4.1.1 At the Clinical Stage

Davunetide: Tau, Microtubules and Neuroprotection

Our research demonstrated that an 8-amino-acid peptide fragment of activity-dependent neuroprotective protein (ADNP) [153–155] was capable of conferring neuroprotection, while inhibiting tau pathology (see recent review [156]) and protecting the microtubule network [157, 158]. This peptide fragment is called davunetide (also known as NAP, intranasal formulation, AL-108) and has the following amino acid sequence:

$$H_2N - Asn - Ala - Pro - Val - Ser - Ile - Pro - Gln - COOH \ [153]$$

Preclinical experiments indicate that davunetide has neuroprotective, cognitive protective, and neurotrophic properties. Importantly, NAP (davunetide), did not affect dividing cells (unlike paclitaxel and other microtubule targeting drugs) [159]. Davunetide has been recently reviewed from the preclinical [2–4, 156] as well as the clinical aspects [5, 160, 161]. In short, davunetide is in Phase 2/3 clinical trial in progressive supranuclear palsy (PSP) a neurodegenerative disease presenting tauopathy, Allon Therapeutics Inc., (www.allontherapeutics.com). As reviewed [5], PSP has emerged as an attractive clinical trial population for three main reasons: (1) It is a debilitating and life-threatening disease for which there is a desire to develop effective treatment options that currently do not exist; (2) it serves to prove whether a tau-directed therapy can affect a primary neurodegenerative disease, thereby reducing the risk of failure in a related condition, namely AD; and (3) it is a rare disease that qualifies as an orphan indication thereby providing certain commercial incentives and allows sponsors to pursue development with more limited resources. As indicated above, davunetide's efficacy and tolerability are being tested in a placebo-controlled study in 300 PSP patients, making it the most advanced drug candidate in this indication, NCT01110720.

Epothilone D (BMS-241027)

Paclitaxel and related compounds have also been suggested to act by protecting microtubules, however, bioavailability, brain specificity and the potential irreversibility of their action suggests that more work is required prior to paclitaxel-like neuroprotective drug candidate [162]. A recent work, however, showed that the cancer drug epothilone D (BMS-241027) protects axonal transport [163]. This drug is now in phase 1 clinical trials, NCT01492374.

Rember (Methylene Blue)

Rember (methylene blue, suggested among others to inhibit tau aggregation) was featured in several AD meetings as per [132, 164], and completed phase 2 studies in AD as reviewed [142], however, updates are still forthcoming.

Lithium

Lithium (inhibition of GSK3) was also reviewed [142]. It was in clinical trials in PSP and CBD, 28 weeks (NCT00703677), showing no safety and no efficacy signs. It was also tested in AD patients (NCT00088387). Disease modifying potential is currently being evaluated (NCT01055392).

Tideglusib (NP12, Nypta®)

NP12 (Nypta®), is a GSK-3 inhibitor [165], that we have previously reviewed [133]. It is a potent thiadiazolidinone derivative, that when injected into the rat hippocampus dramatically reduces kainic acid-induced inflammation, as measured by edema formation using T_2-weighted magnetic resonance imaging and glial activation and has a neuroprotective effect in the damaged areas of the hippocampus [166].

Tideglusib, claimed to be the first non ATP competitive inhibitor of GSK3, has been in three clinical trials, the two recent ones (phase 2) include:

1. Safety, Tolerability, and Efficacy of Two Different Oral Doses of NP031112 Versus Placebo in the Treatment of Patients With Mild-to-Moderate Progressive Supranuclear Palsy (Tauros), NCT01049399 – completed.
2. Efficacy, Safety and Tolerability of Tideglusib to Treat Mild-to-Moderate Alzheimer's Disease Patients (ARGO) – NCT01350362 (active not recruiting). Overall the Tideglusib are not encouraging.

4.1.2 At the Preclinical Stage

Aggregation

At the preclinical stage, a phenylthiazolyl-hydrazide (PTH) compound was suggested as a possible hit in terms of inhibition of tau aggregation and the core of the PTHs crucial for activity was identified, thus representing a putative lead structure [167]. Recently, quantitative high-throughput screening (qHTS) of approximately 292,000 compounds to identify drug-like inhibitors of tau assembly was developed. The fibrillization of a truncated tau fragment that contains four MT-binding domains was monitored in an assay that employed complementary thioflavin T fluorescence and fluorescence polarization methods. Previously described classes of inhibitors as well as new scaffolds were identified, including novel aminothienopyridazines (ATPZs). A number of ATPZ analogues were synthesized, and structure-activity relationships were defined. Further characterization of representative ATPZ compounds showed they do not interfere with tau-mediated MT assembly, and they are significantly more effective at preventing the fibrillization of tau than the Abeta(1–42) peptide which forms AD senile plaques. Thus, the ATPZ molecules are suggested for further development [168]. As a follow-up on these publications, AstraZeneca and the University of Pennsylvania announced a new collaborative research agreement that initially will focus on generating new AD drug candidates for the clinical development pipeline (http://www.astrazeneca.com/research/?itemId=8876304).

Various tau aggregation inhibitory molecules were recently reviewed [132] also taking into consideration that the cellular environment affects tau aggregation. In this respect, it has been suggested that tau fragments across the lysosomal membrane promote formation of tau oligomers at the surface of these organelles which may act as precursors of aggregation and interfere with lysosomal functioning [169].

Kinase Inhibition

Other aspects of tau-related future therapeutics involve targeting tau hyperphosphorylation [170] with inhibitors of the key tau phosphorylating enzymes such as Cdk5/p25 kinase [171], GSK-3 inhibitors, such as lithium [172] (see above) or peptide inhibitors [173] and possible enhancement of protein phosphatase-2A (PP-2A) [174]. Other kinases implicated in tau phosphorylation include CK1, PK1, MARK, and the stress associated kinases, p38MAPK and JNK [170]. ATP competitive inhibitors may present potent drug candidates, but may also have undesirable side effects [143].

Regarding kinase inhibition, internet search of SAR-502250 revealed the patent application: WO/2009/035159 also including Mitsubishi Tanabe Pharma Corporation in the applicants. The patent abstract describes a compound represented by a formula that is disclosed or a pharmaceutically acceptable salt thereof: which is used for preventive and/or therapeutic treatment of a disease caused by abnormal activity of tau protein kinase 1 such as a neurodegenerative diseases (e.g. Alzheimer disease) (http://www.wipo.int/pctdb/en/wo.jsp?WO=2009035159). Furthermore, SRN-003-556 was described as an inhibitor of tau hyperphosphorylation that prevents severe motor impairments in tau transgenic mice [175].

Also at the preclinical stage, several compounds that inhibit MARK kinases, which are known to phosphorylate tau, were reported. The ChemBioNet library, a collection of 17,591 small molecules was screened and four of the identified inhibitors share a similar functional group – the 9-oxo-9H-acridin-10-yl functional group – which may provide a basis for developing inhibitors that can be developed as human therapeutics [78].

Synthesis, Degradation and Neuroprotection

Brunden et al. also review the possibility of increased tau degradation as a therapeutic target also looking at hsp90 [176] as well as macroautophagy [169]. Neuroprotection is another approach to another disease associated with tau deregulation (rasagiline mesylate). A clinical trial with patients with Multiple System Atrophy (MSA), a disease with tau inclusions [177] and with parkinsonism has been completed (NCT00977665).

Yet, another approach to the regulation of tau could be specific inhibition of synthesis, when this seems to lead to accumulation as well as addressing specific tau mutations (please see above and as described in http://www.alzforum.org/res/com/mut/tau/table1.asp). When RNA silencing and targeted in vivo mutagenesis further develop to potential candidate drugs, this would be an interesting avenue to follow. Finally, given the preponderance of 4 repeat (4R) tubulin binding domain tau in the neurodegenerative brain, drugs that target enhancement in the expression of the 3R tau to restore homeostasis are highly desirable, however, keeping in mind that the homeostasis between the 3R and 4R tau isoforms is important and that 3R tau has been associated with the pathology of Pick's disease [53, 88].

Immunotherapy

Another approach to clearing tau aggregates has been tau immunotherapy [178–182]; this approach is still at the preclinical stage, showing potential promise in animal studies [183–185].

Additional Reading

Additional reading on the subject can be found is several reviews including, but not limited to the following literature citations [143, 186] and including insights to the selection of the patient populations [187].

Personalized Medicine

At the current stage of diagnostic and drug development it is difficult to talk about personalized approach. However, addressing patient stratification and dosing will no doubt improve clinical outcome and advance better management of devastating tauopathies.

Acknowledgements I am grateful to the Allon Therapeutics team for valuable input. Studies were supported by Allon Therapeutics Inc.

References

1. Gozes I (2010) Tau pathology and future therapeutics. Curr Alzheimer Res 7:685–696
2. Gozes I (2011) Microtubules, schizophrenia and cognitive behavior: preclinical development of davunetide (NAP) as a peptide-drug candidate. Peptides 32:428–431
3. Gozes I (2011) NAP (davunetide) provides functional and structural neuroprotection. Curr Pharm Des 17:1040–1044
4. Gozes I (2011) Microtubules (tau) as an emerging therapeutic target: NAP (davunetide). Curr Pharm Des 17:3413–3417
5. Gold M, Lorenzl S, Stewart AJ, Morimoto BH, Williams DR, Gozes I (2012) Critical appraisal of the role of davunetide in the treatment of progressive supranuclear palsy. Neuropsychiatr Dis Treat 8:85–93
6. Goedert M, Crowther RA, Spillantini MG (1998) Tau mutations cause frontotemporal dementias. Neuron 21:955–958
7. Gozes I, Littauer UZ (1978) Tubulin microheterogeneity increases with rat brain maturation. Nature 276:411–413
8. Gozes I, Sweadner KJ (1981) Multiple tubulin forms are expressed by a single neurone. Nature 294:477–480
9. Gozes I, Saya D, Littauer UZ (1979) Tubulin microheterogeneity in neuroblastoma and glioma cell lines differs from that of the brain. Brain Res 171:171–175
10. Gao J, Dennis JE, Muzic RF, Lundberg M, Caplan AI (2001) The dynamic in vivo distribution of bone marrow-derived mesenchymal stem cells after infusion. Cells Tissues Organs 169: 12–20
11. Cleveland DW, Hwo SY, Kirschner MW (1977) Purification of tau, a microtubule-associated protein that induces assembly of microtubules from purified tubulin. J Mol Biol 116:207–225
12. Witman GB, Cleveland DW, Weingarten MD, Kirschner MW (1976) Tubulin requires tau for growth onto microtubule initiating sites. Proc Natl Acad Sci U S A 73:4070–4074
13. Bernhardt R, Matus A (1984) Light and electron microscopic studies of the distribution of microtubule-associated protein 2 in rat brain: a difference between dendritic and axonal cytoskeletons. J Comp Neurol 226:203–221
14. Weingarten MD, Lockwood AH, Hwo SY, Kirschner MW (1975) A protein factor essential for microtubule assembly. Proc Natl Acad Sci U S A 72:1858–1862
15. Cleveland DW, Hwo SY, Kirschner MW (1977) Physical and chemical properties of purified tau factor and the role of tau in microtubule assembly. J Mol Biol 116:227–247
16. Drubin DG, Caput D, Kirschner MW (1984) Studies on the expression of the microtubule-associated protein, tau, during mouse brain development, with newly isolated complementary DNA probes. J Cell Biol 98:1090–1097
17. Ebneth A, Godemann R, Stamer K, Illenberger S, Trinczek B, Mandelkow E (1998) Overexpression of tau protein inhibits kinesin-dependent trafficking of vesicles, mitochondria, and endoplasmic reticulum: implications for Alzheimer's disease. J Cell Biol 143:777–794
18. Stamer K, Vogel R, Thies E, Mandelkow E, Mandelkow EM (2002) Tau blocks traffic of organelles, neurofilaments, and APP vesicles in neurons and enhances oxidative stress. J Cell Biol 156:1051–1063
19. Gozes I (1982) Tubulin in the nervous system. Neurochem Int 4:101–120
20. Goedert M, Spillantini MG, Jakes R, Rutherford D, Crowther RA (1989) Multiple isoforms of human microtubule-associated protein tau: sequences and localization in neurofibrillary tangles of Alzheimer's disease. Neuron 3:519–526

21. Himmler A, Drechsel D, Kirschner MW, Martin DW Jr (1989) Tau consists of a set of proteins with repeated C-terminal microtubule-binding domains and variable N-terminal domains. Mol Cell Biol 9:1381–1388
22. Stoothoff W, Jones PB, Spires-Jones TL, Joyner D, Chhabra E, Bercury K, Fan Z, Xie H, Bacskai B, Edd J, Irimia D, Hyman BT (2009) Differential effect of three-repeat and four-repeat tau on mitochondrial axonal transport. J Neurochem 111:417–427
23. Shemesh OA, Erez H, Ginzburg I, Spira ME (2008) Tau-induced traffic jams reflect organelles accumulation at points of microtubule polar mismatching. Traffic 9:458–471
24. Witte H, Neukirchen D, Bradke F (2008) Microtubule stabilization specifies initial neuronal polarization. J Cell Biol 180:619–632
25. Eickholt BJ, Walsh FS, Doherty P (2002) An inactive pool of GSK-3 at the leading edge of growth cones is implicated in Semaphorin 3A signaling. J Cell Biol 157:211–217
26. Hoover BR, Reed MN, Su J, Penrod RD, Kotilinek LA, Grant MK, Pitstick R, Carlson GA, Lanier LM, Yuan LL, Ashe KH, Liao D (2010) Tau mislocalization to dendritic spines mediates synaptic dysfunction independently of neurodegeneration. Neuron 68:1067–1081
27. Lee S, Kim W, Li Z, Hall GF (2012) Accumulation of vesicle-associated human tau in distal dendrites drives degeneration and tau secretion in an in situ cellular tauopathy model. Int J Alzheimers Dis 2012:172837
28. Dickstein DL, Brautigam H, Stockton SD Jr, Schmeidler J, Hof PR (2010) Changes in dendritic complexity and spine morphology in transgenic mice expressing human wild-type tau. Brain Struct Funct 214:161–179
29. Liu L, Drouet V, Wu JW, Witter MP, Small SA, Clelland C, Duff K (2012) Trans-synaptic spread of tau pathology in vivo. PLoS One 7:e31302
30. Goedert M, Clavaguera F, Tolnay M (2010) The propagation of prion-like protein inclusions in neurodegenerative diseases. Trends Neurosci 33:317–325
31. Francon J, Lennon AM, Fellous A, Mareck A, Pierre M, Nunez J (1982) Heterogeneity of microtubule-associated proteins and brain development. Eur J Biochem 129:465–471
32. Hong XP, Peng CX, Wei W, Tian Q, Liu YH, Yao XQ, Zhang Y, Cao FY, Wang Q, Wang JZ (2010) Essential role of tau phosphorylation in adult hippocampal neurogenesis. Hippocampus 20:1339–1349
33. He HJ, Wang XS, Pan R, Wang DL, Liu MN, He RQ (2009) The proline-rich domain of tau plays a role in interactions with actin. BMC Cell Biol 10:81
34. Chambraud B, Sardin E, Giustiniani J, Dounane O, Schumacher M, Goedert M, Baulieu EE (2010) A role for FKBP52 in tau protein function. Proc Natl Acad Sci U S A 107:2658–2663
35. Quinta HR, Galigniana MD (2012) The neuroregenerative mechanism mediated by the Hsp90-binding immunophilin FKBP52 resembles the early steps of neuronal differentiation. Br J Pharmacol 166:637–649
36. Li Y, Chen B (1995) Differential regulation of fyn-associated protein tyrosine kinase activity by macrophage colony-stimulating factor (M-CSF) and granulocyte-macrophage colony-stimulating factor (GM-CSF). J Leukoc Biol 57:484–490
37. Roberson ED, Halabisky B, Yoo JW, Yao J, Chin J, Yan F, Wu T, Hamto P, Devidze N, Yu GQ, Palop JJ, Noebels JL, Mucke L (2011) Amyloid-beta/Fyn-induced synaptic, network, and cognitive impairments depend on tau levels in multiple mouse models of Alzheimer's disease. J Neurosci 31:700–711
38. Usardi A, Pooler AM, Seereeram A, Reynolds CH, Derkinderen P, Anderton B, Hanger DP, Noble W, Williamson R (2011) Tyrosine phosphorylation of tau regulates its interactions with Fyn SH2 domains, but not SH3 domains, altering the cellular localization of tau. FEBS J 278:2927–2937
39. Zempel H, Mandelkow EM (2012) Linking amyloid-beta and tau: amyloid-beta induced synaptic dysfunction via local wreckage of the neuronal cytoskeleton. Neurodegener Dis 10:64–72
40. Ballatore C, Lee VM, Trojanowski JQ (2007) Tau-mediated neurodegeneration in Alzheimer's disease and related disorders. Nat Rev Neurosci 8:663–672

41. Grundke-Iqbal I, Iqbal K, Quinlan M, Tung YC, Zaidi MS, Wisniewski HM (1986) Microtubule-associated protein tau. A component of Alzheimer paired helical filaments. J Biol Chem 261:6084–6089
42. Delacourte A, Defossez A (1986) Alzheimer's disease: tau proteins, the promoting factors of microtubule assembly, are major components of paired helical filaments. J Neurol Sci 76: 173–186
43. Wood JG, Mirra SS, Pollock NJ, Binder LI (1986) Neurofibrillary tangles of Alzheimer disease share antigenic determinants with the axonal microtubule-associated protein tau (tau). Proc Natl Acad Sci U S A 83:4040–4043
44. Kosik KS, Joachim CL, Selkoe DJ (1986) Microtubule-associated protein tau (tau) is a major antigenic component of paired helical filaments in Alzheimer disease. Proc Natl Acad Sci U S A 83:4044–4048
45. Nukina N, Ihara Y (1986) One of the antigenic determinants of paired helical filaments is related to tau protein. J Biochem (Tokyo) 99:1541–1544
46. Goedert M, Wischik CM, Crowther RA, Walker JE, Klug A (1988) Cloning and sequencing of the cDNA encoding a core protein of the paired helical filament of Alzheimer disease: identification as the microtubule-associated protein tau. Proc Natl Acad Sci U S A 85: 4051–4055
47. Wischik CM, Novak M, Thøgersen HC, Edwards PC, Runswick MJ, Jakes R, Walker JE, Milstein C, Roth M, Klug A (1988) Isolation of a fragment of tau derived from the core of the paired helical filament of Alzheimer disease. Proc Natl Acad Sci U S A 85:4506–4510
48. Kidd M (1963) Paired helical filaments in electron microscopy of Alzheimer's disease. Nature 197:192–193
49. Kidd M (1964) Alzheimer's disease – an electron microscopical study. Brain 87:307–320
50. Wisniewski HM, Narang HK, Terry RD (1976) Neurofibrillary tangles of paired helical filaments. J Neurol Sci 27:173–181
51. Crowther RA, Wischik CM (1985) Image reconstruction of the Alzheimer paired helical filament. EMBO J 4:3661–3665
52. Gozes I, Schmitt H, Littauer UZ (1975) Translation in vitro of rat brain messenger RNA coding for tubulin and actin. Proc Natl Acad Sci U S A 72:701–705
53. Robert M, Mathuranath PS (2007) Tau and tauopathies. Neurol India 55:11–16
54. Mason JL, Suzuki K, Chaplin DD, Matsushima GK (2001) Interleukin-1beta promotes repair of the CNS. J Neurosci 21:7046–7052
55. Temporin K, Tanaka H, Kuroda Y, Okada K, Yachi K, Moritomo H, Murase T, Yoshikawa H (2008) IL-1beta promotes neurite outgrowth by deactivating RhoA via p38 MAPK pathway. Biochem Biophys Res Commun 365:375–380
56. Arnett HA, Mason J, Marino M, Suzuki K, Matsushima GK, Ting JP (2001) TNF alpha promotes proliferation of oligodendrocyte progenitors and remyelination. Nat Neurosci 4:1116–1122
57. Platel JC, Dave KA, Gordon V, Lacar B, Rubio ME, Bordey A (2010) NMDA receptors activated by subventricular zone astrocytic glutamate are critical for neuroblast survival prior to entering a synaptic network. Neuron 65:859–872
58. Grundke-Iqbal I, Iqbal K, Tung YC, Quinlan M, Wisniewski HM, Binder LI (1986) Abnormal phosphorylation of the microtubule-associated protein tau (tau) in Alzheimer cytoskeletal pathology. Proc Natl Acad Sci U S A 83:4913–4917
59. Ihara Y, Nukina N, Miura R, Ogawara M (1986) Phosphorylated tau protein is integrated into paired helical filaments in Alzheimer's disease. J Biochem (Tokyo) 99:1807–1810
60. Trojanowski JQ, Lee VM (1995) Phosphorylation of paired helical filament tau in Alzheimer's disease neurofibrillary lesions: focusing on phosphatases. FASEB J 9:1570–1576
61. Liu F, Iqbal K, Grundke-Iqbal I, Hart GW, Gong CX (2004) O-GlcNAcylation regulates phosphorylation of tau: a mechanism involved in Alzheimer's disease. Proc Natl Acad Sci U S A 101:10804–10809
62. Lee G, Thangavel R, Sharma VM, Litersky JM, Bhaskar K, Fang SM, Do LH, Andreadis A, Van Hoesen G, Ksiezak-Reding H (2004) Phosphorylation of tau by fyn: implications for Alzheimer's disease. J Neurosci 24:2304–2312

63. Gong CX, Liu F, Grundke-Iqbal I, Iqbal K (2005) Post-translational modifications of tau protein in Alzheimer's disease. J Neural Transm 112:813–838
64. Gordon-Krajcer W, Yang L, Ksiezak-Reding H (2000) Conformation of paired helical filaments blocks dephosphorylation of epitopes shared with fetal tau except Ser199/202 and Ser202/Thr205. Brain Res 856:163–175
65. Hernandez F, Lucas JJ, Cuadros R, Avila J (2003) GSK-3 dependent phosphoepitopes recognized by PHF-1 and AT-8 antibodies are present in different tau isoforms. Neurobiol Aging 24:1087–1094
66. Terwel D, Dewachter I, Van Leuven F (2002) Axonal transport, tau protein, and neurodegeneration in Alzheimer's disease. Neuromolecular Med 2:151–165
67. Drechsel DN, Hyman AA, Cobb MH, Kirschner MW (1992) Modulation of the dynamic instability of tubulin assembly by the microtubule-associated protein tau. Mol Biol Cell 3:1141–1154
68. Bramblett GT, Goedert M, Jakes R, Merrick SE, Trojanowski JQ, Lee VM (1993) Abnormal tau phosphorylation at Ser396 in Alzheimer's disease recapitulates development and contributes to reduced microtubule binding. Neuron 10:1089–1099
69. Yoshida H, Ihara Y (1993) Tau in paired helical filaments is functionally distinct from fetal tau: assembly incompetence of paired helical filament-tau. J Neurochem 61:1183–1186
70. Biernat J, Gustke N, Drewes G, Mandelkow EM, Mandelkow E (1993) Phosphorylation of Ser262 strongly reduces binding of tau to microtubules: distinction between PHF-like immunoreactivity and microtubule binding. Neuron 11:153–163
71. Jameson L, Frey T, Zeeberg B, Dalldorf F, Caplow M (1980) Inhibition of microtubule assembly by phosphorylation of microtubule-associated proteins. Biochemistry 19:2472–2479
72. Lindwall G, Cole RD (1984) Phosphorylation affects the ability of tau protein to promote microtubule assembly. J Biol Chem 259:5301–5305
73. Alonso A, Zaidi T, Novak M, Grundke-Iqbal I, Iqbal K (2001) Hyperphosphorylation induces self-assembly of tau into tangles of paired helical filaments/straight filaments. Proc Natl Acad Sci U S A 98:6923–6928
74. Haase C, Stieler JT, Arendt T, Holzer M (2004) Pseudophosphorylation of tau protein alters its ability for self-aggregation. J Neurochem 88:1509–1520
75. Litersky JM, Johnson GV (1992) Phosphorylation by cAMP-dependent protein kinase inhibits the degradation of tau by calpain. J Biol Chem 267:1563–1568
76. Eriksson PS, Perfilieva E, Björk-Eriksson T, Alborn AM, Nordborg C, Peterson DA, Gage FH (1998) Neurogenesis in the adult human hippocampus. Nat Med 4:1313–1317
77. Drewes G, Ebneth A, Preuss U, Mandelkow EM, Mandelkow E (1997) MARK, a novel family of protein kinases that phosphorylate microtubule-associated proteins and trigger microtubule disruption. Cell 89:297–308
78. Timm T, von Kries JP, Li X, Zempel H, Mandelkow E, Mandelkow EM (2011) Microtubule affinity regulating kinase activity in living neurons was examined by a genetically encoded fluorescence resonance energy transfer/fluorescence lifetime imaging-based biosensor: inhibitors with therapeutic potential. J Biol Chem 286:41711–41722
79. Liao X, Zhang Y, Wang Y, Wang J (2004) The effect of cdk-5 overexpression on tau phosphorylation and spatial memory of rat. Sci China C Life Sci 47:251–257
80. Ziv Y, Ron N, Butovsky O, Landa G, Sudai E, Greenberg N, Cohen H, Kipnis J, Schwartz M (2006) Immune cells contribute to the maintenance of neurogenesis and spatial learning abilities in adulthood. Nat Neurosci 9:268–275
81. Ohtaki H, Ylostalo JH, Foraker JE, Robinson AP, Reger RL, Shioda S, Prockop DJ (2008) Stem/progenitor cells from bone marrow decrease neuronal death in global ischemia by modulation of inflammatory/immune responses. Proc Natl Acad Sci U S A 105:14638–14643
82. Baker M, Litvan I, Houlden H, Adamson J, Dickson D, Perez-Tur J, Hardy J, Lynch T, Bigio E, Hutton M (1999) Association of an extended haplotype in the tau gene with progressive supranuclear palsy. Hum Mol Genet 8:711–715
83. Pittman AM, Myers AJ, Duckworth J, Bryden L, Hanson M, Abou-Sleiman P, Wood NW, Hardy J, Lees A, de Silva R (2004) The structure of the tau haplotype in controls and in progressive supranuclear palsy. Hum Mol Genet 13:1267–1274

84. Caffrey TM, Wade-Martins R (2007) Functional MAPT haplotypes: bridging the gap between genotype and neuropathology. Neurobiol Dis 27:1–10
85. Pittman AM, Myers AJ, Abou-Sleiman P, Fung HC, Kaleem M, Marlowe L, Duckworth J, Leung D, Williams D, Kilford L, Thomas N, Morris CM, Dickson D, Wood NW, Hardy J, Lees AJ, de Silva R (2005) Linkage disequilibrium fine mapping and haplotype association analysis of the tau gene in progressive supranuclear palsy and corticobasal degeneration. J Med Genet 42:837–846
86. Di Maria E, Cammarata S, Parodi MI, Borghi R, Benussi L, Galli M, Galimberti D, Ghidoni R, Gonella D, Novello C, Pollero V, Perroni L, Odetti P, Scarpini E, Binetti G, Tabaton M (2010) The h1 haplotype of the tau gene (mapt) is associated with mild cognitive impairment. J Alzheimers Dis 19:909–914
87. Borroni B, Grassi M, Agosti C, Premi E, Archetti S, Alberici A, Bellelli G, Caimi L, Di Luca M, Padovani A (2010) Establishing short-term prognosis in frontotemporal lobar degeneration spectrum: Role of genetic background and clinical phenotype. Neurobiol Aging 31:270–279
88. Desai AK, Chand P (2009) Tau-based therapies for Alzheimer's disease: wave of the future? Prim Psychiatry 16:40–46
89. Hauw JJ, Daniel SE, Dickson D, Horoupian DS, Jellinger K, Lantos PL, McKee A, Tabaton M, Litvan I (1994) Preliminary NINDS neuropathologic criteria for Steele-Richardson-Olszewski syndrome (progressive supranuclear palsy). Neurology 44:2015–2019
90. Dickson DW, Rademakers R, Hutton ML (2007) Progressive supranuclear palsy: pathology and genetics. Brain Pathol 17:74–82
91. Spillantini MG, Bird TD, Ghetti B (1998) Frontotemporal dementia and Parkinsonism linked to chromosome 17: a new group of tauopathies. Brain Pathol 8:387–402
92. Litvan I, Agid Y, Jankovic J, Goetz C, Brandel JP, Lai EC, Wenning G, D'Olhaberriague L, Verny M, Chaudhuri KR, McKee A, Jellinger K, Bartko JJ, Mangone CA, Pearce RK (1996) Accuracy of clinical criteria for the diagnosis of progressive supranuclear palsy (Steele-Richardson-Olszewski syndrome). Neurology 46:922–930
93. Pillon B, Dubois B, Ploska A, Agid Y (1991) Severity and specificity of cognitive impairment in Alzheimer's, Huntington's, and Parkinson's diseases and progressive supranuclear palsy. Neurology 41:634–643
94. Litvan I, Agid Y, Calne D, Campbell G, Dubois B, Duvoisin RC, Goetz CG, Golbe LI, Grafman J, Growdon JH, Hallett M, Jankovic J, Quinn NP, Tolosa E, Zee DS (1996) Clinical research criteria for the diagnosis of progressive supranuclear palsy (Steele-Richardson-Olszewski syndrome): report of the NINDS-SPSP international workshop. Neurology 47:1–9
95. Schrag A, Ben-Shlomo Y, Quinn NP (1999) Prevalence of progressive supranuclear palsy and multiple system atrophy: a cross-sectional study. Lancet 354:1771–1775
96. Testa D, Monza D, Ferrarini M, Soliveri P, Girotti F, Filippini G (2001) Comparison of natural histories of progressive supranuclear palsy and multiple system atrophy. Neurol Sci 22:247–251
97. Golbe LI, Ohman-Strickland PA (2007) A clinical rating scale for progressive supranuclear palsy. Brain 130:1552–1565
98. Paviour DC, Price SL, Lees AJ, Fox NC (2007) MRI derived brain atrophy in PSP and MSA-P. Determining sample size to detect treatment effects. J Neurol 254:478–481
99. Whitwell JL, Jack CR Jr, Parisi JE, Knopman DS, Boeve BF, Petersen RC, Ferman TJ, Dickson DW, Josephs KA (2007) Rates of cerebral atrophy differ in different degenerative pathologies. Brain 130:1148–1158
100. Cordato NJ, Duggins AJ, Halliday GM, Morris JG, Pantelis C (2005) Clinical deficits correlate with regional cerebral atrophy in progressive supranuclear palsy. Brain 128:1259–1266
101. Boxer AL, Geschwind MD, Belfor N, Gorno-Tempini ML, Schauer GF, Miller BL, Weiner MW, Rosen HJ (2006) Patterns of brain atrophy that differentiate corticobasal degeneration syndrome from progressive supranuclear palsy. Arch Neurol 63:81–86
102. Hutton M, Lendon CL, Rizzu P, Baker M, Froelich S, Houlden H, Pickering-Brown S, Chakraverty S, Isaacs A, Grover A, Hackett J, Adamson J, Lincoln S, Dickson D, Davies P, Petersen RC, Stevens M, de Graaff E, Wauters E, van Baren J, Hillebrand M, Joosse M,

Kwon JM, Nowotny P, Che LK, Norton J, Morris JC, Reed LA, Trojanowski J, Basun H, Lannfelt L, Neystat M, Fahn S, Dark F, Tannenberg T, Dodd PR, Hayward N, Kwok JB, Schofield PR, Andreadis A, Snowden J, Craufurd D, Neary D, Owen F, Oostra BA, Hardy J, Goate A, van Swieten J, Mann D, Lynch T, Heutink P (1998) Association of missense and 5'-splice-site mutations in tau with the inherited dementia FTDP-17. Nature 393:702–705
103. Rademakers R, Neumann M, Mackenzie IR (2012) Advances in understanding the molecular basis of frontotemporal dementia. Nat Rev Neurol 8:423–434
104. Guillozet-Bongaarts AL, Glajch KE, Libson EG, Cahill ME, Bigio E, Berry RW, Binder LI (2007) Phosphorylation and cleavage of tau in non-AD tauopathies. Acta Neuropathol 113:513–520
105. Lee RH, Pulin AA, Seo MJ, Kota DJ, Ylostalo J, Larson BL, Semprun-Prieto L, Delafontaine P, Prockop DJ (2009) Intravenous hMSCs improve myocardial infarction in mice because cells embolized in lung are activated to secrete the anti-inflammatory protein TSG-6. Cell Stem Cell 5:54–63
106. Nemeth K, Leelahavanichkul A, Yuen PS, Mayer B, Parmelee A, Doi K, Robey PG, Leelahavanichkul K, Koller BH, Brown JM, Hu X, Jelinek I, Star RA, Mezey E (2009) Bone marrow stromal cells attenuate sepsis via prostaglandin E(2)-dependent reprogramming of host macrophages to increase their interleukin-10 production. Nat Med 15:42–49
107. Ginsberg SD, Che S, Counts SE, Mufson EJ (2006) Shift in the ratio of three-repeat tau and four-repeat tau mRNAs in individual cholinergic basal forebrain neurons in mild cognitive impairment and Alzheimer's disease. J Neurochem 96:1401–1408
108. Schrepfer S, Deuse T, Reichenspurner H, Fischbein MP, Robbins RC, Pelletier MP (2007) Stem cell transplantation: the lung barrier. Transplant Proc 39:573–576
109. Shaw LM, Vanderstichele H, Knapik-Czajka M, Clark CM, Aisen PS, Petersen RC, Blennow K, Soares H, Simon A, Lewczuk P, Dean R, Siemers E, Potter W, Lee VM, Trojanowski JQ, Alzheimer's Disease Neuroimaging Initiative (2009) Cerebrospinal fluid biomarker signature in Alzheimer's disease neuroimaging initiative subjects. Ann Neurol 65:403–413
110. Buerger K, Frisoni G, Uspenskaya O, Ewers M, Zetterberg H, Geroldi C, Binetti G, Johannsen P, Rossini PM, Wahlund LO, Vellas B, Blennow K, Hampel H (2009) Validation of Alzheimer's disease CSF and plasma biological markers: the multicentre reliability study of the pilot European Alzheimer's Disease Neuroimaging Initiative (E-ADNI). Exp Gerontol 44:579–585
111. Rissman RA et al (2004) Caspase-cleavage of tau is an early event in Alzheimer disease tangle pathology. J Clin Invest 114:121–130
112. Ohtaki H, Satoh A, Nakamachi T, Yofu S, Dohi K, Mori H, Ohara K, Miyamoto K, Hashimoto H, Shintani N, Baba A, Matsunaga M, Shioda S (2010) Regulation of oxidative stress by pituitary adenylate cyclase-activating polypeptide (PACAP) mediated by PACAP receptor. J Mol Neurosci 42:397–403
113. Mori H, Nakamachi T, Ohtaki H, Yofu S, Sato A, Endo K, Iso Y, Suzuki H, Takeyama Y, Shintani N, Hashimoto H, Baba A, Shioda S (2010) Cardioprotective effect of endogenous pituitary adenylate cyclase-activating polypeptide on doxorubicin-induced cardiomyopathy in mice. Circ J 74:1183–1190
114. Stomrud E, Hansson O, Zetterberg H, Blennow K, Minthon L, Londos E (2010) Correlation of longitudinal cerebrospinal fluid biomarkers with cognitive decline in healthy older adults. Arch Neurol 67:217–223
115. Shiryaev N, Jouroukhin Y, Gozes I (2010) 3R tau expression modifies behavior in transgenic mice. J Neurosci Res 88:2727–2735
116. Borroni B, Hansson O, Zetterberg H, Blennow K, Minthon L, Londos E, Gardoni F, Parnetti L, Magno L, Malinverno M, Saggese E, Calabresi P, Spillantini MG, Padovani A, Di Luca M (2009) Pattern of Tau forms in CSF is altered in progressive supranuclear palsy. Neurobiol Aging 30:34–40
117. Kuiperij HB, Verbeek MM (2012) Detection of tau forms in CSF requires sensitive techniques. Neurobiol Aging 33:1841

118. Constantinescu R, Zetterberg H, Holmberg B, Rosengren L (2009) Levels of brain related proteins in cerebrospinal fluid: an aid in the differential diagnosis of parkinsonian disorders. Parkinsonism Relat Disord 15:205–212
119. Ewers M, Walsh C, Trojanowski JQ, Shaw LM, Petersen RC, Jack CR Jr, Feldman HH, Bokde AL, Alexander GE, Scheltens P, Vellas B, Dubois B, Weiner M, Hampel H, North American Alzheimer's Disease Neuroimaging Initiative (ADNI) (2012) Prediction of conversion from mild cognitive impairment to Alzheimer's disease dementia based upon biomarkers and neuropsychological test performance. Neurobiol Aging 33:1203–1214, e1202
120. Sperling RA, Aisen PS, Beckett LA, Bennett DA, Craft S, Fagan AM, Iwatsubo T, Jack CR Jr, Kaye J, Montine TJ, Park DC, Reiman EM, Rowe CC, Siemers E, Stern Y, Yaffe K, Carrillo MC, Thies B, Morrison-Bogorad M, Wagster MV, Phelps CH (2011) Toward defining the preclinical stages of Alzheimer's disease: recommendations from the National Institute on Aging-Alzheimer's Association workgroups on diagnostic guidelines for Alzheimer's disease. Alzheimers Dement 7:280–292
121. Ray S, Britschgi M, Herbert C, Takeda-Uchimura Y, Boxer A, Blennow K, Friedman LF, Galasko DR, Jutel M, Karydas A, Kaye JA, Leszek J, Miller BL, Minthon L, Quinn JF, Rabinovici GD, Robinson WH, Sabbagh MN, So YT, Sparks DL, Tabaton M, Tinklenberg J, Yesavage JA, Tibshirani R, Wyss-Coray T (2007) Classification and prediction of clinical Alzheimer's diagnosis based on plasma signaling proteins. Nat Med 13:1359–1362
122. Ewers M, Mielke MM, Hampel H (2010) Blood-based biomarkers of microvascular pathology in Alzheimer's disease. Exp Gerontol 45:75–79
123. Rye PD, Booij BB, Grave G, Lindahl T, Kristiansen L, Andersen HM, Horndalsveen PO, Nygaard HA, Naik M, Hoprekstad D, Wetterberg P, Nilsson C, Aarsland D, Sharma P, Lönneborg A (2011) A novel blood test for the early detection of Alzheimer's disease. J Alzheimers Dis 23:121–129
124. Booij BB, Lindahl T, Wetterberg P, Skaane NV, Sæbø S, Feten G, Rye PD, Kristiansen LI, Hagen N, Jensen M, Bårdsen K, Winblad B, Sharma P, Lönneborg A (2011) A gene expression pattern in blood for the early detection of Alzheimer's disease. J Alzheimers Dis 23:109–119
125. Weigand SD, Vemuri P, Wiste HJ, Senjem ML, Pankratz VS, Aisen PS, Weiner MW, Petersen RC, Shaw LM, Trojanowski JQ, Knopman DS, Jack CR Jr, Alzheimer's Disease Neuroimaging Initiative (2011) Transforming cerebrospinal fluid Abeta42 measures into calculated Pittsburgh Compound B units of brain Abeta amyloid. Alzheimers Dement 7: 133–141
126. Braskie MN, Klunder AD, Hayashi KM, Protas H, Kepe V, Miller KJ, Huang SC, Barrio JR, Ercoli LM, Siddarth P, Satyamurthy N, Liu J, Toga AW, Bookheimer SY, Small GW, Thompson PM (2010) Plaque and tangle imaging and cognition in normal aging and Alzheimer's disease. Neurobiol Aging 31:1669–1678
127. Fodero-Tavoletti MT, Okamura N, Furumoto S, Mulligan RS, Connor AR, McLean CA, Cao D, Rigopoulos A, Cartwright GA, O'Keefe G, Gong S, Adlard PA, Barnham KJ, Rowe CC, Masters CL, Kudo Y, Cappai R, Yanai K, Villemagne VL (2011) 18F-THK523: a novel in vivo tau imaging ligand for Alzheimer's disease. Brain 134:1089–1100
128. Arriagada PV, Growdon JH, Hedley-Whyte ET, Hyman BT (1992) Neurofibrillary tangles but not senile plaques parallel duration and severity of Alzheimer's disease. Neurology 42: 631–639
129. Braak H, Braak E (1998) Evolution of neuronal changes in the course of Alzheimer's disease. J Neural Transm Suppl 53:127–140
130. Desikan RS, McEvoy LK, Thompson WK, Holland D, Brewer JB, Aisen PS, Sperling RA, Dale AM, Alzheimer's Disease Neuroimaging Initiative (2012) Amyloid-beta-associated clinical decline occurs only in the presence of elevated P-tau. Arch Neurol 69:709–713
131. Gozes I (2002) Tau as a drug target in Alzheimer's disease. J Mol Neurosci 19:337–338
132. Bulic B, Pickhardt M, Schmidt B, Mandelkow EM, Waldmann H, Mandelkow E (2009) Development of tau aggregation inhibitors for Alzheimer's disease. Angew Chem Int Ed Engl 48:1740–1752

133. Stewart AJ, Fox A, Morimoto BH, Gozes I (2007) Looking for novel ways to treat the hallmarks of Alzheimer's disease. Expert Opin Investig Drugs 16:1183–1196
134. Lemere CA, Masliah E (2010) Can Alzheimer disease be prevented by amyloid-beta immunotherapy? Nat Rev Neurol 6:108–119
135. Winblad B, Andreasen N, Minthon L, Floesser A, Imbert G, Dumortier T, Maguire RP, Blennow K, Lundmark J, Staufenbiel M, Orgogozo JM, Graf A (2012) Safety, tolerability, and antibody response of active Abeta immunotherapy with CAD106 in patients with Alzheimer's disease: randomised, double-blind, placebo-controlled, first-in-human study. Lancet Neurol 11:597–604
136. Voss K, Combs B, Patterson KR, Binder LI, Gamblin TC (2012) Hsp70 alters tau function and aggregation in an isoform specific manner. Biochemistry 51:888–898
137. Yu W, Polepalli J, Wagh D, Rajadas J, Malenka R, Lu B (2012) A critical role for the PAR-1/MARK-tau axis in mediating the toxic effects of Abeta on synapses and dendritic spines. Hum Mol Genet 21:1384–1390
138. Gotz J, Ittner A, Ittner LM (2012) Tau-targeted treatment strategies in Alzheimer's disease. Br J Pharmacol 165:1246–1259
139. Mondragon-Rodriguez S, Perry G, Zhu X, Boehm J (2012) Amyloid Beta and tau proteins as therapeutic targets for Alzheimer's disease treatment: rethinking the current strategy. Int J Alzheimers Dis 2012:630182
140. Hernandez F, Lucas JJ, Avila J (2013) GSK3 and Tau: two convergence points in Alzheimer's disease. J Alzheimers Dis 33:S141–144
141. Schaeffer V, Lavenir I, Ozcelik S, Tolnay M, Winkler DT, Goedert M (2012) Stimulation of autophagy reduces neurodegeneration in a mouse model of human tauopathy. Brain 135(Pt 7):2169–2177
142. Morris M, Maeda S, Vossel K, Mucke L (2011) The many faces of tau. Neuron 70:410–426
143. Brunden KR, Trojanowski JQ, Lee VM (2009) Advances in tau-focused drug discovery for Alzheimer's disease and related tauopathies. Nat Rev Drug Discov 8:783–793
144. Paquet D, Bhat R, Sydow A, Mandelkow EM, Berg S, Hellberg S, Fälting J, Distel M, Köster RW, Schmid B, Haass C (2009) A zebrafish model of tauopathy allows in vivo imaging of neuronal cell death and drug evaluation. J Clin Invest 119:1382–1395
145. Chatterjee S, Sang TK, Lawless GM, Jackson GR (2009) Dissociation of tau toxicity and phosphorylation: role of GSK-3beta, MARK and Cdk5 in a Drosophila model. Hum Mol Genet 18:164–177
146. Mudher A, Shepherd D, Newman TA, Mildren P, Jukes JP, Squire A, Mears A, Drummond JA, Berg S, MacKay D, Asuni AA, Bhat R, Lovestone S (2004) GSK-3beta inhibition reverses axonal transport defects and behavioural phenotypes in Drosophila. Mol Psychiatry 9: 522–530
147. Gotz J, Ittner LM (2008) Animal models of Alzheimer's disease and frontotemporal dementia. Nat Rev Neurosci 9:532–544
148. Ramsden M, Kotilinek L, Forster C, Paulson J, McGowan E, SantaCruz K, Guimaraes A, Yue M, Lewis J, Carlson G, Hutton M, Ashe KH (2005) Age-dependent neurofibrillary tangle formation, neuron loss, and memory impairment in a mouse model of human tauopathy (P301L). J Neurosci 25:10637–10647
149. Le Blanc K, Ringden O (2006) Mesenchymal stem cells: properties and role in clinical bone marrow transplantation. Curr Opin Immunol 18:586–591
150. Prockop DJ, Olson SD (2007) Clinical trials with adult stem/progenitor cells for tissue repair: let's not overlook some essential precautions. Blood 109:3147–3151
151. Giordano A, Galderisi U, Marino IR (2007) From the laboratory bench to the patient's bedside: an update on clinical trials with mesenchymal stem cells. J Cell Physiol 211:27–35
152. Caplan AI (2007) Adult mesenchymal stem cells for tissue engineering versus regenerative medicine. J Cell Physiol 213:341–347
153. Bassan M, Zamostiano R, Davidson A, Pinhasov A, Giladi E, Perl O, Bassan H, Blat C, Gibney G, Glazner G, Brenneman DE, Gozes I (1999) Complete sequence of a novel protein containing a femtomolar-activity-dependent neuroprotective peptide. J Neurochem 72: 1283–1293

154. Zamostiano R, Pinhasov A, Gelber E, Steingart RA, Seroussi E, Giladi E, Bassan M, Wollman Y, Eyre HJ, Mulley JC, Brenneman DE, Gozes I (2001) Cloning and characterization of the human activity-dependent neuroprotective protein. J Biol Chem 276:708–714
155. Gozes I (2007) Activity-dependent neuroprotective protein: from gene to drug candidate. Pharmacol Ther 114:146–154
156. Idan-Feldman A, Ostritsky R, Gozes I (2012) Tau and caspase 3 as targets for neuroprotection. Int J Alzheimers Dis 2012:493670
157. Divinski I, Holtser-Cochav M, Vulih-Schultzman I, Steingart RA, Gozes I (2006) Peptide neuroprotection through specific interaction with brain tubulin. J Neurochem 98:973–984
158. Divinski I, Mittelman L, Gozes I (2004) A femtomolar acting octapeptide interacts with tubulin and protects astrocytes against zinc intoxication. J Biol Chem 279:28531–28538
159. Gozes I, Divinsky I, Pilzer I, Fridkin M, Brenneman DE, Spier AD (2003) From vasoactive intestinal peptide (VIP) through activity-dependent neuroprotective protein (ADNP) to NAP: a view of neuroprotection and cell division. J Mol Neurosci 20:315–322
160. Javitt DC, Buchanan RW, Keefe RS, Kern R, McMahon RP, Green MF, Lieberman J, Goff DC, Csernansky JG, McEvoy JP, Jarskog F, Seidman LJ, Gold JM, Kimhy D, Nolan KS, Barch DS, Ball MP, Robinson J, Marder SR (2012) Effect of the neuroprotective peptide davunetide (AL-108) on cognition and functional capacity in schizophrenia. Schizophr Res 136:25–31
161. Gozes I, Stewart A, Morimoto B, Fox A, Sutherland K, Schmeche D (2009) Addressing Alzheimer's disease tangles: from NAP to AL-108. Curr Alzheimer Res 6:455–460
162. Butler D, Bendiske J, Michaelis ML, Karanian DA, Bahr BA (2007) Microtubule-stabilizing agent prevents protein accumulation-induced loss of synaptic markers. Eur J Pharmacol 562:20–27
163. Zhang B, Carroll J, Trojanowski JQ, Yao Y, Iba M, Potuzak JS, Hogan AM, Xie SX, Ballatore C, Smith AB 3rd, Lee VM, Brunden KR (2012) The microtubule-stabilizing agent, epothilone D, reduces axonal dysfunction, neurotoxicity, cognitive deficits, and Alzheimer-like pathology in an interventional study with aged tau transgenic mice. J Neurosci 32:3601–3611
164. Wischik CM, Edwards PC, Lai RY, Roth M, Harrington CR (1996) Selective inhibition of Alzheimer disease-like tau aggregation by phenothiazines. Proc Natl Acad Sci U S A 93:11213–11218
165. Martinez A, Perez DI (2008) GSK-3 inhibitors: a ray of hope for the treatment of Alzheimer's disease? J Alzheimers Dis 15:181–191
166. Luna-Medina R, Cortes-Canteli M, Sanchez-Galiano S, Morales-Garcia JA, Martinez A, Santos A, Perez-Castillo A (2007) NP031112, a thiadiazolidinone compound, prevents inflammation and neurodegeneration under excitotoxic conditions: potential therapeutic role in brain disorders. J Neurosci 27:5766–5776
167. Pickhardt M, Larbig G, Khlistunova I, Coksezen A, Meyer B, Mandelkow EM, Schmidt B, Mandelkow E (2007) Phenylthiazolyl-hydrazide and its derivatives are potent inhibitors of tau aggregation and toxicity in vitro and in cells. Biochemistry 46:10016–10023
168. Crowe A, Huang W, Ballatore C, Johnson RL, Hogan AM, Huang R, Wichterman J, McCoy J, Huryn D, Auld DS, Smith AB 3rd, Inglese J, Trojanowski JQ, Austin CP, Brunden KR, Lee VM (2009) Identification of aminothienopyridazine inhibitors of tau assembly by quantitative high-throughput screening. Biochemistry 48:7732–7745
169. Wang Y, Martinez-Vicente M, Krüger U, Kaushik S, Wong E, Mandelkow EM, Cuervo AM, Mandelkow E (2009) Tau fragmentation, aggregation and clearance: the dual role of lysosomal processing. Hum Mol Genet 18:4153–4170
170. Hanger DP, Anderton BH, Noble W (2009) Tau phosphorylation: the therapeutic challenge for neurodegenerative disease. Trends Mol Med 15:112–119
171. Liu M, Choi S, Cuny GD, Ding K, Dobson BC, Glicksman MA, Auerbach K, Stein RL (2008) Kinetic studies of Cdk5/p25 kinase: phosphorylation of tau and complex inhibition by two prototype inhibitors. Biochemistry 47:8367–8377
172. Engel T, Goñi-Oliver P, Gómez de Barreda E, Lucas JJ, Hernández F, Avila J (2008) Lithium, a potential protective drug in Alzheimer's disease. Neurodegener Dis 5:247–249

173. Eldar-Finkelman H, Eisenstein M (2009) Peptide inhibitors targeting protein kinases. Curr Pharm Des 15:2463–2470
174. Iqbal K, Grundke-Iqbal I (2008) Alzheimer neurofibrillary degeneration: significance, etiopathogenesis, therapeutics and prevention. J Cell Mol Med 12:38–55
175. Le Corre S, Klafki HW, Plesnila N, Hübinger G, Obermeier A, Sahagún H, Monse B, Seneci P, Lewis J, Eriksen J, Zehr C, Yue M, McGowan E, Dickson DW, Hutton M, Roder HM (2006) An inhibitor of tau hyperphosphorylation prevents severe motor impairments in tau transgenic mice. Proc Natl Acad Sci U S A 103:9673–9678
176. Dickey CA, Koren J, Zhang YJ, Xu YF, Jinwal UK, Birnbaum MJ, Monks B, Sun M, Cheng JQ, Patterson C, Bailey RM, Dunmore J, Soresh S, Leon C, Morgan D, Petrucelli L (2008) Akt and CHIP coregulate tau degradation through coordinated interactions. Proc Natl Acad Sci U S A 105:3622–3627
177. Takeda A, Arai N, Komori T, Iseki E, Kato S, Oda M (1997) Tau immunoreactivity in glial cytoplasmic inclusions in multiple system atrophy. Neurosci Lett 234:63–66
178. Sigurdsson EM (2009) Tau-focused immunotherapy for Alzheimer's disease and related tauopathies. Curr Alzheimer Res 6:446–450
179. Rosenmann H, Meiner Z, Geylis V, Abramsky O, Steinitz M (2006) Detection of circulating antibodies against tau protein in its unphosphorylated and in its neurofibrillary tangles-related phosphorylated state in Alzheimer's disease and healthy subjects. Neurosci Lett 410:90–93
180. Lasagna-Reeves CA, Castillo-Carranza DL, Sengupta U, Sarmiento J, Troncoso J, Jackson GR, Kayed R (2012) Identification of oligomers at early stages of tau aggregation in Alzheimer's disease. FASEB J 26(5):1946–1959
181. Lasagna-Reeves CA, Castillo-Carranza DL, Jackson GR, Kayed R (2011) Tau oligomers as potential targets for immunotherapy for Alzheimer's disease and tauopathies. Curr Alzheimer Res 8:659–665
182. Chai X, Wu S, Murray TK, Kinley R, Cella CV, Sims H, Buckner N, Hanmer J, Davies P, O'Neill MJ, Hutton ML, Citron M (2011) Passive immunization with anti-Tau antibodies in two transgenic models: reduction of Tau pathology and delay of disease progression. J Biol Chem 286:34457–34467
183. Bi M, Ittner A, Ke YD, Gotz J, Ittner LM (2011) Tau-targeted immunization impedes progression of neurofibrillary histopathology in aged P301L tau transgenic mice. PLoS One 6:e26860
184. Boutajangout A, Ingadottir J, Davies P, Sigurdsson EM (2011) Passive immunization targeting pathological phospho-tau protein in a mouse model reduces functional decline and clears tau aggregates from the brain. J Neurochem 118:658–667
185. Boimel M, Grigoriadis N, Lourbopoulos A, Haber E, Abramsky O, Rosenmann H (2010) Efficacy and safety of immunization with phosphorylated tau against neurofibrillary tangles in mice. Exp Neurol 224:472–485
186. Trojanowski JQ, Duff K, Fillit H, Koroshetz W, Kuret J, Murphy D, Refolo L, Frontotemporal Dementia (FTD) Working Group on FTD Drug Discovery (2008) New directions for frontotemporal dementia drug discovery. Alzheimers Dement 4:89–93
187. Iqbal K, Chohan MO, Grundke-Iqbal I (2008) Stratification of patients is the way to go to develop neuroprotective/disease-modifying drugs for Alzheimer's disease. J Alzheimers Dis 15:339–345

Part III

The Use of Biomarkers for Prediction and Prevention of Alzheimer's and Parkinson's Diseases

Silvia Mandel and Amos D. Korczyn

Abstract Among the diverse challenges facing clinicians during the last century, is the staggering increase in the proportion of elderly members, which is directly linked to the prevalence of neurodegenerative diseases like Alzheimer's disease (AD) and Parkinson's disease (PD). AD and PD are considered disorders of multifactorial origin, inevitably progressive and having a long preclinical period. At present, clinical diagnosis of AD and PD is based on a constellation of symptoms and manifestations, although the disease originated several years earlier. Therefore, the availability of biological markers or biomarkers (BMs) for early disease diagnosis will impact the management of AD and PD in several dimensions: (a) help to capture high-risk individuals before symptoms develop, a stage where prevention efforts might be expected to have their greatest impact; (b) provide a measure of disease progression that can be evaluated objectively, while clinical measures are much less accurate; (c) help to discriminate between true AD or PD and other causes presenting with a similar clinical syndrome; (d) delineate pathophysiological processes responsible for the disease; (e) determine the clinical efficacy of novel, disease-modifying (neuroprotective) strategies. In the long run the availability of reliable BMs will significantly advance the research and therapeutics of these neurodegenerative diseases. However, at present, the available BMs are of limited value in all these respects. There is an urgent need to further explore this field in order to have validated BMs on hand once neuroprotective interventions become available.

S. Mandel (✉)
Eve Topf Center for Neurodegenerative Diseases Research and Department of Molecular Pharmacology, Faculty of Medicine, Technion, Institute of Technology, Haifa, Israel
e-mail: mandel@tx.technion.ac.il

A.D. Korczyn (✉)
Tel-Aviv University Medical School, Ramat-Aviv, Israel
e-mail: amoskor@tau.ac.il

S. Mandel (ed.), *Neurodegenerative Diseases: Integrative PPPM Approach as the Medicine of the Future*, Advances in Predictive, Preventive and Personalised Medicine 2, DOI 10.1007/978-94-007-5866-7_6,

Keywords Alzheimer's disease • Apolipoprotein E • Biomarker • Disease modifying therapy • Parkinson's disease • Neuroimaging

Abbreviations

A-beta	Amyloid-beta
AD	Alzheimer's disease
ALS	Amyotrophic lateral sclerosis
apoE	Apolipoprotein E
APP	Beta amyloid precursor protein
BM	Biomarker
CDR	Clinical dementia rating scale
Cho	Choline
CNS	Central nervous system
CSF	Cerebrospinal fluid
DA	Dopamine
DAT	DA transporter
DLB	Dementia with Lewy bodies
DSM	Diagnostic and statistical manual of mental disorders
FDG-PET	2-Fluoro-2-deoxy-D-glucose PET
EEG	Electroencephalogram
EOAD	Early onset Alzheimer's disease
fMRI	Functional magnetic resonance
FTD	Frontotemporal dementia
HFE	Hemochromatosis gene
ICD	International classification of diseases
iTRAQ	Isobaric Tagging for Relative and Absolute protein Quantification
LC	Locus coeruleus-subcoeruleus complex
LLD	Late life dementia
LOAD	Late onset AD
LRRK2	Leucine-rich repeat kinase 2
MALDI TOF/TOF	Matrix-assisted laser desorption/ionization time-of-flight tandem mass spectrometer
MAPT	Microtubule-associated protein tau
MCI	Mild cognitive impairment
MRI	Magnetic resonance
MSA	Multiple system atrophy
NAA	N-acetyl aspartate
NAA/Cr	N-acetyl aspartate/creatine
NFT	Neurofibrillary tangle
PD	Parkinson's disease
PET	Photon emission tomography

PIB-PET	Pittsburgh compound B PET
PPA	Primary progressive aphasia
PSP	Progressive supranuclear palsy
PTGD	Prostaglandin D synthase
REM	Rapid eye movements
RPH	Raphe nuclei caudal
SN	Substantia nigra
SPECT	Single photon emission computerized tomography
STX1	Syntaxin-1
SYN1,2	Synapsin 1,2
SYT1	Synaptotagmin-1
TCS	Transcranial sonography
TH	Tyrosyne-hydroxylase
UCH-L1	Ubiquitin c-terminal hydrolase L1 gene
UPDRS	Unified Parkinson Disease Rating Scale
UPS	Ubiquitin-proteasome system
UPSIT	University of Pennsylvania Smell Identification Test
VaD	Vascular dementia
VMAT	Vesicle monoamine transporter

1 Introduction

Parkinson's disease (PD) and Alzheimer's disease (AD) are neurodegenerative disorders sharing in common a number of characteristics. The more common sporadic forms are considered to be multifactorial in origin, inevitably progressive and have a long preclinical period. Currently, the clinical diagnosis in either disease depends on symptoms and signs. The diagnosis of PD can be made when motor symptoms occur, though the disease has originated several years earlier. Furthermore, at its initial stages PD may be confounded with other diseases, such as essential tremor, multiple system atrophy (MSA) and progressive supranuclear palsy (PSP) [1, 2]. Similarly, the pathological process characteristic of AD begins decades before the first symptoms of cognitive dysfunctions, thus making it difficult to reliably identify pathology based on the clinical phenotype alone. AD may share clinical features with many other dementing disorders including vascular dementia (VaD), dementia with Lewy bodies (DLB), frontotemporal lobe dementia (FTLD) and others [3, 4]. Therefore, the availability of biological markers or biomarkers (BM) for early disease diagnosis will impact the management of AD and PD in several dimensions: first, it will allow to capture high-risk individuals before symptoms develop; second, it will help to discriminate between true AD or PD and other causes of similar clinical syndromes, thus allowing better prognostication. In addition, a BM may help to determine the clinical efficacy of new disease specific neuroprotective therapies. The development of neuroprotective therapies is greatly hindered by the lack of markers capable of serving as surrogate objective endpoints for clinical trials testing these treatments. Finally, a BM may indicate the biological changes underlying the disease process and thus allow the development of new therapeutic modalities.

Among the diverse, significant challenges facing the clinicians, is the improvement of diagnostic measures in order to (1) detect early/subtle symptoms, a phase in which intervention might be expected to have its greatest impact; (2) provide a measure of disease progression that can be evaluated objectively, since clinical measure are much less accurate; (3) provide evidence for relevant biological activity of an experimental drug; (4) delineate pathophysiological processes responsible for the disease. This is particularly important because clinical endpoints such as changes in patient's features, or the time it takes to reach specific disease-associated disabilities may vary drastically among patients or clinical raters and require long trial durations and a large number of subjects before significant effects are seen, thus being inaccurate and expensive measures of disease progression. Also, the current assessment of clinical disease improvement does not always allow for a clear distinction between symptom-masking versus disease-modifying (neuroprotective) therapies.

BMs have been defined as "cellular, biochemical or molecular alterations that are measurable in the biological media such as human tissues, cells or fluids" [5]. Others broaden the definition to include biological characteristics that can be measured objectively and evaluated as indicators of normal or pathological biological processes, or pharmacological responses to therapeutic intervention [6], such as electrophysiological activities (EEG-evoked reactions). BMs for detection of neurodegenerative disorders can be divided into three main categories: genetic, neuroimaging (including structural and functional) and biochemical. Normally, neuroimaging and biochemical BMs mark the presence of pathology, while genetic markers can only serve for risk assessment.

Each BM must be evaluated for its specificity and sensitivity for a given disease. The same applies for BMs that are used to define stages of disease severity [7]. However, increased specificity is usually achieved at the expense of reduced sensitivity and vice versa. Therefore a compromise must be achieved. Whether one attempts to achieve maximal sensitivity or specificity depends on the aims of the study and the 'price' which is involved. For example, if a cheap, safe and effective therapy is available for a condition, a less specific test can be used even if this will eventually result in treatment being given to people who do not have the disease. Conversely, if the treatment is costly or has significant adverse effects, a more specific test will be required.

This chapter underlines the utility of genetic, neuroimaging, and biochemical BMs in the clinic, focusing on their predictive diagnostic value for targeted prevention of AD and PD and their potential use to monitor disease progression and its modification by therapy, as well as to delineate the pathobiological (e.g. molecular) processes underlying the diseases.

2 Biomarkers for AD

During the last century the population has shown a staggering increase in the proportion of elderly members. All epidemiological studies of dementia show a strong correlation between prevalence and age. In the group of 60–64 years of age the prevalence of dementia is below 1%; it then doubles approximately every 5 years reaching a prevalence of about 50% among those aged 90 years or older [8, 9]. This high frequency is important in terms of social policy because it frequently marks the need for permanent nursing care [10].

A variety of neurodegenerative pathologies are associated with the development of late life dementia (LLD), while AD is considered by far the most prevalent. 60–75% of all cases of dementia are diagnosed as AD [8, 11]. However, the clinical diagnosis of demented patients is challenging. At present, diagnosis is based on clinical constellation of symptoms and manifestations. In order to reach a consensus, diagnostic criteria have been formulated, for example the Diagnostic and Statistical Manual of Mental Disorders (DSM) or the International Classification of Diseases (ICD) criteria, and more specific ones like the 1984 McKhann's criteria [12] and recently, the new suggestions of Dubois et al. [13] and McKahann et al. [14] (see also [15]).

AD is associated with progressive memory deterioration and worsening of additional cognitive functions. One factor responsible for cognitive deterioration is a loss of cholinergic transmission in cortical brain regions innervated by neurons arising in the nucleus basalis of Meynert. The presence of extracellular plaques containing deposits of proteins, particularly β-amyloid and α-synuclein, and intracellular neurofibrillary tangles (NFTs) are hallmarks which define the pathology of AD. These changes however, may result from destructive processes causing disruption of microtubule assembly and synaptic loss, rather than its cause. They could contribute to further neuronal damage and disease progression [3]. Depressive symptoms occur in a large proportion of subjects with AD and may occur several years before the typical cognitive and motor symptoms appear. This has been attributed to degeneration of noradrenaline and serotonin neurons innervating the limbic system.

The increasing prevalence of AD and the devastating consequences of LLD motivate the drive to develop diagnostic BMs to reliably identify *in vivo* the pathology associated with this disorder, particularly at an early stage of the pathologic process. It is also quite possible that BMs may also help to identify subclasses of the demented disorders [16]. BMs based on detection of pathology may diagnose the process more definitively and reliably than the traditional cognitive and neurological phenotypes. Strategies to accomplish these include for example, (1) the detection of altered levels of tau and amyloid proteins in the cerebrospinal fluid (CSF), (2) the use of structural magnetic resonance imaging (MRI) to identify disease-specific patterns of regional or generalized atrophy (3) imaging of metabolic or cerebrovascular flow using positron emission tomography (PET) and single photon emission computerized tomography (SPECT), as well as (4) PET imaging using

Table 1 Diagnostic value of Alzheimer's disease BMs

Type of Method	AD biomarker	Diagnostic value
Genetic biomarkers	Genes- mutations in presenilin 1 and 2 and in APP	Identify at-risk family members
	Apolipoprotein E ε4	Possible pharmacogenetic value
Neuroimaging biomarkers	MRI	Useful in diagnosis of presymptomatic and very early clinical symptomatic stages of AD
	Proton MR Spectroscopy (1H MRS)	Helpful in diagnosis of preclinical or MCI stage
	PET	Differentiation of AD from normal aging
	SPECT	Detect functional abnormalities in very mild dementia. Specific ligands particularly hypometabolism can depict amyloid deposits
	fMRI	Sensitive to functional changes before cognitive impairment becomes clinically apparent
Biochemical biomarkers	In serum – plasma Aβ	Elevation of plasma Aβ may be helpful in diagnosis of at risk asymptomatic young individuals from LOAD families. Serum $A\beta_{(1-42)}$: $A\beta_{(1-40)}$ ratio may indicate AD risk
	In cerebrospinal fluid – Tau protein and Aβ-peptides.	May be helpful in detecting incipient AD in early MCI stage May also be helpful in distinguishing AD patients from elderly controls or people with other types of dementia

ligands which indicate the presence of amyloid deposits (ligands for neurofibrillar changes are being developed). At present, diagnostic markers are used extensively in the clinic (Table 1), but mainly as exclusionary markers, for example elevated levels of 14-3-3 or tau protein in the CSF, characteristic EEG discharges and MRI signals which suggest the existence of prion diseases, "exclude" the diagnosis of AD [17].

There are several reasons for the emerging need to incorporate BMs into diagnostic criteria used by clinicians to indentify the pathology responsible for the patient's cognitive failure. These include the fact that a single type of pathology can produce a variety of different types of cognitive and behavioral symptoms and conversely, given clinical phenotypes can result from different pathological processes. This phenotypic convergence makes it difficult to reliably identify the pathology based solely on the clinical phenotype. Even more important is the recognition that a variety of neurodegenerative and vascular pathologies can coexist, and in fact it seems that most patients with LLD have mixed pathologies underlying the cognitive deterioration [18]. A particular problem is whether AD can be detected at a very early stage, much before clinical criteria sufficient to diagnose dementia, appear. This early clinical stage, when cognitive deterioration just manifests itself and before the more significant features of dementia appears, called mild cognitive impairment (MCI) may be particularly important once early intervention becomes available [15]. In fact, the updated guidelines delivered by the National Institute on Aging and the Alzheimer's Association [14], consider AD a three-phase illness: a preclinical phase (asymptomatic, with molecular evidence of AD); a MCI phase, preceding the progression to overt clinical dementia, which is the last phase of the new diagnostic criteria for AD. This strategy has been made possible thanks to the development of imaging and molecular markers that may identify pathological processes long before dementia manifests.

2.1 Conceptual Issues

One central problem is the concept of AD as a single disease entity. At present, the definition of AD is based on the final result, the accumulation of β-amyloid deposits and NFTs. It is quite clear that similar, or identical, pathological lesions can results from different causes, e.g. specific genetic mutations cause the disease in a minority of subjects, while in the majority AD is sporadic, and presumably is the consequence of multiple environmental factors. It is thus not clear to what extent the initial biological processes are similar. This of course has important implications on the issue at hand, i.e. whether the same BM will be reliable in each etiologic subtype. Thus, preliminary results from the DIAN (Dominantly Inherited Alzheimer's Network) study (presented at the Alzheimer's Association International Conference in Paris, 2011) following preclinical subjects with dominantly inherited AD, showed that the typical CSF low-levels of Aβ are preceded by elevated CSF Aβ.

A practical question is what are the BMs used for. In reality, an accurate and sensitive diagnostic BM which will distinguish normal people from patients with advanced AD is not very useful, since at that stage the difference is obvious. The helpful BM is one that is informative in situations where clinical data are not sufficient to allow a clear diagnosis [19].

BMs can therefore be used for different purposes:

1. Identify 'at risk' individuals prior to disease onset.
2. Identify people with early AD, prior to the development of the full clinical phenotype, for example at the MCI stage. 19–50% of all sufferers from MCI will develop to dementia over 4 years [20].
3. Differentiate AD from other neurodegenerative processes in demented individuals.
4. To monitor effects of intervention on the evolution of the disease.
5. To gain insight into processes underlying the pathological changes in brain areas known to be linked with AD in its various degrees of severity, i.e. temporal, parietal, and frontal lobes cortex. Renewed attention to changes in the brain-stem nuclei such as the locus coeruleus, may suggest exploration of these sites, as well as the nucleus basalis of Meynert and the olfactory bulb in early stages of the disease [21, 22].

As will be seen, most if not all BMs which have been suggested reveal statistical differences between groups (e.g. normals versus early AD, or AD versus other LLD), but none of them has been shown to be sufficiently robust to allow a definite diagnosis in an individual person. This is because of the heterogeneity which exists even at baseline, for example in hippocampal size, among different individuals. However, repeated studies may show a change such that a more abnormal result (e.g. a hippocampus which is smaller than at baseline, although still within normal limits) might be helpful in detecting a disease process [23, 24].

2.1.1 Genetic Markers of AD

Molecular genetic studies of neurodegenerative disorders have been useful both in identifying genes that may be causative or associated with specific diseases and in uncovering functional mechanisms subserved by products of those genes that may be markers for diagnosis or early detection and targets for intervention. Genetics may provide information to assess who is at risk, but except for the case of rare genetic disorders, it does not provide complete information when or whether those 'at risk' may expect the onset of neurodegeneration.

Genes Causing AD

In the case of AD, mutations in three genes are now known to cause the disorder in families with an autosomal dominant pattern of inheritance: β-amyloid precursor protein (APP), presenilin 1 and presenilin 2 (PS1 and PS2, respectively). Detection of the mutations in these genes can identify at-risk family members, but it is not generalizable to sporadic AD [25, 26]. The identification of mutations in APP has been the basis of the amyloid hypothesis and identified potential sites for pharmacological interventions. However, changes in the gene, or in PS1 and 2, have not been identified in people with LLD. AD is different in genetic and sporadic cases, thus these cannot all be regarded as a single disease entity. This suggests a different etiopathogenic mechanism of disease, at least in the early stages, which should lead to different therapeutic approaches.

An important issue is the genetic association with the increased risk of developing AD in families that carry the apolipoprotein E (apoE) ε4 allele. ApoE is a protein involved in the transport of cholesterol. Apart from its presence in the plasma, it is also produced by astrocytes and neurons in the CNS to support and repair neurons. The *apoE* gene is localized on chromosome 19 and three alleles are described in humans (ε2, ε3 and ε4). A large volume of evidence has reported an association of *APOE ε4* and late-onset familial and sporadic AD [27]. As many as 40–50% of AD patients possess ε4 allele compared to 15–25% of controls [28]. Subjects homozygous for ε4 are reported to have six to eightfold increased risk of developing AD compared to the risk of heterozygotic subjects (increased by two to fourfold)

[27]. However, only ~40% of all patients with LOAD carry *APOE ε4*. At present, the existence of ε4 and particularly ε4/ε4 genotype in an asymptomatic individual, merely increases the risk of developing AD [29], and its existence in a demented individual cannot establish the cause as being AD. Recent data, discovered using phylogenetic analysis, suggest that a variable length poly-T sequence polymorphism in the translocase of outer mitochondrial membrane 40 homolog (*TOMM40*) gene, adjacent to the *apoE* gene, is associated with age-of-onset of AD. When linked to APOE ε3, the longer TOMM40 poly-T repeats are associated with earlier age at onset and the shorter with later age at onset, questioning the alleged 'neutral' role of the ε3 allele in AD risk [30]. More detailed information on potential genetic risk factors for late-onset AD (LOAD), is provided by Grossman et al. [31].

One important PET study [32] found that cognitively normal 50–65 years-old ε4 carriers have low rate of cerebral glucose metabolism in brain regions associated with AD (posterior cingulate, parietal, temporal and prefrontal cortex). The last findings motivated the authors [33] to consider whether ε4 carriers have these regional brain abnormalities even when still young. For this purpose apolipoprotein E genotype was established from healthy young adults (20–39 years of age). Clinical rating, neuropsychological tests and PET where preformed in 12 ε4 heterozygotes, all with ε3/ε4 genotype and 15 noncarriers of ε4 allele. The results of this study indicated that even cognitively healthy ε4 carriers have functional brain abnormalities in regions that are associated with AD several decades later. These abnormalities were detected even though the young ε4 carriers were restricted to persons with one copy of this allele who tend to have lower risk of AD than those with two ε4 allele copies.

In a recent fMRI study, Bookheimer and coworkers [34] tried to determine the relationship between brain responses to tasks requiring memory and the genetic risk of AD by performing APOE genotyping and MRI of the brain in older persons with intact cognition. Both the magnitude and the extent of brain activation during memory-activation tasks in regions affected by AD, including the left hippocampal, parietal, and prefrontal regions, were greater among the carriers of the apoE ε4 allele than among non-carriers. During periods of recall, the carriers of the apoE ε4 allele had a greater average increase in signal intensity in the hippocampal region and a greater mean number of activated regions throughout the brain than did non-carriers. This suggests that the patterns of brain activation during tasks requiring memory differ depending on the genetic risk of AD. It is still unknown however, whether these may predict subsequent decline in memory. This can only be verified by prospective studies.

Obviously, genetic markers are by themselves not sufficient to make a clinical diagnosis of AD. The mutation, if it exists, has been there since conception and birth of the individual. Thus, a person carrying a disease-causing mutation is not immune to the development of other diseases impairing cognition, and the diagnosis must be supported by clinical evidence.

Transgenic animal models have been constructed that mimic some aspects of the AD pathology [35] and proof of principle studies have been able to reduce the transgene-induced pathology with strategies that use anti-inflammatory medications, antibody stimulation by immunization, or other disruptions of pathobiological cascades [36–38]. However, the relevance of these models to human sporadic AD has been contested [39].

Ultimately, sensitive methods for identifying genetic background may provide a better means for characterizing a statistical risk before disease manifestation. Nevertheless, it seems logical to assume that effective identification of the disease must combine genetic characteristics with identifying a clinical phenotype or/and additional BMs, either for early detection or for the more correct diagnosis of an already existing disease [40].

Genome Wide-Association Studies (GWAS) in AD

The vast majority of AD cases does not carry the disease-causing mutations or risk gene alleles discussed in the previous section. Instead, these "sporadic" forms are believed to be the result of various genetic and non-genetic factors acting in concert, to alter the risk to develop AD [41]. During the past decade, genome-wide association studies (GWAS) have provided several additional AD susceptibility loci that are common in the general population [42]. The mainstream of this genetic approach is aimed at the identification of common DNA variants (called "polymorphisms") that can modify susceptibility to AD. Systematic evaluation of GWAS revealed about 40 candidate AD loci showing significant risk effects in meta-analyses combining all available published data (see the AlzGene database at http://www.alzgene.org and [41, 43]). The findings point to a role of the innate immune system, cholesterol metabolism, and protein trafficking, as potential mechanisms linking GWAS findings to AD [43], recently supported by two new Alzheimer's GWAS reports from the NIA-ADGC [44] and European consortium [45]. Of all susceptibility genes described, three loci: CLU, PICALM and CR1 have been replicated in independent samples [46–48] and are now considered to be true susceptibility loci for AD (for an extensive review see [41]).

It should be taken into consideration that the gene loci revealed by GWAS exert only very small risk effects and thus, the majority of heritable cases of AD can not presently be accounted for by the known disease genes. Rather, as suggested [43], they may be the result of inherited rare sequence variants in genes that predispose to both early- and late-onset forms of AD.

2.1.2 Neuroimaging Markers

During the past decade neuroimaging has emerged as a strategy to attempt defining neurodegenerative disorders phenotype in both preclinical and early clinical disease stages. Imaging provides a window to the underlying processes in these disorders since loss of brain volume is one of the earliest consequences of neuron death. When extreme, it can distinguish from age related changes. It has long been appreciated that AD produces regionally specific neurodegeneration starting in entorhinal cortex and the hippocampus [49–51], corresponding to Braak staging [52, 53]. Using standardized qualitative visual rating scale, it is possible to detect these changes on MRI images [54].

Retrospective and prospective neuropathological studies have shown neuritic plaques and neurofibrillary tangles, characteristic of AD in approximately 60% of MCI patients that have come to autopsy [55]. These new data prove, as long suspected, that significant pathological changes occur in the presymptomatic and very early clinical symptomatic stages of AD and have further encouraged efforts to find clinical or biological markers of impending symptomatic disease [56, 57]. However, MCI is so heterogeneous in its presentation, underlying pathology and evolution [58] that it cannot be used as a surrogate marker.

Structural Imaging Using MRI

While helpful in supporting the diagnosis of AD, structural imaging using MRI is confounded by the presence of atrophy due to other pathological processes and the generalized atrophy that occurs as part of normal aging [59, 60]. One way to confront this problem is to use measurements of multiple anatomically specific brain volumes other then the hippocampus and entorhinal cortex that are affected by AD. Examples include: the amygdalae, superior temporal sulcus, cingulate gyrus, temporal-parietal regions and the frontal lobe. This provides the potential to increase the diagnostic power of MRI-derived atrophy measurements by additional anatomically defined areas [61]. While being diagnostically helpful, volumetric MRI is less likely to assist in monitoring diseases progression. Thus, two potential diseases modifying studies aiming at reducing amyloid load showed a paradoxical reduction in hippocampal volumes in treated patients, perhaps partly due to elimination of amyloid [62, 63].

Although very high resolution computed tomography can be used to image neuritic plaques in experimental animals [64], it is currently not possible to accomplish this in a clinical setting. However, by modifying existing clinical MRI acquisition protocols to incorporate measures of T_1 relaxation time, it is possible to quantify the accumulation of abnormal protein aggregates in anatomically defined brain regions. This opens the possibility of developing an MRI-based molecular imaging BMs for AD related pathology that could be used in healthcare facilities [65].

Proton MR Spectroscopy (^{1}H MRS) is a diagnostic imaging technique that is sensitive to changes at the cellular level. With ^{1}H MRS, several of the major proton-containing metabolites in the brain can be measured during a common data acquisition period. The metabolite N-acetyl aspartate (NAA) is a marker for neuronal integrity. NAA decreases in a variety of neurological disorders including clinically diagnosed AD [66, 67]. The decrease of NAA or the NAA/creatine (Cr) ratio was observed in the posterior cingulate of AD patients [68, 69].

In patients with mild to moderate AD, NAA/Cr levels are lower than normal in the medial occipital lobe, including the visual cortex. This regional pattern is in agreement with the distribution of the neurofibrillary pathology, and associated neuron loss in people with mild to moderate AD, indicating that NAA/Cr levels are potential surrogates for early detection of AD and for disease progression [70]. Lower NAA/Choline has also been observed at the preclinical or MCI stage [71].

Another metabolite signal of interest in the ^{1}H MRS of the brain in AD is the choline peak. Some studies identified elevated choline and choline/Cr ratios, while some reported similar levels in people with AD compared to normal [72]. The largest amount of choline in the brain is in membrane phospholipids that are precursors of choline and acetylcholine synthesis. It has been postulated that the choline peak is the consequence of membrane phosphatidylcholine catabolism to provide free choline for the chronically deficient acetylcholine production in AD [73]. This appears to be useful in detection of presumed pathologic changes in people with clinical diagnosis of AD.

An important new addition to the imaging tools is resting-state functional MRI (fMRI). In reality, the correct term should not be "resting-state" but rather "task-free fMRI" (TF-fMRI). This method is useful in identifying baseline connectivity between brain areas, based on their metabolic activities using temporal resolution. It has been applied to neurodegenerative diseases, and AD may have a specific signature, although this is still not widely agreed [74].

PET and SPECT

AD is associated with metabolic impairments that often have a 'typical' pattern, initially affecting the parietal and temporal lobes as well as posterior cingulate cortex before progressing to the frontal lobes and other brain structures. This metabolic activity can be visualized using [^{18}F]-2-fluoro-deoxy-D-glucose (FDG)-PET and the pattern of involvement is often useful in distinguishing AD from other LLD [75, 76] (Fig. 1).

The impaired metabolic activity may, in part, be due to a loss of neurons in these areas and in part may be a sign of reduced activation, for example reflecting synaptic loss [78]. In any case, it is not the mere loss but rather its regional distribution which characterize AD. However, while the distribution of metabolic changes or even atrophic changes may have typical distributions, these are rarely highly specific. The observed changes, e.g. frontotemporal atrophy or hypo-metabolism associated with primary progressive aphasia (PPA), for example, may reflect a large number of underlying pathological processes (e.g. AD, FTLD).

Evidence exists that the pattern of temporal and parietal hypometabolism and hypoperfusion are reasonably sensitive and specific to AD. Comparing AD patients to controls showed relatively high sensitivity and specificity (85–90% respectively) in differentiating AD patients from normal controls [79, 80]. Although differentiation of AD from normal aging seems clinically irrelevant, the evidence for the ability of PET and SPECT to detect functional abnormalities in very mild dementia has considerable importance in the early diagnosis of dementia [79–81]. A recent European multi center PET study that used quantitative FDG-PET methodology

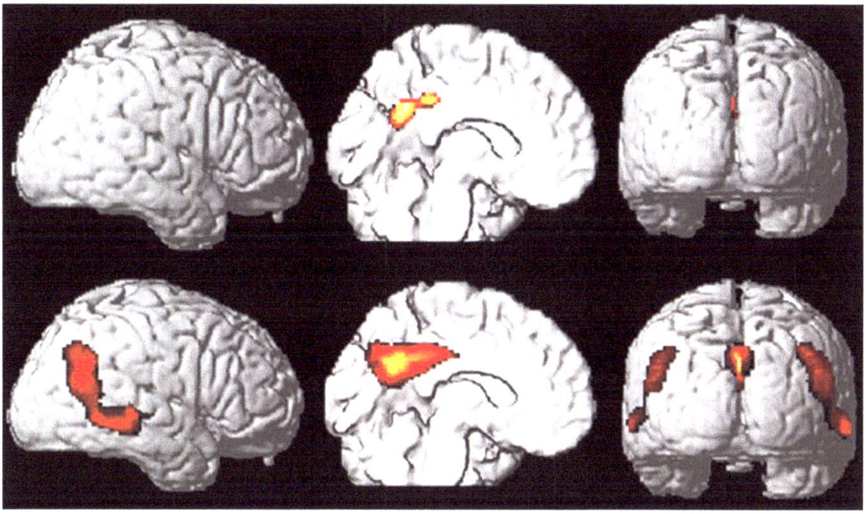

Fig. 1 Brain hypometabolism in MCI and AD. Brain regions showing reduced glucose metabolic rate (MRglc) in amnestic MCI (*top row*) and AD (*bottom row*) in comparison with matched normal controls. The cortical MRglc reductions in an amnestic MCI group (n = 37) were found to be restricted to the posterior cingulate cortex, while they were spatially more extended and involved several regions of the association cortex in an AD group (n = 87). The distribution of MRglc reductions in MCI suggests that these patients share some common feature with AD patients. Statistical parametric maps are superimposed onto the right lateral (*left*), medial (*middle*) and posterior (*right*) views of a volume-rendered spatially normalized MRI study ($p < 0.001$, uncorrected for multiple comparisons) (taken with permission from Mosconi [77])

found 93% sensitivity and 93% specificity in differentiating AD patients from controls when patients with Mini Mental State Examination (MMSE) score of 24 or greater were included [82].

Routine SPECT images provide a qualitative visual map of cerebral perfusion and have been used for many years in evaluation of patients with cognitive impairment as a surrogate marker of impaired neuronal metabolism [83]. Recently, compatible amyloid ligands have been developed to visualize aggregation of amyloid-beta (Aβ) peptides. These ligands can be readily radiolabeled for in vivo imaging using PET. The benzonthiazole derived [^{11}C]-Pittsburgh compound B PET (PIB-PET) ligand is an early example. It is retained in areas of the brain known to harbor fibrillar amyloid and increased levels can be demonstrated in some individuals with MCI. However, abnormal retention can also occur in individuals with normal cognition, and longitudinal studies to date have not demonstrated progressively increased PIB retention in patients with AD [77, 84–86].

A major disadvantage of [^{11}C]-based compounds is their short half-life, limiting their use to sites near cyclotrons. [^{18}F]-based imaging markers have a longer half-life and are therefore more easy to use. Recently, florbetapir has received approval as a diagnostic aid for AD-related amyloidisis [87]. One of the limitations of both compounds is that they only bind to fibrillar Aβ, whereas the presumed relevant species are oligomeric molecules [87].

However, regarding the use of PET and SPECT for clinical diagnosis one must consider several problems. First, while clinical diagnosis is needed in community populations, imaging studies more often are drawn from academic centers. Second, an imaging diagnosis has not yet been shown to supplement a clinical evaluation in any circumstance. Another approach is therefore to compare clinical and imaging diagnosis in the same sample, asking whether imaging provides diagnostically useful information in addition to the clinical data [78]. In this regard it is worth mentioning the Alzheimer's Disease Neuroimaging Initiative (ADNI) whose primary goal is to determine whether brain imaging, biochemical markers, and clinical and neuropsychological assessment can accurately measure the progression of MCI and early AD. The ADNI cross-sectional and longitudinal study has successfully recruited more than 800 individuals, normal controls, subjects with MCI, and subjects with mild AD, followed for 12 months, that were very similar to those seen in MCI and mild AD clinical trials [88]. The data from this clinical trial, which can be accessed in a public website (www.adni-info.org/), are of utmost relevance to researchers and clinicians for helping diagnose AD, develop new treatments and assess their effectiveness in the quest to improve safety and decrease the time and cost of clinical trials. Importantly, the associations of the genetic data with the neuroimaging, neurological and biochemical patterns detected in the ADNI data will allow the survey of genetic factors related to the rate of progression of AD.

Lastly, the diagnosis needs a cheap and fast BM, and none of the imaging methods mentioned above can do that.

2.1.3 Biochemical Markers – Serum and CSF

Most of the CSF is produced at the plexus choroideus. The CSF then flows from the ventricular space to the subarachnoid space and absorbed back to the serum through the Paccioni granulations. A further fraction of the CSF flows along the spinal cord. Here CSF proteins can reach the serum along the nerve sheaths by passive diffusion [55]. Thus, potentially promising BMs can be measured in the CSF. Lumbar puncture is a relatively safe and uncomplicated procedure for sampling of CSF, and only a small number of AD patients develop post-puncture complications [19].

In the past several years numerous blood and CSF tests have been proposed for early detection of AD and related disorders. These tests are based on disease pathology. Most common measurements of the CSF regarding AD focus on the tau protein and Aβ-peptides. Aβ-peptides result from enzymatic breakdown of APP by α-, β- and γ-secretases and make up the main part of amyloid deposits in AD [89].

Neurofibrillary protein aggregates containing phosphorylated tau are among the major hallmarks of AD. In normal cells, tau stabilizes axonal microtubules, which are the tracks for intracellular traffic. In AD, tau becomes abnormally phosphorylated, aggregates into paired helical filaments and loses its ability to maintain the microtubule tracks. Quantification of tau and Aβ in the CSF represents the most intensively studied BM of AD [90]. Approximately 80% of patients who meet clinical criteria for AD have elevated levels of CSF tau [91]. Autopsy studies confirmed this association and validated the relationship between high levels of CSF tau and a pathologic diagnosis of AD [92].

The study by Arai and coworkers [93] addressed the question whether CSF BMs could detect incipient AD among MCI patients. Results presented by these authors seem to indicate that high CSF total tau (t-tau) could discriminate between memory impaired individuals that later progressed to AD and those that did not convert. According to the data presented in the manuscript the test demonstrated 90% sensitivity and 100% specificity. The results have been recently confirmed [94]. However, since the values of these BMs do not change significantly throughout the course of the disease, they cannot be of utility as biochemical predictors in AD-modifying treatments with anti-acetyl cholinesterase agents, inhibitors of β and γ secretases or vaccination regimens.

The development of specific phosphate-dependent tau antibodies raised the possibility that AD could be uniquely identified by quantifying tau phosphorylated in one of several locations: threonine-181 (p-tau181), serine-199 (p-tau199) or threonine-231 (p-tau231). Boban and coworkers [95] tried to determine the diagnostic accuracy of the level of CSF t-tau, p-tau181 and p-tau199 as BMs in a group of AD and VaD patients. The results showed that CSF t-tau and p-tau199 were useful markers for differentiating between AD and VaD. However, while elevated levels of t-tau in the CSF supported the diagnosis of AD, failure to find elevated CSF t-tau reduced, but did not eliminate, the possibility of AD pathology.

Another line of recent evidence suggests that CSF p-tau231 levels decline during the natural course of AD. Seventeen pharmacologically untreated patients with AD were followed up to 6 years with repeated serial CSF measurements. CSF p-tau231 concentration, but not t-tau, decreased over time in AD, independent of age. Rate of change was inversely correlated to the MMSE score at baseline and provided objective measure for the cognitive decline [96]. These results suggest that CSF p-tau231 may have the potential to track AD progression and may be a valuable tool to map effects of disease modifying drugs on AD specific neurodegeneration.

Neuritic plaques associated with AD contain β-pleated sheets of amyloid. The 42-amino-acid fragment of $A\beta_{(1-42)}$ is especially prone to fibrillization and disproportionally accumulates in extracellular lesions in AD brains and was found to be reduced in the CSF of AD patients [92, 97]. However similar reductions occur in a variety of other conditions and as a diagnostic BM CSF $A\beta_{(1-42)}$ provides only a small addition of information beyond what is obtained from the CSF t-tau level [93].

The reduction in CSF $A\beta_{(1-42)}$ in AD has been hypothesized to reflect the deposition of the peptide in senile plaques, with lower levels diffusing to the CSF. Thus, according to Hulstaert and coworkers [98], the decrease in $A\beta_{(1-42)}$ is probably the most consistent BM finding in AD. $A\beta_{(1-42)}$ alone showed a sensitivity of 78% and specificity of 81% in distinguishing AD patients from normal elderly controls. Studies correlating CSF A$\beta_{(1-42)}$ protein concentrations with cognitive performance in AD were partly contradictory. Cross-sectionally, the concentrations of A$\beta_{(1-42)}$ protein and cognitive measures were either inversely correlated [99] or no significant correlation was found [98]. In a longitudinal study, a decrease of CSF A$\beta_{(1-42)}$ protein has been found within a 3 year follow-up [100]. A highly significant correlation between low CSF concentrations at baseline and 1 year serial follow-up was found. There was no correlation between CSF levels and disease duration or severity [19]. The potential value of A$\beta_{(1-42)}$ measurements during the course of AD progression should be further evaluated.

Others [101] criticized that such differences might be the result of differences in patients and control material. Furthermore, low CSF levels of A$\beta_{(1-42)}$ were also found in Creutzfeldt-Jakob disease [102, 103], DLB [104] and amyotrophic lateral sclerosis (ALS) [103], thus making the ability of A$\beta_{(1-42)}$ to distinguish between AD and other neurological disorders relatively limited. At present, however, it seems that Aβ can at best contribute to diagnostic accuracy but is unhelpful in staging disease severity. Furthermore, it is unknown whether interventions, such as reducing Aβ formation, will further reduce CSF Aβ levels or change towards normality, by increasing the levels.

The frequently used calculation of Aβ:tau ratio, has never been properly validated. It is unknown whether this ratio is superior to diagnosing AD based on reduced Aβ and elevated tau, or any other method of incorporating these values. The use of such parameters requires availability of positive predictive and negative predictive values in each. Moreover, since these values differ according to age, separate values may be required in each age group. Finally, although the Aβ and p-tau indicate AD pathology as either present or absent, they are of much more limited value in patients with mixed pathology. Obviously, patients with mixed pathology underlying their dementia will benefit less than those with significant vascular coexistent lesions.

In addition to CSF BMs, studies by others have focused on plasma BMs. In one study [105] plasma Aβ levels were found to be elevated in EOAD (Early Onset AD) as well as in sporadic LOAD and in first degree relatives of patients with LOAD. This finding provides evidence for novel and yet unknown factors that lead to elevation of plasma Aβ in asymptomatic young individuals from LOAD families. The study by Ringman and coworkers [106] provided preliminary support for a model in which BMs can predict disease progression in presymptomatic familial AD persons. Participants underwent evaluations for presenilin 1 or APP mutations and cognitive evaluations with Clinical Dementia Rating Scale (CDR). The data obtained in this study indicated that plasma A$\beta_{(1-42)}$ is elevated in familial AD mutations carriers and that this level may decrease with the cognitive decline of disease progression prior to the development of dementia. These results are in conflict with those found by Hulstaert and coworkers [98].

CSF BMs based on Aβ and tau derive from the known pathology of the disease. However, modern technology allows for non-biased exploration of CSF constituents using proteomics methodology. Preliminary unconfirmed results have been reported [107].

Promising novel BMs for AD have been recently documented by Ray and coworkers [108] focusing on the expression of 18 plasma proteins, out of 120 known signaling and inflammatory proteins, that predicted the progression of patients who had MCI to AD 2–6 years later. The protein signature pointed to systemic dysregulation of hematopoiesis, immune responses, apoptosis and neuronal support. In another proteomic approach using two-dimensional gel electrophoresis coupled with mass spectrometry [109], Hye and coworkers characterized a number of proteins in plasma previously implicated in the disease pathology of AD, *versus* control patients, including complement factor H (CFH) precursor and alpha-2-macroglobulin (alpha-2 M), whose expression correlated with disease severity. Other BMs (e.g. isoprostanes) have also been studied in the plasma [110, 111]. However, further studies aimed at investigating plasma BMs would be necessary to estimate sensitivity, specificity and replication in independent institutions, to evaluate their diagnostic value.

2.1.4 Transcriptomics Signatures as Potential Markers of AD

Currently, there is no blood test that can diagnose AD, making the detection of individuals at risk or at earliest stages of the disease practically impossible. For a BM to be clinically useful, non-invasive detection in accessible peripheral tissue is desirable. Blood lymphocytes may offer valuable surrogate markers for neuropsychiatric disorders, as they share significant gene expression similarities to less accessible CNS tissues [112]. Despite the differences in embryonic lineage and biological function between blood and brain, this correlation could be explained, in part, by the fact that blood cells circulate through many different body tissues and thus, may be exposed to the same toxic atmosphere surrounding. Under these circumstances, the blood cells can alter their transcriptome in response to pathological perturbations, acting as "sentinels", as proposed recently [113].

A number of studies have indicated that peripheral blood-based transcriptomic tests could be used for diagnostic profiling of AD. A large-scale transcriptomic study from the blood of 80 AD patients and 70 non-demented control individuals identified a RNA signature composed of 133 genes that correctly distinguished AD patients from control with high sensitivity and specificity. However it remains to be investigated whether it can differentiate AD from others types of dementia [114]. In a similar approach using an array of 96 selected gene probes, gene expression analysis was performed on blood RNA from a cohort of 203 probable AD and 209 cognitively healthy age matched controls. Although the gene signature could distinguish between mild to moderated AD and healthy individuals, the performance was modest reaching a specificity value of $\sim$72% [115]. It is possible, as discussed in the paper, that non-AD patients were included in the cohort, or that some of the control subjects have a mild cognitive impairment and thus possibly very early AD despite the normal history.

3 Biomarkers for Parkinson's Disease

3.1 Etiology and Pathogenetic Mechanisms of PD

PD may be the second most common neurodegenerative disease after AD, featuring an incidence that increases with age and a higher prevalence throughout the male population. Like AD, PD is also a multifactorial disease and several factors related to genes, age, sex, and environment may increase the risk of the disease. A small minority of all PD cases is of genetic origin [116, 117]. Mutations in alpha (α)-synuclein, parkin, ubiquitin c-terminal hydrolase *(UCH-L1)* and DJ-1 have been linked to earlier onset forms of PD, with onset before the age of 40 years. For details see Table 3 and Sect. 3.1.2 below. Several pesticides such as paraquat, maneb, dieldrin, eptaclor and more recently rotenone have been associated with an increased prevalence of PD among farmers. Moreover, a positive association between exposure to heavy metals including manganese, mercury, lead and PD has been suggested [118].

The classical cardinal symptoms of PD include bradykinesia, hypokinesia and akinesia, muscular rigidity and resting tremor, often progressing to postural instability, gait problems and freezing. More recently, considerable attention is given to a myriad of non-motor symptoms related to autonomic disturbances, such as hypotension, constipation, bladder and thermoregulatory dysfunctions and sleep disturbances [119]. PD may also be associated with olfactory dysfunction, depression and anxiety together with the onset of cognitive deficits and dementia as the disease progresses [120] (see Table 2).

The pathological hallmark of PD is represented by degeneration of the substantia nigra (SN) pars compacta dopaminergic neurons associated to the presence of intra-cytoplasmatic inclusions of ubiquitin and α-synuclein denominated Lewy bodies (Fig. 2). It is interesting to note that the motor symptoms characterizing the disease are manifested once degeneration of the dopaminergic nigro-striatal pathway has reached at least 50–60% [121]. This observation is in accordance with the notion that there is a prolonged period that precedes motor symptoms manifestation and is characterized by an extensive non-dopaminergic pathology, which involves several other neurotransmitters, including acetylcholine, noradrenaline, serotonin, glutamate and adenosine [122]. These abnormalities are responsible for the non-motor pathology of PD that may precede the onset of motor symptoms, sometimes by years and are often considered more debilitating to the afflicted patients. Common manifestations are freezing, falling, cognitive decline and the aforementioned dysfunctions such as anosmia (loss of sense of smell), anxiety, and depression [123]. This receives support from the post-mortem study of Braak et al. [124] who mapped the sequence of disease progression from brainstem to the basal ganglia and cortical regions, according to sequential accumulation of α-synuclein-immunopositive deposits in different brain structures. Further data suggest that PD pathology may even originate in peripheral tissues as the myenteric plexus in the gut and perhaps cardiac neurons [125, 126].

Table 2 Clinical and biological markers for Parkinson's disease

Clinical markers			
Autonomic system	**Neuropsychiatric manifestations**	**Sleep disorders**	**Sensory symptoms**
Gastrointestinal dysfunction	Depression	Impaired sleep initiation and maintenance	Pain
Cardiovascular dysfunction with orthostatic hypotension	Anxiety	Rapid eye movement sleep behavior disorder (RSBD)	Hyposmia
Urinary and sexual dysfunction	Panic attacks	Excessive daytime sleepiness	Visual dysfunction
Constipation	Dementia		
Nausea	Psychosis		
Increased sweating			
Biological markers (biomarkers)			
Biochemical CSF	**Imaging SPECT and PET**	**Genetic familial genes**	**Transcriptomic blood**
α-synuclein,	DA turnover (^{18}F-dopa);	*e.g., LRKK2*	e.g., HSP70-interacting protein ST13;
DJ-1	DAT (β-CIT, β-CFT);	*α-synuclein*	SKP1, PSMC4, HIP2
$A\beta_{1-42}$	VMAT2 (^{11}C-DTBZ);	*Pink1*	
Tau	Metabolic (FDG)	*DJ-1* *Parkin*	
Blood	**MRI and transcranial sonography**	**GWAS**	**Brain**
e.g., Serum urate; Complement C3c; factor H and B; α-synuclein; Fractalkine	Brain volume Hyperechogenicity	e.g., *MAPT, SNCA, HLA-DRB5*	e.g., SKP1, ALDH1A1, STX1, SYT1, PGC1-alpha

It is apparent that the damage to SN pars compacta occurs later in the course of the disease, while other brain areas and peripheral tissues are initially affected at the pre-symptomatic phase of the disease. Thus, similar to AD, there is a pre-PD state (preclinical/prodromal) before the disease can be formally diagnosed [127]. Clearly, the identification of suitable valid biological markers that can be monitored from preclinical stages in asymptomatic 'at-risk' subjects, will allow the capturing of high-risk individuals in the valuable therapeutic window of 5 years or more before the appearance of clinical signs. The early detection could dramatically modify our current approaches to research and treatment.

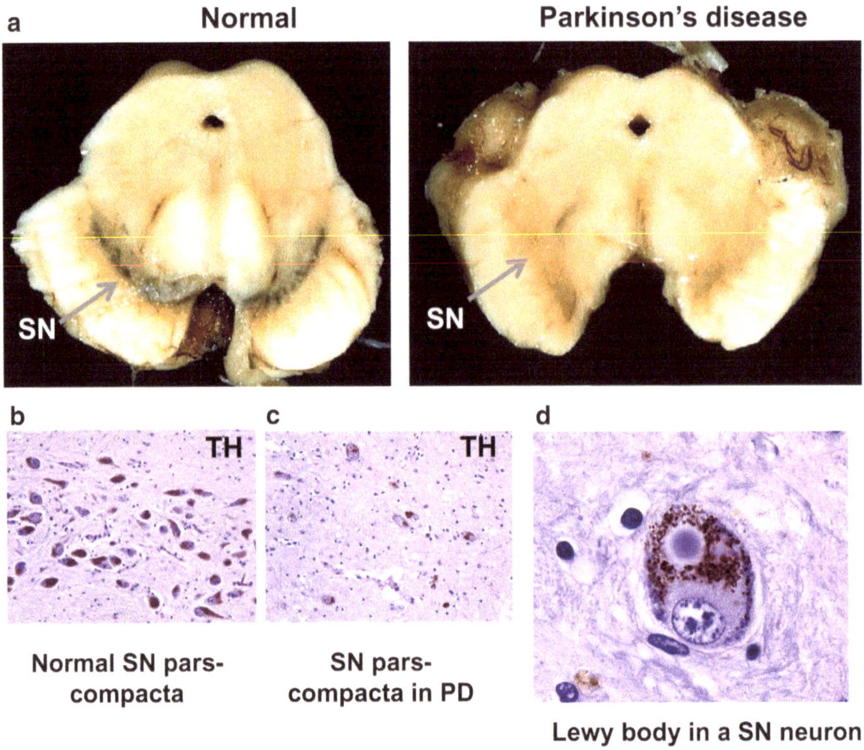

Fig. 2 Substantia nigra degeneration in PD and dementia with Lewy bodies. The core pathology of PD affects the dopamine (*DA*)-producing neurons of the substantia nigra (*SN*). DA is produced by SN neurons and transported along the axons of these neurons to the striatum. The triad of rigidity, bradykinesia and tremor at rest, correlates with degeneration of the dopaminergic nigrostriatal pathway and DA depletion in the striatum. In advanced PD, loss of these neurons results in depigmentation of the SN (**a**) and in loss of the DA synthesizing enzyme tyrosine-hydroxylase (*TH*) (**b**, **c**). PD is a synucleinopathy. Fibrils made of insoluble polymers of α-synuclein are deposited in the neuronal body, forming round lamellated eosinophilic cytoplasmic inclusions, the Lewy bodies (**d**). α-synuclein is also deposited in neuronal processes as Lewy neurites

Currently, the diagnosis and outcome measures of PD rest on the physician's physical examination scored with the Unified Parkinson's Disease Rating Scale (UPDRS) [128] and the Hoehn and Yahr (H&Y) staging scale [129] which, however do not always correlate. Although a diagnosis of PD can be relatively accurate in patients with a typical presentation of cardinal signs and response to levodopa treatment, the differential diagnosis versus different forms of parkinsonism may have greater overlap and thus misdiagnosis can occur in 25% of the patients or more, particularly at the earlier stages [130]. Thus, there is hope that a combination of different BMs (imaging, biochemical and genetic) will help diagnose presymptomatic disease stages, and more accurately when symptomatic, and complement the clinical outcome in clinical studies of disease modifying therapies. Obviously, misdiagnosis can lead to inappropriate therapy, for example prescribing monoamine inhibitors to patients with another underlying pathology, such as PSP.

3.1.1 Clinical Predictive Behavioral Tests

Symptoms including sleep disturbances, mood disorders and olfactory and autonomic dysfunctions (e.g. constipation) have been identified during both the pre-motor and motor symptomatic phases of the disease. Accordingly, the presence of these dysfunctions may contribute towards the clinical diagnosis of PD.

Presence of sleep disturbances and prolonged waking throughout the night are frequent and commonly associated with PD. Rapid eye movements (REM) sleep behavior disorder in patients affected by PD is closely related to incipient cognitive impairment, whereas quantitative motor testing, olfactory deficits and autonomic symptoms are not [131, 132]. Reduced REM sleep latency (time from sleep onset to the first REM episode) was found in a high proportion in PD patients, however its incidence was significantly higher in depressed patients suggesting that differences in latency may be related to depression rather than PD.

Depressed mood is often associated with PD and several recent studies have confirmed that depression in the PD patient is a major determinant of quality of life [133]. Depression in PD might be associated with a specific loss of dopamine (DA) and noradrenaline innervation in the limbic system [134]. Interestingly, several studies have suggested that depression in the PD patient is associated with a more rapid deterioration in cognitive functions [133].

Olfactory deficits reported in PD are thought to represent a sensitive (but not specific) marker of the disease which could allow an early diagnosis and help in differentiating PD from other diseases with features of parkinsonism. The diagnostic accuracy of olfactory testing in differentiating PD from other disorders is however insufficient to justify routine clinical use [135]. Olfactory dysfunctions are often measured by means of the University of Pennsylvania Smell Identification Test (UPSIT). Major deficit in olfactory judgement is present in PD, suggesting that this deficit is associated with disruption of olfactory areas situated in the temporal lobes and the prefrontal cortex [136], although olfactory bulb changes may have an important role.

Diagnostic test battery. Montgomery et al. [137] developed a behavioral diagnostic test battery in order to provide with early diagnosis of PD. The test battery incorporates tests for motor, olfaction and mood evaluation. As mentioned above, symptoms of olfactory dysfunction and depression often antedate the diagnosis of PD; addition of the wrist flexion and extension movement test provided a test battery featuring high specificity (88%) and sensitivity rate (69%), not previously achieved by any test alone. The latter battery test has been successfully applied in the prospective identification of as yet undiagnosed patients with signs suggestive of PD who have been subsequently diagnosed as affected by the disease [138].

MCI. Based on findings reported in several studies on PD featuring dementia and DLB, Troster [139] performed a critical evaluation to determine whether MCI may constitute an early symptom of PD with dementia and whether it may prove possible to differentiate between PD with dementia or DLB. The conclusions drawn in the review underline how, at present, the data available do not allow an adequate evaluation of the utility of MCI, which for PD mainly concerns executive functions, as a predictive test for PD with dementia. The author acknowledged that to date scarce attention has been devoted to cognitive impairment and the role of depression on the latter. There is an ongoing debate whether the concept of MCI is applicable to PD [140–142]. Moreover, PD with dementia and DLB are viewed by some as separate disease entities [143], while others consider them to be extremes in a continuum [139].

Autonomic dysfunctions including orthostatic dizziness, constipation, bladder dysfunction, pain, erectile dysfunction and hyperhydrosis are associated to PD and appear as early symptoms prior to the onset of motor disorders in some patients [144, 145]. A new diagnostic strategy utilizing skin biopsy developed by Sommer et al. [146] and applied to PD patients by Rossi et al. [147] suggests that small peripheral fibers may be reduced in PD patients. Analysis of small caliber sensory nerves might be used in the characterization of early autonomic dysfunctions in PD, but this needs to be confirmed.

3.1.2 Genetic Markers of PD

Genes Causing PD

Parkinsonism has been long viewed as a ‘sporadic’ disorder although a familial component was noted over a century ago by the studies of Leroux in 1880 [148] and Gowers in 1900 [149], who first suggested that heritable factors might increase disease susceptibility. It was not until the last decade that the contribution of genetics to familial forms of PD became much greater acknowledged. At present, 15 genomic chromosomal regions have been described and 5 pathogenic mutations in genes unequivocally linked to familial PD have been identified: the autosomal dominant *LRRK2* and *SNCA* and autosomal recessive *Parkin*, *PINK1* and *DJ1* [116, 150] (Table 3).

Dominantly-Inherited PD

Genetic studies of PD began with the discovery of mutations in the gene *SNCA* that encodes α-synuclein. Three point mutations have been identified that segregate with familial PD, Ala53Thr [151], Ala30Pro [152] and Glu46Lys [153]. Patients carrying *SNCA* mutations clinically respond to levodopa, although disease onset is earlier than in patients with idiopathic PD, and progression appears to be more rapid. Neuropathologically, *SNCA* carriers display similar features as seen in the idiopathic disease including cell degeneration, Lewy bodies, and neuritic staining. Notably, not

Table 3 Identified genes linked to familial PD

Locus	Gene/Protein	Chromosome location	Inheritance	Suggested function of protein
PARK 1/4	SNCA/ α-synuclein	4q21–q23	AD	Synaptic function/vesicle trafficking
PARK 2	parkin	6q25–q27	AR	E3 ubiquitin ligase
PARK 5	UCH-L1	4p14	AD	Ubiquitin C-terminal
PARK 6	PINK1	1p35–p36	AR	Mitochondrial serine/threonine kinase
PARK 7	DJ1	1p36	AR	Chaperone, oxidative stress response
PARK 8	LRRK2/dardarin	12p11.2–q13.1	AD	Protein kinase
PARK 9	ATP13 A2	1p36	AR	Lysosomal type 5 ATPase
PARK 13	OMI/HTRA2	2p12	AD	Serine protease
PARK 14	PLA2G6	22q13.1	AR	A2 phospholipase
PARK 15	FBXO7	22q12–q13	AR	SCF E3 ubiquitin ligase component
PARK 11 (provisional)	GIGYF2	2q36–q37	AD	Insulin signaling
PARK 17	VPS35	16:q12	AD	Component of the retromer cargo-recognition complex
PARK 18	EIF4G1	3q27	AD	mRNA translation initiation
Not determined	NR4A2/NURR1	2q22–q23	Unknown	Probable nuclear receptor
Not determined	Synphilin-1	5q23.1–q23.3	Unknown	Synaptic function, protein degradation

AD autosomal dominant, *AR* autosomal recessive

only mis-sense mutations, but also multiplications (duplication and triplication) of *SNCA* cause PD with a clear effect of dosage on phenotypic severity in terms of age of onset, disease duration and severity [154–156].

A mutation in the *ubiquitin carboxy-terminal hydrolase L1 gene (UCH-L1,* PARK5*)* was detected in a single sibling pair in a German family resembling typical PD with disease onset around 50 years of age [157]. UCH-L1 has ubiquitin hydrolase and ligase activities thus playing a role in the ubiquitin proteasomal system (UPS) which degrades damaged proteins system. This and its reported localization in Lewy bodies [158] support the suggested link between the UPS, endosomal-lysosomal pathways and Lewy bodies in late-onset PD, posing UCH-L1 a convincing candidate gene for PD.

Mutations in the *leucine-rich repeat kinase 2 (LRRK2)* gene (PARK8) are the most common genetic determinant of PD identified to date and have also been found in late-onset PD patients without a known family history of PD, suggesting a significant genetic component but also incomplete penetrance [116, 117]. *LRRK2* mutant PD is estimated to account for approximately 7% of familial PD cases and up to 3% of apparently sporadic disease, although significant ethnic differences exist. Accumulating evidence indicates that the G2019S mutation in *LRRK2* and homozygous or compound heterozygous mutations in the lysosomal enzyme β-glucocerebrosidase (*GBA*), the enzyme that is deficient in Gaucher's disease, are among the most prevalent genetic risk factors for PD. These mutations are most common among the Ashkenazi-Jewish population [159, 160]. *LRRK2* encodes a putative protein kinase and is predicted to be involved in multiple functions in view of the diverse protein domains which include a GTPase and kinase domain, as well leucine-rich and WD40 repeats protein interaction domains [150]. Other monogenic parkinsonisms caused by dominant mutations include the vacuolar protein sorting 35 (*VPS35*) and eukaryotic translation initiation factor 4 gamma 1 (*EIF4G1*) gene [161], which are more similar to the common, late-onset form of idiopathic PD (Table 3).

Recessively-Inherited PD

Four recessive forms of PD/parkinsonism have been identified with mutations in *Parkin (PRKN,* PARK2*)* [162], *PTEN-induced kinase 1 (PINK1,* PARK6*)* [163], *DJ1* (PARK7) [164] and *ATPase type 13A2 (ATP13A2,* PARK13*)* [165]. These mutations are relatively rare, with probable loss-of-function mechanisms, resulting in early age of onset (age <45 years) (Table 3).

Mutations in the *Parkin* gene were originally identified in Japanese families with autosomal recessive, juvenile parkinsonism [162]. Parkin has been accredited to function as an E3 ubiquitin-ligase, targeting proteins for degradation by the proteasome [166]. While several parkin-interacting proteins have been identified as putative targets for Parkin ubiquitination, their pathological role in the death of the SN pars compacta neurons is still elusive. PD patients carrying *Parkin* mutations have different pathological changes than sporadic cases; α-synuclein deposits and Lewy bodies are not seen in the SN, whereas changes in the lower brainstem and elsewhere have not been reported consistently.

PINK1 contains a serine/threonine protein kinase domain [163], and an N-terminal mitochondrial targeting sequence that localizes the protein to mitochondria. Recently, PINK1 has been demonstrated to regulate phosphorylation of another PD-linked gene, *Omi/HtrA2* (PARK13) modulating the response to cellular oxidative stress [167] and to act in a common pathway with Parkin in maintaining mitochondrial integrity and function in both muscles and dopaminergic neurons [168].

DJ1 encodes a mitochondrial protein putatively operating as an oxidative stress sensor within cell. DJ-1 mutations are rare, accounting for less than 1% of early-onset PD. The phenotype of an early-onset Parkinsonism is comparable with *Parkin*- and *PINK1*-linked forms of the disease [116].

Mutations in the *ATP13A2* (PARK9), encoding a protein with lysosomal localization, have been recently linked to autosomal recessive parkinsonism in families with Kufor-Rakeb syndrome [165]. Patients have juvenile onset of atypical parkinsonism, accompanied by pyramidal cell degeneration and dementia. Mutations in other genes, including, *PLA2G6* (PARK14), and *FBXO7* (PARK15), cause more rare forms of recessive parkinsonism with very early-onset (<30 years) and usually additional, atypical features (pyramidal, dystonic, ocular movement, and cognitive disturbances) [161].

Collectively, the discovery of genes linked to heritable forms of PD provides potential targets for therapeutic intervention. *PINK1, DJ1, parkin* and *OMI/HTRA2* are all associated with the mitochondria and have been implicated in protection against oxidative stress, alluding to a common pathway in their cell mechanism of action. The malfunction of these proteins may compromise mitochondrial integrity with subsequent increased levels of free radicals, reduced cellular energy and eventual impaired UPS function and protein aggregation, further supporting a link between mitochondrial impairment and the pathogenesis of PD. Thus, modulation of the function of Parkin, PINK1 or DJ-1 or interfering with the function of LRRK2 or with the expression and accumulation of α-synuclein may hold promise for neuroprotection in PD.

GWAS in PD

In addition to the familial PD-linked genes, several genomic loci have been designated in GWAS as either being linked or influencing the age of onset of PD, such as microtubule-associated protein tau (*MAPT*) and *SNCA* [169, 170] (for an overview of all published large-scale and GWAS in PD go to http://www.pdgene.org/largescale.asp). Recently, a meta-analysis of datasets from five PD GWAS from the USA and Europe has recognized 11 loci that surpassed the threshold for genome-wide significance ($p < 5 \times 10^{-8}$). Six were previously identified loci (MAPT, SNCA, HLA-DRB5, BST1, GAK and LRRK2) and five were newly identified loci (ACMSD, STK39, MCCC1/LAMP3, SYT11, and CCDC62/HIP1R) [171]. Moreover, a sequence variation in the putative promoter region of RAB7L1 localized to PARK16 locus was associated with reduced risk for PD in a genetically homogeneous Ashkenazi Jewish population [172].

GWAS promise to extend genome coverage in order to reveal common PD susceptibility genes and shed light on the (likely) myriad genetic factors involved in this complex disease.

3.1.3 Neuroimaging Markers

PET and SPECT

Imaging involves either detecting alterations in brain structure or examining functional changes in brain metabolic systems. In PD, degeneration of the dopaminergic system is accompanied by cholinergic, noradrenergic, and serotonergic dysfunction. Function of these dopaminergic and nondopaminergic systems can be imaged with PET and SPECT and correlated with motor and nonmotor disturbances. Dopa decarboxylase activity at DA terminals and DA turnover can both be measured with ^{18}F-dopa PET. Presynaptic dopamine transporters (DATs) can be identified *in vivo* with PET and SPECT tracers such as ^{11}C-2-β-carbomethoxy-3β-(4-fluorophenyl) tropane (β-CFT) and ^{123}I-2β-carbomethoxy-3β-(4-iodophenyl)-

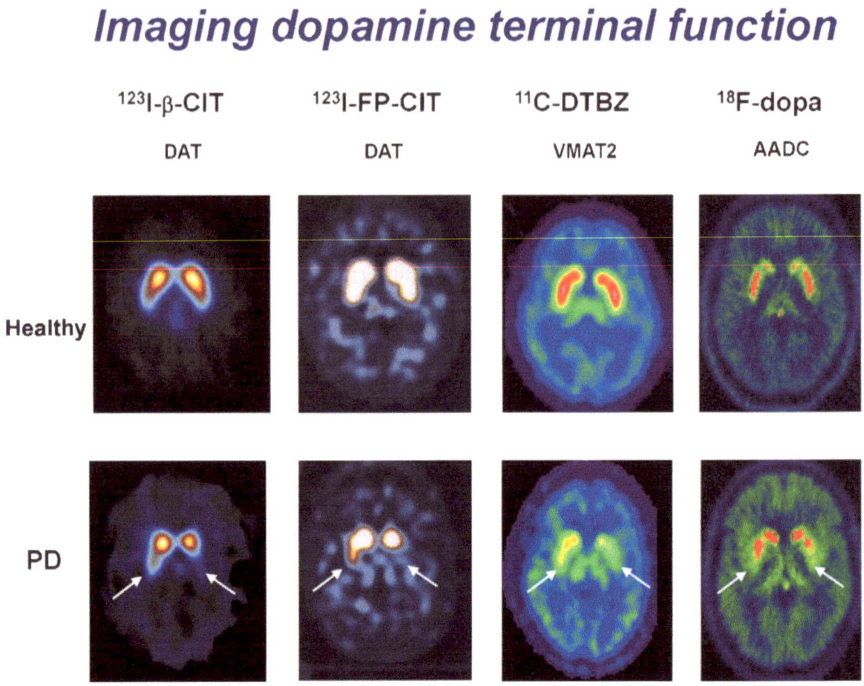

Fig. 3 Images of striatal beta-CIT SPECT (DAT), FP-CIT SPECT (DAT), 11C-DTBZ PET (VMAT2), and 18F-dopa PET (measuring striatal aromatic amino acid decarboxylase (AADC)) uptake in healthy volunteers and early PD. It can be seen that the four imaging modalities all show asymmetrically reduced posterior putamen dopaminergic function in PD (taken with permission from Brooks and Pavese [173])

N-(3-iodophenyl)tropane (β-CIT; Fig. 3), while vesicle monoamine transporter (VMAT) density in DA terminals can be examined with ^{11}C-dihydrotetrabenazine (^{11}C-DTBZ) PET. Extensive microglial activation can be detected in PD with 18 kDa translocator protein (TSPO) markers such as ^{11}C-PK11195 PET.

Measurements of DA terminal function can sensitively detect DA deficiency in both symptomatic patients and individuals at risk for parkinsonian syndromes (e.g. carriers of genetic mutations known to be associated with parkinsonism, elderly subjects with idiopathic hyposmia) [174], and are very helpful in differentiating PD from other disorders such as essential tremor or dystonic related tremor, but have poor specificity for discriminating between typical (idiopathic) and atypical PD (e.g. head trauma, drug-induced parkinsonism, PSP and MSA). On the other hand, measurements of glucose metabolism with FDG-PET can be helpful as normal or raised levels were found in the lentiform nucleus of PD but levels were reduced in MSA and PSP [175]. These markers are sensitive and able to demonstrate changes even in presymptomatic cases, or bilaterally in patients whose clinical manifestations are still unilateral. However, they are not sensitive to monitor disease progression, as evidenced by two prospective studies using ropirinole and pramipexole (CALM-PD-CIT and REAL-PET studies), where results of the imaging did not parallel the clinical changes [176, 177].

A relatively new imaging approach is demonstration of cardiac sympathetic innervations. The use of ^{123}I-metaiodobenzylguanidine (^{123}I-MIBG)-SPECT can demonstrate reduced or absent noradrenergic supply in PD, but not in MSA [173, 178].

MRI

MRI and transcranial sonography (TCS) can reveal brain structural changes such as volumetric reduction and hyper echogenicity of certain midbrain and striatal areas in patients with PD [179]. TCS and MRI and particularly diffusion MRI, can be helpful in differentiating PD from MSA [180] and might be particularly valuable for revealing a susceptibility to PD, although they correlate less well with either clinical status or loss of DA terminal function in the striatum. The sensitivity and specificity of these markers is still being examined, but it is clear that neither can monitor disease progression.

It is likely that newer MRI methods will be developed that will allow better diagnosis. One such potential is in higher field MRI. For example, 7 tesla (T) MRI has been shown to visualize anatomical alterations occurring in the SN of PD patients [181]. Similarly, 7 T proton magnetic resonance spectroscopy (MRS) was able to demonstrate elevated levels of pontine and putamenal GABA levels in PD [182]. Both of these studies were performed on a small number of subjects and the comparisons were made with normal controls, rather than patients with other neurological diseases.

Although the available BMs for PD are promising, none can yet predict the disease with a high percent of certainty, thus it seems that the best diagnosis in the near future will require a combination of markers and clinical assessment. Currently, a comprehensive multicenter study (Parkinson's Progression Markers Initiative (PPMI)) is under way in North America and Europe, designed to identify PD progression biomarkers in a cohort of recently diagnosed patients with PD and healthy subjects using advanced imaging, biological sampling (longitudinal collection of blood, cerebrospinal fluid (CSF) and urine), and clinical and behavioral assessments (http://www.ppmi-info.org) will be available. These biomarkers may aid in detecting asymptomatic individuals with early disease who may benefit from early therapeutic intervention.

3.2 Imaging of Iron: A Potential Biomarker in Neurodegenerative Diseases and Aging

There is increasing evidence that iron is involved in the processes that underlie many neurodegenerative diseases [183, 184]. Iron-related neurodegenerative disorders can be separated into those that result from iron accumulation in specific brain regions, and those that result from defects in iron metabolism and/or homeostasis. Normally, the total iron level in the brain of newborns is quite low, but increases rapidly until stable levels are reached at an age of 20–30 years [185]. Non-heme iron (mostly ferritin) progressively accumulates particularly in regions that are affected by AD and PD such as the putamen, motor cortex, prefrontal cortex and thalamus during the first 30–35 years of life, and variable changes are observed in older individuals [184, 186]. Furthermore, subcortical brain areas responsible for motor functions (extrapyramidal regions) appear to have more iron than non-motor-related regions, which might explain why movement disorders are commonly associated with iron imbalance [183, 185].

A quantitative evaluation study has revealed deposits of iron (Fe^{3+}) in microglia, oligodendrocytes, astrocytes located close to neurons, non-pigmented neurons and in the rim of Lewy bodies in the SN pars compacta of patients with PD. A similar picture of iron accumulation in glia and neurons was found in the putamen and pallidum of the same patients [187, 188]. Significant amounts of iron are sequestered in neuromelanin granules of dopaminergic and noradrenergic neurons of the human SN and locus coeruleus [189]. In light of the aforementioned findings there is a growing conviction that iron distribution and levels in certain brain areas might indicate a degree of vulnerability to PD.

Recent structural neuroimaging studies applying non-invasive TCS have shown that more than 90% of PD-patients show hyperechogenicity of the SN compared to healthy individuals [190] and this could be related in part to an increased iron content of the SN [191]. Even more, in young healthy adults who presented with increased echogenicity of the SN, a functional impairment of the nigrostriatal system was evident [191]. Thus, TCS technique may identify a susceptibility marker that can be detected prior to the clinical manifestations of PD.

3.2.1 Biochemical Markers

As mentioned above, no blood or CSF BM is available that correlate with sporadic forms of PD. We present here a series of biochemical tests, the clinical application of which needs to be investigated, together with other tests already applied in clinical settings, for the above purpose.

CSF is used in proteomic search strategy as the best biological fluid for detecting disease BMs, reflecting the state of substances present in the brain. Pioneer studies [192, 193] described differences in CSF protein pattern in PD patients compared to healthy controls such as variations in prostaglandin D synthase isoforms, though none of them have reached the phase of clinical screening. To date, the two markers that have been most broadly investigated in CSF of PD affected individuals are DJ-1 and α-synuclein, two proteins intimately involved in familial and sporadic PD pathogenesis. Some studies have described interesting methods for use in the quantification of α-synuclein in diseases characterized by synucleinopathy. Lower CSF levels of α-synuclein have been observed in neurological diseases with primary synucleinopathy, such as DLB, MSA and PD [194], whereas plasma levels were found increased [195]. It can be speculated that the reduction of CSF α-synuclein concentrations reflects intraneuronal accumulation (and deposition) of aggregated α-synuclein, whereas its increase probably indicates widespread neurodegeneration. However, quantification of total CSF α-synuclein did not enable to separate individual patients with α-synuclein-associated PD from other synucleopathies such as MSA and DLB [196]. The results from this and other studies indicate that the performance of CSF α-synuclein concentrations as a clinical laboratory test is insufficient as a single measure, and thus additional markers are needed for improved laboratory testing of individuals with PD or related diseases. In this respect, an interesting finding in clinically diagnosed PD subjects relates to the reported increase in CSF oligomers of α-synuclein or oligomers/total α-synuclein ratio, compared to patients with PSP, AD or control subjects [197], which may be considered for joint measurements with other markers.

Using CSF of PD patients, Zhang et al. [198] reported on the first large scale proteomic experiment, trying to identify the presence of additional BMs for PD and AD. Of these BMs, however, only $A\beta_{1-42}$ and tau proteins displayed different levels among controls, AD and PD and between AD and PD, as expected. While brain derived neurotrophic factor, interleukin-8, vitamin D binding protein, β_2 microglobulin, haptoglobin, apolipoprotein A-I, A-II and E were significantly different from healthy controls, they did not differ between PD and AD. Similarly, Shi et al. [199] have identified a panel of seven CSF proteins that could aid in PD diagnosis and differential diagnosis. Among these, an increase in CSF fractalkine, along with decreased $A\beta_{1-42}$ levels, correlated with a higher UPDRS score in cross-sectional samples and in a set of longitudinally collected PD samples from the DATATOP study. More recently, Ishigami and coworkers reported a CSF proteomic patterns that discriminates PD from MSA using mass spectrometry analysis [200], though the number of participants in each cohort was relatively small.

New potential BMs in blood are another focus of interest in PD research. Pioneer studies performed on lymphocytes or platelets of PD patients have detected subtle abnormalities in DA signaling [201, 202] or mitochondrial functions [203–206], which constitute hallmarks of PD. None of the latter abnormalities, however, has reached the phase of clinical screening. At present, the most studied marker in clinical settings linked to a reduced risk of PD, is the concentration of serum urate, the main end product of purine metabolism [207, 208]. Uric acid appears to be the first molecular factor directly linked to the progression of typical PD as revealed by a prospective trial showing an inverse correlation of urate levels with clinical and radiographic progression of PD [209]. Urate exerts an antioxidant efficacy comparable to that displayed by ascorbate; accordingly, the high levels registered may act as a defense against oxidative damage, one main acknowledged feature of PD. Urate is implicated in cerebrovascular diseases, therefore, rather than being utilized as a neuroprotective agent, its blood levels may serve as an indirect marker of neuroprotection in neurodegenerative diseases such as PD. More recently, a metabolomic profiling of blood from PD patients has shown a differential expression of analytes related to oxidative damage, lower uric acid and increased glutathione

[210] Thus, urate may be of use in neuroprotective trials as an index of disease progression [209, 211]. However, this association appears to be sex-dependent, as no significant lower rates of PD were detected among women [212]. Presence of adenosine A2A receptor alterations in putamen of PD patients that mirrors up-regulation in human peripheral blood elements has been also reported. A statistically significant linear correlation among the A2A receptor density or tumor necrosis factor-alpha (TNF-alpha) levels in blood lymphocytes and plasma respectively and UPDRS motor score was described by Varani and coworkers [213]. The correlation found in the basal ganglia and UPDRS motor score highlights the central role of A2A receptor antagonists in the pharmacological treatment of PD [214].

Using quantitative 2D gel electrophoresis, 34 different serum proteins have been identified in PD and ALS, 9 of which belong to the complement system [215–217]. Indeed, differences have been reported between the components of complement C3c and complement factor H and B in affected subjects compared to controls. These findings lend further support to evidence suggesting that inflammatory processes involving complement cascade elements may be involved in the onset of PD.

Interestingly, plasma α-synuclein was found elevated early in the development of PD and the levels were not changed after DA agonists treatment [218]. Yet, serum levels of alpha-synuclein are unlikely to be a diagnostic marker since in addition to the brain, it is produced in peripheral organs, blood cells and skin [219, 220]. Regarding DJ-1, no differences were reported in its serum levels measured in PD patients and healthy controls in a study performed by Maita et al. [221], whereas other studies performed in CSF or plasma, reported higher DJ-1 levels in PD patients than controls [222, 223]. Moreover, while upregulation of CSF DJ-1 levels in early stages of PD was more marked compared to levels observed in advanced stages and in controls, plasma DJ-1 levels appeared to be higher in advanced disease. Likewise, these studies revealed similar increases in PD and in neurological diseases with primary synucleinopathy [222, 223]. Thus, how DJ-1 and α-synuclein in biological fluids may be of use as BMs in differentiating PD from other neurological disorders remains to be established.

For the scope of BMs in neurodegenerative diseases, it should be borne in mind how clinical proteomic and consequently predictive BMs are strongly influenced by pre-analytical factors such as time or condition of storage [224]. Accumulation of altered proteins and impaired protein clearance may underlie the onset of PD. Recent advances made in methodological procedures used in the study of PD-associated proteins, involving the systematic combined use of BMs may lead to findings capable of providing a crucial contribution to enhancing diagnosis of PD.

3.2.2 Transcriptomics Signatures as Potential Markers of PD

The next section will describe potential BMs based on gene expression studies performed in post-mortem brain tissue and blood-derived RNA from sporadic PD.

Gene Expression Profiling of Sporadic PD SN

The current unsatisfactory situation with early diagnosis and available measures of neuroprotection in PD has prompted researchers to look for new approaches using smart advanced biotechnology. During the course of the past decade, we are witnessing the advent of high throughput gene and protein techniques which are becoming especially relevant in the neuroprotective field for aging and neurodegenerative diseases, since they can provide important targets to develop disease modifying therapies that may have an impact on the disease. Recent microarray-assisted large-scale transcriptomic studies in human *post-mortem* SN from sporadic PD have revealed significant dysregulation of genes from biological processes linked to previously established neurodegenerative mechanisms both in sporadic and hereditary PD [225]. These include protein aggregation, mitochondrial dysfunction, oxidative stress, cell cycle, vesicle trafficking, synaptic transmission, DA metabolism and cell adhesion/cytoskeleton maintenance. The selected number of candidate genes identified by the various research groups may offer clues for potential gene intersections or cross-talks along the neurodegenerative cascade.

Interestingly, all the gene expression profiling studies conducted so far in the SN of PD patients [225], reported a dysfunction in vesicle trafficking and synaptic transmission functional group such as syntaxin-1 (*STX1*), synaptotagmin-1 (*SYT1*) and synapsin 1 and 2 (*SYN1,2*). Also, all the studies revealed a consistent elevation in heat-shock proteins/ stress genes, such as the endoplasmic reticulum protein homocysteine-inducible, endoplasmic reticulum stress-inducible, ubiquitin-like domain member 1 (*HERPUD1*) and DnaJ (Hsp40) homologue subfamily B, members 1 and 5 (*DNAJB1/5*), heat shock proteins A1A and A1B (*HSPA1A, HSPA1B*) and heat shock 27 kDa protein 1 (*HSPB1*) [225].

More recently, a global meta-analysis of nine genome-wide expression studies (meta-GWES) integrating transcriptomic data from post-mortem SN (including the above described studies) and laser-captured DA neurons of PD patients [226], revealed decreases in expression of genes related to several biological processes associated with PD, such as nuclear-encoded mitochondrial electron transport and glucose and protein metabolism. In particular, under expression of peroxisome proliferator gene, PGC-1alpha in PD patients was observed, a master regulator of genes involved in mitochondrial biogenesis and function [226].

The main outcome of the above studies is to provide a view into potential disease mechanistic pathways underlying pathophysiology of PD. Considering the multifactorial nature of PD, reflected in the phenotypic and clinical manifestation variability and the heterogeneity in dopaminergic marker imaging, the information obtained from the functional genomics studies will be useful to assess whether the BM would be relevant to the disease process and consequently, to drug specific treatment.

Gene Expression Profiling of Sporadic PD Blood

In a transcriptome-wide scan study in whole blood tissue from early-staged PD individuals and age-matched controls, Scherzer et al. [227, 228] have found a panel of genes that may modify PD risk. The authors reported under-expression of the HSP70-interacting protein ST13 and coexpression and coinduction of *SNCA* gene with critical enzymes of heme metabolism, in patients with PD versus healthy individuals. In another study, using a Human Exon array platform, a down regulation of genes related to B cell activity was identified in peripheral blood leukocytes from 30 female PD-patients [229].

Regardless of which disease is investigated, the most precise results will be obtained for genes that show similar expression profiles across blood and CNS tissues [112]. Supporting the rationale of identifying molecular changes in peripheral blood that may respond to the pathology in the brain of sporadic PD, Grünblatt et al. [230] reported a cluster signature of four genes in blood tissue that discriminated between PD and healthy controls. These transcripts have been primarily found by the same authors to be altered in PD SNpc [231]. More recently, our group has identified a five-gene panel that optimally discriminates between early PD (newly diagnosed, before or after receiving DA replacing medication) from healthy controls, based on stepwise multivariate logistic regression analysis of genes that were found significantly altered in sporadic PD SN tissue [232].

These findings suggest that blood and neuronal cells might have a common regulatory mechanism for gene expression. Additional cohort studies are warranted that would include larger PD population size, adequately selected on the basis of disease onset, date of medication initiation and follow-up program. In this study paradigm, PD patients could be followed during the course of the disease for correlation in their gene expression profile in blood to identify a PD biomarker panel phenotype that can potentially be used for diagnosis of early PD (risk markers) and monitor disease progression and its modification by therapy.

4 Future Perspective

The rapid growth of molecular genetic, imaging and laboratory technology has expanded to the point at which the application of technically advanced BMs becomes widely available and applicable. Such BMs can be useful in mapping disease pathogenesis and identifying subjects at risk for AD and PD.

BMs derived from genes mutated in familial PD/AD, or from transcriptomics of sporadic brain/blood tissue studies as well as polymorphisms in sporadic cases, may assist to identify individuals "at-risk" during the asymptomatic period. Early screening, detection, and diagnosis of AD and PD will allow intervention with disease modifying therapies at earlier stages, once such drugs become available, and thus may potentially improve clinical outcome. Neurochemical BMs measured in the periphery by means of proteomics/metabolomics in blood plasma, CSF and other tissues may be useful adjuncts to imaging and clinical assessment tools for AD and PD. They may also provide valuable information about pathogenic mechanisms during clinical testing of neuroprotective/disease modifying drugs, which is especially relevant to personalized treatment.

In spite of the intensive research in the field of BMs for neurodegenerative diseases, there is some frustration with the progress being made. Many of the BMs discussed above show group differences (e.g. between normals and diseased), but none so far has been able to reliably diagnose AD or PD in their preclinical stages in an individual subject. At present the diagnosis depends on clinical or pathological phenotypes, which suffer however from heterogeneity. This, added to the phenotypic convergence, indicate the limits of present-day understanding of these diseases. Since both AD and PD have a long preclinical period, future efforts should be focused on this time window which allows neuroprotective treatment. Large multicenter studies are warranted to evaluate the diagnostic value of the combined application of BMs with multiple modalities for prediction, targeted prevention and evaluation of disease modifying therapies in AD and PD.

References

1. Weiner WJ (2005) A differential diagnosis of Parkinsonism. Rev Neurol Dis 2:124–131
2. Poewe W, Wenning G (2002) The differential diagnosis of Parkinson's disease. Eur J Neurol 9(Suppl 3):23–30
3. Beach TG, Monsell SE, Phillips LE, Kukull W (2012) Accuracy of the clinical diagnosis of Alzheimer disease at National Institute on Aging Alzheimer Disease Centers, 2005–2010. J Neuropathol Exp Neurol 71:266–273
4. Hulette C, Nochlin D, Mckeel D, Morris JC, Mirra SS, Sumi SM, Heyman A (1997) Clinical-neuropathologic findings in multi-infarct dementia: a report of six autopsied cases. Neurology 48:668–672
5. Hulka BS, Wilcosky T (1988) Biological markers in epidemiologic research. Arch Environ Health 43:83–89
6. Naylor S (2003) Biomarkers: current perspectives and future prospects. Expert Rev Mol Diagn 3:525–529
7. Masterman DL, Mendez MF, Fairbanks LA, Cummings JL (1997) Sensitivity, specificity, and positive predictive value of technetium 99-HMPAO SPECT in discriminating Alzheimer's disease from other dementias. J Geriatr Psychiatry Neurol 10:15–21
8. Fratiglioni L, Launer LJ, Andersen K, Breteler MM, Copeland JR, Dartigues JF, Lobo A, Martinez-Lage J, Soininen H, Hofman A (2000) Incidence of dementia and major subtypes in Europe: a collaborative study of population-based cohorts. Neurologic Diseases in the Elderly Research Group. Neurology 54:S10–S15
9. Lobo A, Launer LJ, Fratiglioni L, Andersen K, Di Carlo A, Breteler MM, Copeland JR, Jagger C, Martinez-Lage J, Soininen H, Hofman A (2000) Prevalence of dementia and major subtypes in Europe: a collaborative study of population-based cohorts. Neurologic Diseases in the Elderly Research Group. Neurology 54:S4–S9
10. Graham JE, Rockwood K, Beattie BL, Eastwood R, Gauthier S, Tuokko H, McDowell I (1997) Prevalence and severity of cognitive impairment with and without dementia in an elderly population. Lancet 349:1793–1796
11. Kawas C, Gray S, Brookmeyer R, Fozard J, Zonderman A (2000) Age-specific incidence rates of Alzheimer's disease: the Baltimore longitudinal study of aging. Neurology 54:2072–2077
12. Mckhann G, Drachman D, Folstein M, Katzman R, Price D, Stadlan EM (1984) Clinical diagnosis of Alzheimer's disease: report of the NINCDS-ADRDA work group under the auspices of Department of Health and Human Services Task Force on Alzheimer's Disease. Neurology 34:939–944

13. Dubois B, Feldman HH, Jacova C, Dekosky ST, Barberger-Gateau P, Cummings J, Delacourte A, Galasko D, Gauthier S, Jicha G, Meguro K, O'brien J, Pasquier F, Robert P, Rossor M, Salloway S, Stern Y, Visser PJ, Scheltens P (2007) Research criteria for the diagnosis of Alzheimer's disease: revising the NINCDS-ADRDA criteria. Lancet Neurol 6:734–746
14. Mckhann GM, Knopman DS, Chertkow H, Hyman BT, Jack CR Jr, Kawas CH, Klunk WE et al (2011) The diagnosis of dementia due to Alzheimer's disease: recommendations from the National Institute on Aging-Alzheimer's Association workgroups on diagnostic guidelines for Alzheimer's disease. Alzheimers Dement 7:263–269
15. Korczyn AD (2011) Commentary on "Recommendations from the National Institute on Aging-Alzheimer's Association workgroups on diagnostic guidelines for Alzheimer's disease". Alzheimers Dement 7:333–334
16. Lyketsos CG, Szekely CA, Mielke MM, Rosenberg PB, Zandi PP (2008) Developing new treatments for Alzheimer's disease: the who, what, when, and how of biomarker-guided therapies. Int Psychogeriatr 20:871–889
17. Sunderland T, Hampel H, Takeda M, Putnam KT, Cohen RM (2006) Biomarkers in the diagnosis of Alzheimer's disease: are we ready? J Geriatr Psychiatry Neurol 19:172–179
18. Korczyn AD (2002) Mixed dementia–the most common cause of dementia. Ann N Y Acad Sci 977:129–134
19. Andreasen N, Minthon L, Davidsson P, Vanmechelen E, Vanderstichele H, Winblad B, Blennow K (2001) Evaluation of CSF-tau and CSF-Abeta42 as diagnostic markers for Alzheimer disease in clinical practice. Arch Neurol 58:373–379
20. Hanninen T, Hallikainen M, Tuomainen S, Vanhanen M, Soininen H (2002) Prevalence of mild cognitive impairment: a population-based study in elderly subjects. Acta Neurol Scand 106:148–154
21. Middleton LE, Grinberg LT, Miller B, Kawas C, Yaffe K (2011) Neuropathologic features associated with Alzheimer disease diagnosis: age matters. Neurology 77:1737–1744
22. Grinberg LT, Rueb U, Heinsen H (2011) Brainstem: neglected locus in neurodegenerative diseases. Front Neurol 2:42
23. Jack CR Jr, Petersen RC, Xu Y, O'brien PC, Smith GE, Ivnik RJ, Boeve BF, Tangalos EG, Kokmen E (2000) Rates of hippocampal atrophy correlate with change in clinical status in aging and AD. Neurology 55:484–489
24. Henneman WJ, Sluimer JD, Barnes J, Van Der Flier WM, Sluimer IC, Fox NC, Scheltens P, Vrenken H, Barkhof F (2009) Hippocampal atrophy rates in Alzheimer disease: added value over whole brain volume measures. Neurology 72:999–1007
25. Gasser T (2003) Overview of the genetics of parkinsonism. Adv Neurol 91:143–152
26. Bertram L, Tanzi RE (2008) Thirty years of Alzheimer's disease genetics: the implications of systematic meta-analyses. Nat Rev Neurosci 9:768–778
27. Mulder C, Scheltens P, Visser JJ, Van Kamp GJ, Schutgens RB (2000) Genetic and biochemical markers for Alzheimer's disease: recent developments. Ann Clin Biochem 37(Pt 5):593–607
28. Strittmatter WJ, Weisgraber KH, Huang DY, Dong LM, Salvesen GS, Pericak-Vance M, Schmechel D, Saunders AM, Goldgaber D, Roses AD (1993) Binding of human apolipoprotein E to synthetic amyloid beta peptide: isoform-specific effects and implications for late-onset Alzheimer disease. Proc Natl Acad Sci U S A 90:8098–8102
29. Tanzi RE, Bertram L (2005) Twenty years of the Alzheimer's disease amyloid hypothesis: a genetic perspective. Cell 120:545–555
30. Roses AD, Lutz MW, Amrine-Madsen H, Saunders AM, Crenshaw DG, Sundseth SS, Huentelman MJ, Welsh-Bohmer KA, Reiman EM (2010) A TOMM40 variable-length polymorphism predicts the age of late-onset Alzheimer's disease. Pharmacogenomics J 10:375–384
31. Grossman I, Lutz MW, Crenshaw DG, Saunders AM, Burns DK, Roses AD (2010) Alzheimer's disease: diagnostics, prognostics and the road to prevention. EPMA J 1:293–303

32. Reiman EM, Caselli RJ, Chen K, Alexander GE, Bandy D, Frost J (2001) Declining brain activity in cognitively normal apolipoprotein E epsilon 4 heterozygotes: a foundation for using positron emission tomography to efficiently test treatments to prevent Alzheimer's disease. Proc Natl Acad Sci U S A 98:3334–3339
33. Reiman EM, Chen K, Alexander GE, Caselli RJ, Bandy D, Osborne D, Saunders AM, Hardy J (2004) Functional brain abnormalities in young adults at genetic risk for late-onset Alzheimer's dementia. Proc Natl Acad Sci U S A 101:284–289
34. Bookheimer SY, Strojwas MH, Cohen MS, Saunders AM, Pericak-Vance MA, Mazziotta JC, Small GW (2000) Patterns of brain activation in people at risk for Alzheimer's disease. N Engl J Med 343:450–456
35. Lindsten K, Menendez-Benito V, Masucci MG, Dantuma NP (2003) A transgenic mouse model of the ubiquitin/proteasome system. Nat Biotechnol 21:897–902
36. Morgan D, Diamond DM, Gottschall PE, Ugen KE, Dickey C, Hardy J, Duff K, Jantzen P, DiCarlo G, Wilcock D, Connor K, Hatcher J, Hope C, Gordon M, Arendash GW (2000) A beta peptide vaccination prevents memory loss in an animal model of Alzheimer's disease. Nature 408:982–985
37. Schenk D, Barbour R, Dunn W, Gordon G, Grajeda H, Guido T, Hu K, Huang J, Johnson-Wood K, Khan K, Kholodenko D, Lee M, Liao Z, Lieberburg I, Motter R, Mutter L, Soriano F, Shopp G, Vasquez N, Vandevert C, Walker S, Wogulis M, Yednock T, Games D, Seubert P (1999) Immunization with amyloid-beta attenuates Alzheimer-disease-like pathology in the PDAPP mouse. Nature 400:173–177
38. Rockenstein E, Crews L, Masliah E (2007) Transgenic animal models of neurodegenerative diseases and their application to treatment development. Adv Drug Deliv Rev 59:1093–1102
39. Korczyn AD (2008) The amyloid cascade hypothesis. Alzheimers Dement 4:176–178
40. Corder EH, Saunders AM, Strittmatter WJ, Schmechel DE, Gaskell PC, Small GW, Roses AD, Haines JL, Pericak-Vance MA (1993) Gene dose of apolipoprotein E type 4 allele and the risk of Alzheimer's disease in late onset families. Science 261:921–923
41. Guerreiro RJ, Hardy J (2011) Alzheimer's disease genetics: lessons to improve disease modelling. Biochem Soc Trans 39:910–916
42. Bertram L (2011) Alzheimer's genetics in the GWAS era: a continuing story of 'replications and refutations'. Curr Neurol Neurosci Rep 11:246–253
43. Bertram L, Lill CM, Tanzi RE (2010) The genetics of Alzheimer disease: back to the future. Neuron 68:270–281
44. Naj AC, Jun G, Beecham GW et al (2011) Common variants at MS4A4/MS4A6E, CD2AP, CD33 and EPHA1 are associated with late-onset Alzheimer's disease. Nat Genet 43:436–441
45. Hollingworth P, Harold D, Sims R et al (2011) Common variants at ABCA7, MS4A6A/MS4A4E, EPHA1, CD33 and CD2AP are associated with Alzheimer's disease. Nat Genet 43:429–435
46. Lambert JC, Heath S, Even G, Campion D, Sleegers K, Hiltunen M, Combarros O et al (2009) Genome-wide association study identifies variants at CLU and CR1 associated with Alzheimer's disease. Nat Genet 41:1094–1099
47. Harold D, Abraham R, Hollingworth P et al (2009) Genome-wide association study identifies variants at CLU and PICALM associated with Alzheimer's disease. Nat Genet 41:1088–1093
48. Seshadri S, Fitzpatrick AL, Ikram MA et al (2010) Genome-wide analysis of genetic loci associated with Alzheimer disease. JAMA 303:1832–1840
49. Jack CR Jr (2011) Alliance for aging research AD biomarkers work group: structural MRI. Neurobiol Aging 32(Suppl 1):S48–S57
50. Csernansky JG, Wang L, Joshi S, Miller JP, Gado M, Kido D, Mckeel D, Morris JC, Miller MI (2000) Early DAT is distinguished from aging by high-dimensional mapping of the hippocampus. Dementia of the Alzheimer type. Neurology 55:1636–1643
51. Dickerson BC, Goncharova I, Sullivan MP, Forchetti C, Wilson RS, Bennett DA, Beckett LA, deToledo-Morrell L (2001) MRI-derived entorhinal and hippocampal atrophy in incipient and very mild Alzheimer's disease. Neurobiol Aging 22:747–754

52. Braak H, Alafuzoff I, Arzberger T, Kretzschmar H, Del Tredici K (2006) Staging of Alzheimer disease-associated neurofibrillary pathology using paraffin sections and immunocytochemistry. Acta Neuropathol 112:389–404
53. Braak H, Braak E (1991) Neuropathological stageing of Alzheimer-related changes. Acta Neuropathol 82:239–259
54. Leung KK, Barnes J, Ridgway GR, Bartlett JW, Clarkson MJ, Macdonald K, Schuff N, Fox NC, Ourselin S; Alzheimer's Disease Neuroimaging Initiative (2010) Automated cross-sectional and longitudinal hippocampal volume measurement in mild cognitive impairment and Alzheimer's disease. Neuroimage 51:1345–1359
55. Reiber H, Peter JB (2001) Cerebrospinal fluid analysis: disease-related data patterns and evaluation programs. J Neurol Sci 184:101–122
56. Dekosky ST, Ikonomovic MD, Styren SD, Beckett L, Wisniewski S, Bennett DA, Cochran EJ, Kordower JH, Mufson EJ (2002) Upregulation of choline acetyltransferase activity in hippocampus and frontal cortex of elderly subjects with mild cognitive impairment. Ann Neurol 51:145–155
57. Davis KL, Mohs RC, Marin D, Purohit DP, Perl DP, Lantz M, Austin G, Haroutunian V (1999) Cholinergic markers in elderly patients with early signs of Alzheimer disease. JAMA 281:1401–1406
58. Werner P, Korczyn AD (2008) Mild cognitive impairment: conceptual, assessment, ethical, and social issues. Clin Interv Aging 3:413–420
59. Resnick SM, Goldszal AF, Davatzikos C, Golski S, Kraut MA, Metter EJ, Bryan N, Zonderman AB (2000) One-year age changes in MRI brain volumes in older adults. Cereb Cortex 10:464–472
60. Sullivan EV, Pfefferbaum A, Adalsteinsson E, Swan GE, Carmelli D (2002) Differential rates of regional brain change in callosal and ventricular size: a 4-year longitudinal MRI study of elderly men. Cereb Cortex 12:438–445
61. Chetelat G, Baron JC (2003) Early diagnosis of Alzheimer's disease: contribution of structural neuroimaging. Neuroimage 18:525–541
62. Imbimbo BP (2009) Why did tarenflurbil fail in Alzheimer's disease? J Alzheimers Dis 17:757–760
63. Musiek ES, Chen Y, Korczykowski M, Saboury B, Martinez PM, Reddin JS, Alavi A, Kimberg DY, Wolk DA, Julin P, Newberg AB, Arnold SE, Detre JA (2012) Direct comparison of fluorodeoxyglucose positron emission tomography and arterial spin labeling magnetic resonance imaging in Alzheimer's disease. Alzheimers Dement 8:51–59
64. Connor DM, Benveniste H, Dilmanian FA, Kritzer MF, Miller LM, Zhong Z (2009) Computed tomography of amyloid plaques in a mouse model of Alzheimer's disease using diffraction enhanced imaging. Neuroimage 46:908–914
65. Borthakur A, Wheaton AJ, Gougoutas AJ, Akella SV, Regatte RR, Charagundla SR, Reddy R (2004) In vivo measurement of T1rho dispersion in the human brain at 1.5 tesla. J Magn Reson Imaging 19:403–409
66. Klunk WE, Panchalingam K, Moossy J, Mcclure RJ, Pettegrew JW (1992) N-acetyl-L-aspartate and other amino acid metabolites in Alzheimer's disease brain: a preliminary proton nuclear magnetic resonance study. Neurology 42:1578–1585
67. Kwo-on-Yuen PF, Newmark RD, Budinger TF, Kaye JA, Ball MJ, Jagust WJ (1994) Brain N-acetyl-L-aspartic acid in Alzheimer's disease: a proton magnetic resonance spectroscopy study. Brain Res 667:167–174
68. Jessen F, Block W, Traber F, Keller E, Flacke S, Papassotiropoulos A, Lamerichs R, Heun R, Schild HH (2000) Proton MR spectroscopy detects a relative decrease of N-acetylaspartate in the medial temporal lobe of patients with AD. Neurology 55:684–688
69. Schuff N, Capizzano AA, Du AT, Amend DL, O'neill J, Norman D, Kramer J, Jagust W, Miller B, Wolkowitz OM, Yaffe K, Weiner MW (2002) Selective reduction of N-acetylaspartate in medial temporal and parietal lobes in AD. Neurology 58:928–935

70. Kantarci K, Jack CR Jr, Xu YC, Campeau NG, O'Brien PC, Smith GE, Ivnik RJ, Boeve BF, Kokmen E, Tangalos EG, Petersen RC (2000) Regional metabolic patterns in mild cognitive impairment and Alzheimer's disease: a 1H MRS study. Neurology 55:210–217
71. Leinonen V, Alafuzoff I, Aalto S, Suotunen T, Savolainen S, Nagren K, Tapiola T, Pirttilä T, Rinne J, Jääskeläinen JE, Soininen H, Rinne JO (2008) Assessment of beta-amyloid in a frontal cortical brain biopsy specimen and by positron emission tomography with carbon 11-labeled Pittsburgh Compound B. Arch Neurol 65:1304–1309
72. Valenzuela MJ, Sachdev PS, Wen W, Shnier R, Brodaty H, Gillies D (2000) Dual voxel proton magnetic resonance spectroscopy in the healthy elderly: subcortical-frontal axonal N-acetylaspartate levels are correlated with fluid cognitive abilities independent of structural brain changes. Neuroimage 12:747–756
73. Mackay S, Ezekiel F, Di Sclafani V, Meyerhoff DJ, Gerson J, Norman D, Fein G, Weiner MW (1996) Alzheimer disease and subcortical ischemic vascular dementia: evaluation by combining MR imaging segmentation and H-1 MR spectroscopic imaging. Radiology 198:537–545
74. Vemuri P, Jones DT, Jack CR Jr (2012) Resting state functional MRI in Alzheimer's disease. Alzheimers Res Ther 4:2
75. Foster NL, Heidebrink JL, Clark CM, Jagust WJ, Arnold SE, Barbas NR, Decarli CS, Turner RS, Koeppe RA, Higdon R, Minoshima S (2007) FDG-PET improves accuracy in distinguishing frontotemporal dementia and Alzheimer's disease. Brain 130:2616–2635
76. Silverman DH, Small GW, Chang CY, Lu CS, Kung De Aburto MA, Chen W, Czernin J, Rapoport SI, Pietrini P, Alexander GE, Schapiro MB, Jagust WJ, Hoffman JM, Welsh-Bohmer KA, Alavi A, Clark CM, Salmon E, de Leon MJ, Mielke R, Cummings JL, Kowell AP, Gambhir SS, Hoh CK, Phelps ME (2001) Positron emission tomography in evaluation of dementia: regional brain metabolism and long-term outcome. JAMA 286:2120–2127
77. Mosconi L (2005) Brain glucose metabolism in the early and specific diagnosis of Alzheimer's disease. FDG-PET studies in MCI and AD. Eur J Nucl Med Mol Imaging 32:486–510
78. Jagust W (2004) Molecular neuroimaging in Alzheimer's disease. NeuroRx 1:206–212
79. Johnson KA, Holman BL, Rosen TJ, Nagel JS, English RJ, Growdon JH (1990) Iofetamine I 123 single photon emission computed tomography is accurate in the diagnosis of Alzheimer's disease. Arch Intern Med 150:752–756
80. Eberling JL, Jagust WJ, Reed BR, Baker MG (1992) Reduced temporal lobe blood flow in Alzheimer's disease. Neurobiol Aging 13:483–491
81. Johnson KA, Jones K, Holman BL, Becker JA, Spiers PA, Satlin A, Albert MS (1998) Preclinical prediction of Alzheimer's disease using SPECT. Neurology 50:1563–1571
82. Herholz K, Salmon E, Perani D, Baron JC, Holthoff V, Frolich L, Schonknecht P, Ito K, Mielke R, Kalbe E, Zündorf G, Delbeuck X, Pelati O, Anchisi D, Fazio F, Kerrouche N, Desgranges B, Eustache F, Beuthien-Baumann B, Menzel C, Schröder J, Kato T, Arahata Y, Henze M, Heiss WD (2002) Discrimination between Alzheimer dementia and controls by automated analysis of multicenter FDG PET. Neuroimage 17:302–316
83. Jagust W, Thisted R, Devous MD Sr, Van Heertum R, Mayberg H, Jobst K, Smith AD, Borys N (2001) SPECT perfusion imaging in the diagnosis of Alzheimer's disease: a clinical-pathologic study. Neurology 56:950–956
84. Archer HA, Edison P, Brooks DJ, Barnes J, Frost C, Yeatman T, Fox NC, Rossor MN (2006) Amyloid load and cerebral atrophy in Alzheimer's disease: an 11C-PIB positron emission tomography study. Ann Neurol 60:145–147
85. Engler H, Forsberg A, Almkvist O, Blomquist G, Larsson E, Savitcheva I, Wall A, Ringheim A, Långström B, Nordberg A (2006) Two-year follow-up of amyloid deposition in patients with Alzheimer's disease. Brain 129:2856–2866
86. Kemppainen NM, Aalto S, Wilson IA, Nagren K, Helin S, Bruck A, Oikonen V, Kailajärvi M, Scheinin M, Viitanen M, Parkkola R, Rinne JO (2006) Voxel-based analysis of PET amyloid ligand [11C]PIB uptake in Alzheimer disease. Neurology 67:1575–1580

87. Choi SR, Schneider JA, Bennett DA, Beach TG, Bedell BJ, Zehntner SP, Krautkramer MJ, Kung HF, Skovronsky DM, Hefti F, Clark CM (2012) Correlation of amyloid PET ligand florbetapir F 18 binding with Abeta aggregation and neuritic plaque deposition in postmortem brain tissue. Alzheimer Dis Assoc Disord 26:8–16
88. Petersen RC, Aisen PS, Beckett LA, Donohue MC, Gamst AC, Harvey DJ, Jack CR Jr, Jagust WJ, Shaw LM, Toga AW, Trojanowski JQ, Weiner MW (2010) Alzheimer's Disease Neuroimaging Initiative (ADNI): clinical characterization. Neurology 74:201–209
89. Shoji M, Golde TE, Ghiso J, Cheung TT, Estus S, Shaffer LM, Cai XD, McKay DM, Tintner R, Frangione B, Younkin SG (1992) Production of the Alzheimer amyloid beta protein by normal proteolytic processing. Science 258:126–129
90. Blennow K (2004) Cerebrospinal fluid protein biomarkers for Alzheimer's disease. NeuroRx 1:213–225
91. Blennow K, Vanmechelen E, Hampel H (2001) CSF total tau, Abeta42 and phosphorylated tau protein as biomarkers for Alzheimer's disease. Mol Neurobiol 24:87–97
92. Clark CM, Xie S, Chittams J, Ewbank D, Peskind E, Galasko D, Morris JC, McKeel DW Jr, Farlow M, Weitlauf SL, Quinn J, Kaye J, Knopman D, Arai H, Doody RS, DeCarli C, Leight S, Lee VM, Trojanowski JQ (2003) Cerebrospinal fluid tau and beta-amyloid: how well do these biomarkers reflect autopsy-confirmed dementia diagnoses? Arch Neurol 60: 1696–1702
93. Arai H, Nakagawa T, Kosaka Y, Higuchi M, Matsui T, Okamura N, Tashiro M, Sasaki H (1997) Elevated cerebrospinal fluid tau protein level as a predictor of dementia in memory-impaired patients. Alzheimer's Res 3:211–213
94. Visser PJ, Verhey F, Knol DL, Scheltens P, Wahlund LO, Freund-Levi Y, Tsolaki M, Minthon L, Wallin AK, Hampel H, Bürger K, Pirttila T, Soininen H, Rikkert MO, Verbeek MM, Spiru L, Blennow K (2009) Prevalence and prognostic value of CSF markers of Alzheimer's disease pathology in patients with subjective cognitive impairment or mild cognitive impairment in the DESCRIPA study: a prospective cohort study. Lancet Neurol 8:619–627
95. Boban M, Grbic K, Mladinov M, Hof PR, Sussmair C, Ackl N, Stanic G, Bader B, Danek A, Simić G (2008) Cerebrospinal fluid markers in differential diagnosis of Alzheimer's disease and vascular dementia. Coll Antropol 32(Suppl 1):31–36
96. Hampel H, Buerger K, Kohnken R, Teipel SJ, Zinkowski R, Moeller HJ, Rapoport SI, Davies P (2001) Tracking of Alzheimer's disease progression with cerebrospinal fluid tau protein phosphorylated at threonine 231. Ann Neurol 49:545–546
97. Buerger K, Teipel SJ, Zinkowski R, Blennow K, Arai H, Engel R, Hofmann-Kiefer K, McCulloch C, Ptok U, Heun R, Andreasen N, DeBernardis J, Kerkman D, Moeller H, Davies P, Hampel H (2002) CSF tau protein phosphorylated at threonine 231 correlates with cognitive decline in MCI subjects. Neurology 59:627–629
98. Hulstaert F, Blennow K, Ivanoiu A, Schoonderwaldt HC, Riemenschneider M, De Deyn PP, Bancher C, Cras P, Wiltfang J, Mehta PD, Iqbal K, Pottel H, Vanmechelen E, Vanderstichele H (1999) Improved discrimination of AD patients using beta-amyloid(1–42) and tau levels in CSF. Neurology 52:1555–1562
99. Samuels SC, Silverman JM, Marin DB, Peskind ER, Younki SG, Greenberg DA, Schnur E, Santoro J, Davis KL (1999) CSF beta-amyloid, cognition, and APOE genotype in Alzheimer's disease. Neurology 52:547–551
100. Tapiola T, Pirttila T, Mikkonen M, Mehta PD, Alafuzoff I, Koivisto K, Soininen H (2000) Three-year follow-up of cerebrospinal fluid tau, beta-amyloid 42 and 40 concentrations in Alzheimer's disease. Neurosci Lett 280:119–122
101. Jensen M, Schroder J, Blomberg M, Engvall B, Pantel J, Ida N, Basun H, Wahlund LO, Werle E, Jauss M, Beyreuther K, Lannfelt L, Hartmann T (1999) Cerebrospinal fluid A beta42 is increased early in sporadic Alzheimer's disease and declines with disease progression. Ann Neurol 45:504–511
102. Van Everbroeck B, Green AJ, Pals P, Martin JJ, Cras P (1999) Decreased levels of amyloid-beta 1–42 in cerebrospinal fluid of Creutzfeldt-Jakob disease patients. J Alzheimers Dis 1:419–424

103. Otto M, Esselmann H, Schulz-Shaeffer W, Neumann M, Schroter A, Ratzka P, Cepek L, Zerr I, Steinacker P, Windl O, Kornhuber J, Kretzschmar HA, Poser S, Wiltfang J (2000) Decreased beta-amyloid1-42 in cerebrospinal fluid of patients with Creutzfeldt-Jakob disease. Neurology 54:1099–1102
104. Kanemaru K, Kameda N, Yamanouchi H (2000) Decreased CSF amyloid beta42 and normal tau levels in dementia with Lewy bodies. Neurology 54:1875–1876
105. Crystal HA, Davies P (2008) Toward a plasma marker for Alzheimer disease: some progress, but still a long way to go. Neurology 70:586–587
106. Ringman JM, Younkin SG, Pratico D, Seltzer W, Cole GM, Geschwind DH, Rodriguez-Agudelo Y, Schaffer B, Fein J, Sokolow S, Rosario ER, Gylys KH, Varpetian A, Medina LD, Cummings JL (2008) Biochemical markers in persons with preclinical familial Alzheimer disease. Neurology 71:85–92
107. Ghidoni R, Benussi L, Paterlini A, Albertini V, Binetti G, Emanuele E (2011) Cerebrospinal fluid biomarkers for Alzheimer's disease: the present and the future. Neurodegener Dis 8:413–420
108. Ray S, Britschgi M, Herbert C, Takeda-Uchimura Y, Boxer A, Blennow K, Friedman LF, Galasko DR, Jutel M, Karydas A, Kaye JA, Leszek J, Miller BL, Minthon L, Quinn JF, Rabinovici GD, Robinson WH, Sabbagh MN, So YT, Sparks DL, Tabaton M, Tinklenberg J, Yesavage JA, Tibshirani R, Wyss-Coray T (2007) Classification and prediction of clinical Alzheimer's diagnosis based on plasma signaling proteins. Nat Med 13:1359–1362
109. Hye A, Lynham S, Thambisetty M, Causevic M, Campbell J, Byers HL, Hooper C, Rijsdijk F, Tabrizi SJ, Banner S, Shaw CE, Foy C, Poppe M, Archer N, Hamilton G, Powell J, Brown RG, Sham P, Ward M, Lovestone S (2006) Proteome-based plasma biomarkers for Alzheimer's disease. Brain 129:3042–3050
110. Montine TJ, Beal MF, Cudkowicz ME, O'donnell H, Margolin RA, Mcfarland L, Bachrach AF, Zackert WE, Roberts LJ, Morrow JD (1999) Increased CSF F2-isoprostane concentration in probable AD. Neurology 52:562–565
111. Irizarry MC, Hyman BT (2003) Brain isoprostanes: a marker of lipid peroxidation and oxidative stress in AD. Neurology 61:436–437
112. Sullivan PF, Fan C, Perou CM (2006) Evaluating the comparability of gene expression in blood and brain. Am J Med Genet B Neuropsychiatr Genet 141B:261–268
113. Burczynski ME, Dorner AJ (2006) Transcriptional profiling of peripheral blood cells in clinical pharmacogenomic studies. Pharmacogenomics 7:187–202
114. Fehlbaum-Beurdeley P, Jarrige-Le Prado AC, Pallares D, Carriere J, Guihal C, Soucaille C, Rouet F, Drouin D, Sol O, Jordan H, Wu D, Lei L, Einstein R, Schweighoffer F, Bracco L (2010) Toward an Alzheimer's disease diagnosis via high-resolution blood gene expression. Alzheimers Dement 6:25–38
115. Rye PD, Booij BB, Grave G, Lindahl T, Kristiansen L, Andersen HM, Horndalsveen PO, Nygaard HA, Naik M, Hoprekstad D, Wetterberg P, Nilsson C, Aarsland D, Sharma P, Lönneborg A (2011) A novel blood test for the early detection of Alzheimer's disease. J Alzheimers Dis 23:121–129
116. Rosner S, Giladi N, Orr-Urtreger A (2008) Advances in the genetics of Parkinson's disease. Acta Pharmacol Sin 29:21–34
117. Farrer MJ (2006) Genetics of Parkinson disease: paradigm shifts and future prospects. Nat Rev Genet 7:306–318
118. Greenamyre JT, Betarbet R, Sherer TB (2003) The rotenone model of Parkinson's disease: genes, environment and mitochondria. Parkinsonism Relat Disord 9(Suppl 2):S59–S64
119. Korczyn AD, Gurevich T (2010) Parkinson's disease: before the motor symptoms and beyond. J Neurol Sci 289:2–6
120. Obeso JA, Rodriguez-Oroz MC, Rodriguez M, Lanciego JL, Artieda J, Gonzalo N, Olanow CW (2000) Pathophysiology of the basal ganglia in Parkinson's disease. Trends Neurosci 23:S8–S19
121. Fearnley JM, Lees AJ (1991) Ageing and Parkinson's disease: substantia nigra regional selectivity. Brain 114:2283–2301

122. Schapira AH, Bezard E, Brotchie J, Calon F, Collingridge GL, Ferger B, Hengerer B, Hirsch E, Jenner P, Le Novère N, Obeso JA, Schwarzschild MA, Spampinato U, Davidai G (2006) Novel pharmacological targets for the treatment of Parkinson's disease. Nat Rev Drug Discov 5:845–854
123. Langston JW (2006) The Parkinson's complex: parkinsonism is just the tip of the iceberg. Ann Neurol 59:591–596
124. Braak H, Del Tredici K, Rub U, De Vos RA, Jansen Steur EN, Braak E (2003) Staging of brain pathology related to sporadic Parkinson's disease. Neurobiol Aging 24:197–211
125. Braak H, Muller CM, Rub U, Ackermann H, Bratzke H, De Vos RA, Del Tredici K (2006) Pathology associated with sporadic Parkinson's disease–where does it end? J Neural Transm Suppl 70:89–97
126. Bloch A, Probst A, Bissig H, Adams H, Tolnay M (2006) Alpha-synuclein pathology of the spinal and peripheral autonomic nervous system in neurologically unimpaired elderly subjects. Neuropathol Appl Neurobiol 32:284–295
127. Olanow CW, Obeso JA (2012) The significance of defining preclinical or prodromal Parkinson's disease. Mov Disord 27:666–669
128. Fahn S, Elton R, Committee UD (1987) Unified Parkinson's disease rating scale. In: Fahn S, Marsden C, Calne DB, Goldstein M (eds) Recent developments in Parkinson's disease. Macmillan Health Care Information, Florham Park, pp 153–163
129. Hoehn MM, Yahr MD (1967) Parkinsonism: onset, progression and mortality. Neurology 17:427–442
130. Tolosa E, Wenning G, Poewe W (2006) The diagnosis of Parkinson's disease. Lancet Neurol 5:75–86
131. Vendette M, Gagnon JF, Decary A, Massicotte-Marquez J, Postuma RB, Doyon J, Panisset M, Montplaisir J (2007) REM sleep behavior disorder predicts cognitive impairment in Parkinson disease without dementia. Neurology 69:1843–1849
132. Postuma RB, Gagnon JF, Vendette M, Charland K, Montplaisir J (2008) REM sleep behaviour disorder in Parkinson's disease is associated with specific motor features. J Neurol Neurosurg Psychiatry 79:1117–1121
133. Burn DJ (2002) Beyond the iron mask: towards better recognition and treatment of depression associated with Parkinson's disease. Mov Disord 17:445–454
134. Remy P, Doder M, Lees A, Turjanski N, Brooks D (2005) Depression in Parkinson's disease: loss of dopamine and noradrenaline innervation in the limbic system. Brain 128:1314–1322
135. Mckinnon JH, Demaerschalk BM, Caviness JN, Wellik KE, Adler CH, Wingerchuk DM (2007) Sniffing out Parkinson disease: can olfactory testing differentiate parkinsonian disorders? Neurologist 13:382–385
136. Hudry J, Thobois S, Broussolle E, Adeleine P, Royet JP (2003) Evidence for deficiencies in perceptual and semantic olfactory processes in Parkinson's disease. Chem Senses 28: 537–543
137. Montgomery EB Jr, Koller WC, Lamantia TJ, Newman MC, Swanson-Hyland E, Kaszniak AW, Lyons K (2000) Early detection of probable idiopathic Parkinson's disease: I. Development of a diagnostic test battery. Mov Disord 15:467–473
138. Montgomery EB Jr, Lyons K, Koller WC (2000) Early detection of probable idiopathic Parkinson's disease: II. A prospective application of a diagnostic test battery. Mov Disord 15:474–478
139. Troster AI (2008) Neuropsychological characteristics of dementia with Lewy bodies and Parkinson's disease with dementia: differentiation, early detection, and implications for "mild cognitive impairment" and biomarkers. Neuropsychol Rev 18:103–119
140. Barone P, Aarsland D, Burn D, Emre M, Kulisevsky J, Weintraub D (2011) Cognitive impairment in nondemented Parkinson's disease. Mov Disord 26:2483–2495
141. Poletti M, Emre M, Bonuccelli U (2011) Mild cognitive impairment and cognitive reserve in Parkinson's disease. Parkinsonism Relat Disord 17:579–586
142. Korczyn AD (2011) Comment: should we diagnose MCI in Parkinson disease? J Neural Transm 118:1177

143. Emre M, Aarsland D, Brown R, Burn DJ, Duyckaerts C, Mizuno Y, Broe GA, Cummings J, Dickson DW, Gauthier S, Goldman J, Goetz C, Korczyn A, Lees A, Levy R, Litvan I, McKeith I, Olanow W, Poewe W, Quinn N, Sampaio C, Tolosa E, Dubois B (2007) Clinical diagnostic criteria for dementia associated with Parkinson's disease. Mov Disord 22:1689–1707, quiz 1837
144. Korczyn AD (1990) Autonomic nervous system disturbances in Parkinson's disease. Adv Neurol 53:463–468
145. Wolters E, Braak H (2006) Parkinson's disease: premotor clinico-pathological correlations. J Neural Transm Suppl 70:309–319
146. Sommer C, Lauria G (2007) Skin biopsy in the management of peripheral neuropathy. Lancet Neurol 6:632–642
147. Rossi A, Giovenali P, Benvenuti M, Di Iorio W, Calabresi P (2007) Skin biopsy: a new diagnostic tool for autonomic dysfunctions in Parkinson's disease? Lancet Neurol 6:848–849
148. Leroux P-D (1880) Contribution à l'Étude des Causes de la Paralysie Agitante. (in French). Thèse de Paris, Imprimeur de la Faculté de Médecine
149. Gowers WR (1900) A manual of diseases of the nervous system. Blakiston, Philadelphia
150. Wider C, Foroud T, Wszolek ZK (2010) Clinical implications of gene discovery in Parkinson's disease and parkinsonism. Mov Disord 25(Suppl 1):S15–S20
151. Polymeropoulos MH, Lavedan C, Leroy E, Ide SE, Dehejia A, Dutra A, Pike B, Root H, Rubenstein J, Boyer R, Stenroos ES, Chandrasekharappa S, Athanassiadou A, Papapetropoulos T, Johnson WG, Lazzarini AM, Duvoisin RC, Di Iorio G, Golbe LI, Nussbaum RL (1997) Mutation in the alpha-synuclein gene identified in families with Parkinson's disease. Science 276:2045–2047
152. Kruger R, Kuhn W, Muller T, Woitalla D, Graeber M, Kosel S, Przuntek H, Epplen JT, Schöls L, Riess O (1998) Ala30Pro mutation in the gene encoding alpha-synuclein in Parkinson's disease. Nat Genet 18:106–108
153. Zarranz JJ, Alegre J, Gomez-Esteban JC, Lezcano E, Ros R, Ampuero I, Vidal L, Hoenicka J, Rodriguez O, Atarés B, Llorens V, Gomez Tortosa E, del Ser T, Muñoz DG, de Yebenes JG (2004) The new mutation, E46K, of alpha-synuclein causes Parkinson and Lewy body dementia. Ann Neurol 55:164–173
154. Singleton AB, Farrer M, Johnson J, Singleton A, Hague S, Kachergus J, Hulihan M, Peuralinna T, Dutra A, Nussbaum R, Lincoln S, Crawley A, Hanson M, Maraganore D, Adler C, Cookson MR, Muenter M, Baptista M, Miller D, Blancato J, Hardy J, Gwinn-Hardy K (2003) Alpha-Synuclein locus triplication causes Parkinson's disease. Science 302:841
155. Ibanez P, Bonnet AM, Debarges B, Lohmann E, Tison F, Pollak P, Agid Y, Dürr A, Brice A (2004) Causal relation between alpha-synuclein gene duplication and familial Parkinson's disease. Lancet 364:1169–1171
156. Chartier-Harlin MC, Kachergus J, Roumier C, Mouroux V, Douay X, Lincoln S, Levecque C, Larvor L, Andrieux J, Hulihan M, Waucquier N, Defebvre L, Amouyel P, Farrer M, Destée A (2004) Alpha-synuclein locus duplication as a cause of familial Parkinson's disease. Lancet 364:1167–1169
157. Leroy E, Boyer R, Auburger G, Leube B, Ulm G, Mezey E, Harta G, Brownstein MJ, Jonnalagada S, Chernova T, Dehejia A, Lavedan C, Gasser T, Steinbach PJ, Wilkinson KD, Polymeropoulos MH (1998) The ubiquitin pathway in Parkinson's disease. Nature 395: 451–452
158. Lowe J, McDermott H, Landon M, Mayer RJ, Wilkinson KD (1990) Ubiquitin carboxyl-terminal hydrolase (PGP 9.5) is selectively present in ubiquitinated inclusion bodies characteristic of human neurodegenerative diseases. J Pathol 161:153–160
159. Aharon-Peretz J, Rosenbaum H, Gershoni-Baruch R (2004) Mutations in the glucocerebrosidase gene and Parkinson's disease in Ashkenazi Jews. N Engl J Med 351:1972–1977
160. Ozelius LJ, Senthil G, Saunders-Pullman R, Ohmann E, Deligtisch A, Tagliati M, Hunt AL, Klein C, Henick B, Hailpern SM, Lipton RB, Soto-Valencia J, Risch N, Bressman SB (2006) LRRK2 G2019S as a cause of Parkinson's disease in Ashkenazi Jews. N Engl J Med 354: 424–425

161. Bonifati V (2011) Autosomal recessive parkinsonism. Parkinsonism Relat Disord 18(Suppl 1):S4–S6
162. Kitada T, Asakawa S, Hattori N, Matsumine H, Yamamura Y, Minoshima S, Yokochi M, Mizuno Y, Shimizu N (1998) Mutations in the parkin gene cause autosomal recessive juvenile parkinsonism. Nature 392:605–608
163. Valente EM, Abou-Sleiman PM, Caputo V, Muqit MM, Harvey K, Gispert S, Ali Z, Del Turco D, Bentivoglio AR, Healy DG, Albanese A, Nussbaum R, González-Maldonado R, Deller T, Salvi S, Cortelli P, Gilks WP, Latchman DS, Harvey RJ, Dallapiccola B, Auburger G, Wood NW (2004) Hereditary early-onset Parkinson's disease caused by mutations in PINK1. Science 304:1158–1160
164. Bonifati V, Rizzu P, Van Baren MJ, Schaap O, Breedveld GJ, Krieger E, Dekker MC, Squitieri F, Ibanez P, Joosse M, van Dongen JW, Vanacore N, van Swieten JC, Brice A, Meco G, van Duijn CM, Oostra BA, Heutink P (2003) Mutations in the DJ-1 gene associated with autosomal recessive early-onset parkinsonism. Science 299:256–259
165. Ramirez A, Heimbach A, Grundemann J, Stiller B, Hampshire D, Cid LP, Goebel I, Mubaidin AF, Wriekat AL, Roeper J, Al-Din A, Hillmer AM, Karsak M, Liss B, Woods CG, Behrens MI, Kubisch C (2006) Hereditary parkinsonism with dementia is caused by mutations in ATP13A2, encoding a lysosomal type 5 P-type ATPase. Nat Genet 38:1184–1191
166. Shimura H, Hattori N, Kubo S, Mizuno Y, Asakawa S, Minoshima S, Shimizu N, Iwai K, Chiba T, Tanaka K, Suzuki T (2000) Familial Parkinson disease gene product, parkin, is a ubiquitin-protein ligase. Nat Genet 25:302–305
167. Plun-Favreau H, Klupsch K, Moisoi N, Gandhi S, Kjaer S, Frith D, Harvey K, Deas E, Harvey RJ, McDonald N, Wood NW, Martins LM, Downward J (2007) The mitochondrial protease HtrA2 is regulated by Parkinson's disease-associated kinase PINK1. Nat Cell Biol 9: 1243–1252
168. Park J, Lee SB, Lee S, Kim Y, Song S, Kim S, Bae E, Kim J, Shong M, Kim JM, Chung J (2006) Mitochondrial dysfunction in Drosophila PINK1 mutants is complemented by parkin. Nature 441:1157–1161
169. Hamza TH, Zabetian CP, Tenesa A, Laederach A, Montimurro J, Yearout D, Kay DM, Doheny KF, Paschall J, Pugh E, Kusel VI, Collura R, Roberts J, Griffith A, Samii A, Scott WK, Nutt J, Factor SA, Payami H (2010) Common genetic variation in the HLA region is associated with late-onset sporadic Parkinson's disease. Nat Genet 42:781–785
170. Satake W, Nakabayashi Y, Mizuta I, Hirota Y, Ito C, Kubo M, Kawaguchi T, Tsunoda T, Watanabe M, Takeda A, Tomiyama H, Nakashima K, Hasegawa K, Obata F, Yoshikawa T, Kawakami H, Sakoda S, Yamamoto M, Hattori N, Murata M, Nakamura Y, Toda T (2009) Genome-wide association study identifies common variants at four loci as genetic risk factors for Parkinson's disease. Nat Genet 41:1303–1307
171. Nalls MA, Plagnol V, Hernandez DG, Sharma M, Sheerin UM, Saad M, Simon-Sanchez J, Schulte C, Lesage S, Sveinbjörnsdóttir S, Stefánsson K, Martinez M, Hardy J, Heutink P, Brice A, Gasser T, Singleton AB, Wood NW (2011) Imputation of sequence variants for identification of genetic risks for Parkinson's disease: a meta-analysis of genome-wide association studies. Lancet 377:641–649
172. Gan-Or Z, Bar-Shira A, Dahary D, Mirelman A, Kedmi M, Gurevich T, Giladi N, Orr-Urtreger A (2012) Association of sequence alterations in the putative promoter of RAB7L1 with a reduced parkinson disease risk. Arch Neurol 69:105–110
173. Brooks DJ, Pavese N (2012) Imaging biomarkers in Parkinson's disease. Prog Neurobiol 95:614–628
174. Brooks DJ (2008) Technology insight: imaging neurodegeneration in Parkinson's disease. Nat Clin Pract Neurol 4:267–277
175. Eckert T, Tang C, Eidelberg D (2007) Assessment of the progression of Parkinson's disease: a metabolic network approach. Lancet Neurol 6:926–932
176. Group PS (2002) Dopamine transporter brain imaging to assess the effects of pramipexole vs levodopa on Parkinson disease progression. JAMA 287:1653–1661

177. Whone AL, Watts RL, Stoessl AJ, Davis M, Reske S, Nahmias C, Lang AE, Rascol O, Ribeiro MJ, Remy P, Poewe WH, Hauser RA, Brooks DJ; REAL-PET Study Group (2003) Slower progression of Parkinson's disease with ropinirole versus levodopa: the REAL-PET study group. Ann Neurol 54:93–101
178. Goldstein DS, Holmes C, Bentho O, Sato T, Moak J, Sharabi Y, Imrich R, Conant S, Eldadah BA (2008) Biomarkers to detect central dopamine deficiency and distinguish Parkinson disease from multiple system atrophy. Parkinsonism Relat Disord 14:600–607
179. Berg D (2008) Biomarkers for the early detection of Parkinson's and Alzheimer's disease. Neurodegener Dis 5:133–136
180. Nicoletti G, Lodi R, Condino F, Tonon C, Fera F, Malucelli E, Manners D, Zappia M, Morgante L, Barone P, Barbiroli B, Quattrone A (2006) Apparent diffusion coefficient measurements of the middle cerebellar peduncle differentiate the Parkinson variant of MSA from Parkinson's disease and progressive supranuclear palsy. Brain 129:2679–2687
181. Kwon DH, Kim JM, Oh SH, Jeong HJ, Park SY, Oh ES, Chi JG, Kim YB, Jeon BS, Cho ZH (2012) Seven-Tesla magnetic resonance images of the substantia nigra in Parkinson disease. Ann Neurol 71:267–277
182. Emir UE, Tuite PJ, Oz G (2012) Elevated pontine and putamenal GABA levels in mild-moderate Parkinson disease detected by 7 tesla proton MRS. PLoS One 7:e30918
183. Zecca L, Youdim MB, Riederer P, Connor JR, Crichton RR (2004) Iron, brain ageing and neurodegenerative disorders. Nat Rev Neurosci 5:863–873
184. Gerlach M, Double KL, Youdim MB, Riederer P (2006) Potential sources of increased iron in the substantia nigra of parkinsonian patients. J Neural Transm Suppl 70:133–142
185. Koeppen AH (1995) The history of iron in the brain. J Neurol Sci 134(Suppl):1–9
186. Hallgren B, Sourander P (1958) The effect of age on the nonhemin iron in the human brain. J Neurochem 3:41
187. Jellinger K, Paulus W, Grundke-Iqbal I, Riederer P, Youdim MB (1990) Brain iron and ferritin in Parkinson's and Alzheimer's diseases. J Neural Transm Park Dis Dement Sect 2: 327–340
188. Sofic E, Paulus W, Jellinger K, Riederer P, Youdim MB (1991) Selective increase of iron in substantia nigra zona compacta of parkinsonian brains. J Neurochem 56:978–982
189. Zecca L, Shima T, Stroppolo A, Goj C, Battiston GA, Gerbasi R, Sarna T, Swartz HM (1996) Interaction of neuromelanin and iron in substantia nigra and other areas of human brain. Neuroscience 73:407–415
190. Berg D, Siefker C, Becker G (2001) Echogenicity of the substantia nigra in Parkinson's disease and its relation to clinical findings. J Neurol 248:684–689
191. Berg D, Roggendorf W, Schroder U, Klein R, Tatschner T, Benz P, Tucha O, Preier M, Lange KW, Reiners K, Gerlach M, Becker G (2002) Echogenicity of the substantia nigra: association with increased iron content and marker for susceptibility to nigrostriatal injury. Arch Neurol 59:999–1005
192. Harrington MG, Merril CR (1984) Two-dimensional electrophoresis and "ultrasensitive" silver staining of cerebrospinal fluid proteins in neurological diseases. Clin Chem 30: 1933–1937
193. Harrington MG, Fonteh AN, Biringer RG, Hühmer AF, Cowan RP (2006) Prostaglandin D synthase isoforms from cerebrospinal fluid vary with brain pathology. Dis Markers 22: 73–81
194. Mollenhauer B, Cullen V, Kahn I, Krastins B, Outeiro TF, Pepivani I, Ng J, Schulz-Schaeffer W, Kretzschmar HA, McLean PJ, Trenkwalder C, Sarracino DA, Vonsattel JP, Locascio JJ, El-Agnaf OM, Schlossmacher MG (2008) Direct quantification of CSF alpha-synuclein by ELISA and first cross-sectional study in patients with neurodegeneration. Exp Neurol 213:315–325
195. El-Agnaf OM, Salem SA, Paleologou KE, Curran MD, Gibson MJ, Court JA, Schlossmacher MG, Allsop D (2006) Detection of oligomeric forms of alpha-synuclein protein in human plasma as a potential biomarker for Parkinson's disease. FASEB J 20:419–425

196. Mollenhauer B, Locascio JJ, Schulz-Schaeffer W, Sixel-Doring F, Trenkwalder C, Schlossmacher MG (2011) Alpha-Synuclein and tau concentrations in cerebrospinal fluid of patients presenting with parkinsonism: a cohort study. Lancet Neurol 10:230–240
197. Tokuda T, Qureshi MM, Ardah MT, Varghese S, Shehab SA, Kasai T, Ishigami N, Tamaoka A, Nakagawa M, El-Agnaf OM (2010) Detection of elevated levels of alpha-synuclein oligomers in CSF from patients with Parkinson disease. Neurology 75:1766–1772
198. Zhang J, Sokal I, Peskind ER, Quinn JF, Jankovic J, Kenney C, Chung KA, Millard SP, Nutt JG, Montine TJ (2008) CSF multianalyte profile distinguishes Alzheimer and Parkinson diseases. Am J Clin Pathol 129:526–529
199. Shi M, Bradner J, Hancock AM, Chung KA, Quinn JF, Peskind ER, Galasko D, Jankovic J, Zabetian CP, Kim HM, Leverenz JB, Montine TJ, Ginghina C, Kang UJ, Cain KC, Wang Y, Aasly J, Goldstein D, Zhang J (2010) Cerebrospinal fluid biomarkers for Parkinson disease diagnosis and progression. Ann Neurol 69:570–580
200. Ishigami N, Tokuda T, Ikegawa M, Komori M, Kasai T, Kondo T, Matsuyama Y, Nirasawa T, Thiele H, Tashiro K, Nakagawa M (2012) Cerebrospinal fluid proteomic patterns discriminate Parkinson's disease and multiple system atrophy. Mov Disord 27:851–857
201. Nagai Y, Ueno S, Saeki Y, Soga F, Hirano M, Yanagihara T (1996) Decrease of the D3 dopamine receptor mRNA expression in lymphocytes from patients with Parkinson's disease. Neurology 46:791–795
202. Caronti B, Antonini G, Calderaro C, Ruggieri S, Palladini G, Pontieri FE, Colosimo C (2001) Dopamine transporter immunoreactivity in peripheral blood lymphocytes in Parkinson's disease. J Neural Transm 108:803–807
203. Parker WD Jr, Boyson SJ, Parks JK (1989) Abnormalities of the electron transport chain in idiopathic Parkinson's disease. Ann Neurol 26:719–723
204. Benecke R, Strumper P, Weiss H (1993) Electron transfer complexes I and IV of platelets are abnormal in Parkinson's disease but normal in Parkinson-plus syndromes. Brain 116(Pt 6):1451–1463
205. Haas RH, Nasirian F, Nakano K, Ward D, Pay M, Hill R, Shults CW (1995) Low platelet mitochondrial complex I and complex II/III activity in early untreated Parkinson's disease. Ann Neurol 37:714–722
206. Migliore L, Petrozzi L, Lucetti C, Gambaccini G, Bernardini S, Scarpato R, Trippi F, Barale R, Frenzilli G, Rodilla V, Bonuccelli U (2002) Oxidative damage and cytogenetic analysis in leukocytes of Parkinson's disease patients. Neurology 58:1809–1815
207. Davis JW, Grandinetti A, Waslien CI, Ross GW, White LR, Morens DM (1996) Observations on serum uric acid levels and the risk of idiopathic Parkinson's disease. Am J Epidemiol 144:480–484
208. Weisskopf MG, O'Reilly E, Chen H, Schwarzschild MA, Ascherio A (2007) Plasma urate and risk of Parkinson's disease. Am J Epidemiol 166:561–567
209. Schwarzschild MA, Schwid SR, Marek K, Watts A, Lang AE, Oakes D, Shoulson I, Ascherio A; Parkinson Study Group PRECEPT Investigators, Hyson C, Gorbold E, Rudolph A, Kieburtz K, Fahn S, Gauger L, Goetz C, Seibyl J, Forrest M, Ondrasik J (2008) Serum urate as a predictor of clinical and radiographic progression in Parkinson disease. Arch Neurol 65:716–723
210. Bogdanov M, Matson WR, Wang L, Matson T, Saunders-Pullman R, Bressman SS, Flint Beal M (2008) Metabolomic profiling to develop blood biomarkers for Parkinson's disease. Brain 131:389–396
211. Ascherio A, Lewitt PA, Xu K, Eberly S, Watts A, Matson WR, Marras C, Kieburtz K, Rudolph A, Bogdanov MB, Schwid SR, Tennis M, Tanner CM, Beal MF, Lang AE, Oakes D, Fahn S, Shoulson I, Schwarzschild MA; Parkinson Study Group DATATOP Investigators (2009) Urate as a predictor of the rate of clinical decline in Parkinson disease, Parkinson study group DATATOP investigators. Arch Neurol 66:1460–1468
212. O'Reilly EJ, Gao X, Weisskopf MG, Chen H, Schwarzschild MA, Spiegelman D, Ascherio A (2010) Plasma urate and Parkinson's disease in women. Am J Epidemiol 172:666–670

213. Varani K, Vincenzi F, Tosi A, Gessi S, Casetta I, Granieri G, Fazio P, Leung E, MacLennan S, Granieri E, Borea PA (2010) A2A adenosine receptor overexpression and functionality, as well as TNF-alpha levels, correlate with motor symptoms in Parkinson's disease. FASEB J 24:587–598
214. Morelli M, Di Paolo T, Wardas J, Calon F, Xiao D, Schwarzschild MA (2007) Role of adenosine A2A receptors in parkinsonian motor impairment and l-DOPA-induced motor complications. Prog Neurobiol 83:293–309
215. Goldknopf IL, Sheta EA, Bryson J, Folsom B, Wilson C, Duty J, Yen AA, Appel SH (2006) Complement C3c and related protein biomarkers in amyotrophic lateral sclerosis and Parkinson's disease. Biochem Biophys Res Commun 342:1034–1039
216. Sheta EA, Appel SH, Goldknopf IL (2006) 2D gel blood serum biomarkers reveal differential clinical proteomics of the neurodegenerative diseases. Expert Rev Proteomics 3:45–62
217. Goldknopf IL (2008) Blood-based proteomics for personalized medicine: examples from neurodegenerative disease. Expert Rev Proteomics 5:1–8
218. Duran R, Barrero FJ, Morales B, Luna JD, Ramirez M, Vives F (2010) Plasma alpha-synuclein in patients with Parkinson's disease with and without treatment. Mov Disord 25:489–493
219. Barbour R, Kling K, Anderson JP, Banducci K, Cole T, Diep L, Fox M, Goldstein JM, Soriano F, Seubert P, Chilcote TJ (2008) Red blood cells are the major source of alpha-synuclein in blood. Neurodegener Dis 5:55–59
220. Michell AW, Luheshi LM, Barker RA (2005) Skin and platelet alpha-synuclein as peripheral biomarkers of Parkinson's disease. Neurosci Lett 381:294–298
221. Maita C, Tsuji S, Yabe I, Hamada S, Ogata A, Maita H, Iguchi-Ariga SM, Sasaki H, Ariga H (2008) Secretion of DJ-1 into the serum of patients with Parkinson's disease. Neurosci Lett 431:86–89
222. Waragai M, Wei J, Fujita M, Nakai M, Ho GJ, Masliah E, Akatsu H, Yamada T, Hashimoto M (2006) Increased level of DJ-1 in the cerebrospinal fluids of sporadic Parkinson's disease. Biochem Biophys Res Commun 345:967–972
223. Waragai M, Nakai M, Wei J, Fujita M, Mizuno H, Ho G, Masliah E, Akatsu H, Yokochi F, Hashimoto M (2007) Plasma levels of DJ-1 as a possible marker for progression of sporadic Parkinson's disease. Neurosci Lett 425:18–22
224. Zetterberg H, Ruetschi U, Portelius E, Brinkmalm G, Andreasson U, Blennow K, Brinkmalm A (2008) Clinical proteomics in neurodegenerative disorders. Acta Neurol Scand 118:1–11
225. Mandel S, Amit T, Kalfon L, Youdim MB (2007) Applying transcriptomic and proteomic knowledge to Parkinson's disease drug discovery. Expert opin drug disc 2:1225–1240
226. Zheng B, Liao Z, Locascio JJ, Lesniak KA, Roderick SS, Watt ML, Eklund AC, Zhang-James Y, Kim PD, Hauser MA, Grünblatt E, Moran LB, Mandel SA, Riederer P, Miller RM, Federoff HJ, Wüllner U, Papapetropoulos S, Youdim MB, Cantuti-Castelvetri I, Young AB, Vance JM, Davis RL, Hedreen JC, Adler CH, Beach TG, Graeber MB, Middleton FA, Rochet JC, Scherzer CR; Global PD Gene Expression (GPEX) Consortium (2010) PGC-1alpha, a potential therapeutic target for early intervention in Parkinson's disease, Global PD Gene Expression (GPEX) consortium. Sci Transl Med 2:52ra73
227. Scherzer CR, Eklund AC, Morse LJ, Liao Z, Locascio JJ, Fefer D, Schwarzschild MA, Schlossmacher MG, Hauser MA, Vance JM, Sudarsky LR, Standaert DG, Growdon JH, Jensen RV, Gullans SR (2007) Molecular markers of early Parkinson's disease based on gene expression in blood. Proc Natl Acad Sci U S A 104:955–960
228. Scherzer CR, Grass JA, Liao Z, Pepivani I, Zheng B, Eklund AC, Ney PA, Ng J, McGoldrick M, Mollenhauer B, Bresnick EH, Schlossmacher MG (2008) GATA transcription factors directly regulate the Parkinson's disease-linked gene alpha-synuclein. Proc Natl Acad Sci U S A 105:10907–10912
229. Kedmi M, Bar-Shira A, Gurevich T, Giladi N, Orr-Urtreger A (2011) Decreased expression of B cell related genes in leukocytes of women with Parkinson's disease. Mol Neurodegener 6:66

230. Grunblatt E, Zehetmayer S, Jacob CP, Muller T, Jost WH, Riederer P (2010) Pilot study: peripheral biomarkers for diagnosing sporadic Parkinson's disease. J Neural Transm 117:1387–1393
231. Grunblatt E, Mandel S, Jacob-Hirsch J, Zeligson S, Amariglo N, Rechavi G, Li J, Ravid R, Roggendorf W, Riederer P, Youdim MB (2004) Gene expression profiling of parkinsonian substantia nigra pars compacta; alterations in ubiquitin-proteasome, heat shock protein, iron and oxidative stress regulated proteins, cell adhesion/cellular matrix and vesicle trafficking genes. J Neural Transm 111:1543–1573
232. Molochnikov L, Rabey JM, Dobronevsky E, Bonucelli U, Ceravolo R, Frosini D, Grunblatt E, Riederer P, Jacob C, Aharon-Peretz J, Bashenko Y, Youdim MB, Mandel SA (2012) A molecular signature in blood identifies early Parkinson's disease. Mol Neurodegener 7:26

Clinical Trials of AD Delay of Onset: Enrichment by a Prognostic Genetic Biomarker

Iris Grossman, Michael W. Lutz, Donna G. Crenshaw, Ann M. Saunders, Ornit Chiba-Falek, William K. Gottschalk, and Allen D. Roses

Abstract Modern medicine is shifting its focus to predictive and preventive healthcare utilization. To support this change, early intervention clinical trials that delay the onset of symptomatic disease must be conducted in a safe and cost-effective manner, and within a reasonable timeframe. Biomarkers in general, and pharmacogenetic determinants in particular, hold great promise for enabling the conduct of early intervention studies by allowing enrichment of the study population with *high risk* individuals while also minimizing participation of *low risk* subjects. In this chapter, we review one possible approach for the co-development of a therapeutic and companion prognostic. In the example cited, at the conclusion of the clinical trial, the biomarker will be evaluated for its utility in identifying individuals at risk of developing mild cognitive impairment (MCI) due to Alzheimer's disease (AD) in the next 5 years (i.e. prospective data will be generated to determine the biomarker's qualification as a prognostic). Simultaneously, the therapeutic will be evaluated for its efficacy in delaying the onset of MCI due to AD in high risk subjects with normal cognition at study entry. This clinical trial in planning is projected to launch internationally in 2013. The strategic and operational aspects of this development program are expected to pave the way to the application of similar approaches for other early intervention indications, particularly within the context of neurodegenerative diseases.

I. Grossman
IsraGene, Ltd., Rosh HaAyin 48092, Israel

M.W. Lutz • D.G. Crenshaw • A.M. Saunders • O. Chiba-Falek • W.K. Gottschalk
Duke University, 90344, Durham, NC 27708-0120, USA

A.D. Roses (✉)
Zinfandel Pharmaceuticals Inc., 110395, Research Triangle Park, Chapel Hill, NC 27709, USA
e-mail: allen.roses@zinfandelpharma.com

S. Mandel (ed.), *Neurodegenerative Diseases: Integrative PPPM Approach as the Medicine of the Future*, Advances in Predictive, Preventive and Personalised Medicine 2, DOI 10.1007/978-94-007-5866-7_7,

1 Introduction

Early intervention for the purpose of delaying symptomatic disease is gradually becoming the focus of attention in the healthcare sector, since it provides a sound foundation for minimizing healthcare costs and disease burden, while maximizing quality of life. Stakeholders are advocating the shift to predictive and preventive healthcare utilization [1], with researchers, care providers, and insurers actively pursuing early intervention strategies, investing their resources to either improve outcome of currently available treatments, or to developing efficacious and cost-effective novel therapies. However, there are only a few examples where non-vaccine, active intervention strategies have been successfully and systematically developed. One of the primary reasons for the limited number of examples is the medical relevance that must be demonstrated for active early intervention therapies in order for regulators to approve their administration to healthy individuals. In the case of early intervention clinical trials, which require administration of a drug to subjects without symptoms, careful consideration of the benefit-risk equation is warranted. Additionally, a prerequisite to development of an active early intervention is the availability of a validated way of assigning risk profiles to currently healthy subjects and predicting their future likelihood of health status conversion within a specific timeline. It is therefore critical to develop prognostic biomarkers that can reliably define the indicated subpopulation for which the benefit-risk profile is clearly favorable and active intervention is justified. In order to establish these attributes, developers need to demonstrate high analytical and clinical specificity and sensitivity of the assay. Next, the clinical utility of the therapy in conjunction with qualification of the prognostic biomarker must be prospectively evaluated to provide definitive evidence supporting registration, marketing and clinical adoption, and eventual reimbursement by payers. Alzheimer's disease (AD) is a specific and timely example of where advances in preventive care, disease prognostics, and disease diagnostics are desperately needed, given that the global burden of the disease is estimated to reach 1 in 85 persons worldwide by 2050 [2] and development programs for disease treatments have faltered.

This chapter first evaluates the characteristics of clinically useful biomarkers in general, and then focuses on genetic markers and their utility in early intervention for AD. We use the example of a novel genetic marker for age of onset of mild cognitive impairment (MCI) due to AD and discuss its application in the design of a delay of disease onset clinical trial that may directly impact medical practice. This chapter is an extension of the Grossman et al. publication [3], featuring overview of basic concepts of the approach and a case study, as well as a set of questions and answers that apply to the development and application of enrichment biomarkers for use in the conduct of registration clinical trials.

2 Properties of Clinically Useful Biomarkers

Evidence-based medicine principles guide the assessment of biomarker utility and indicate the properties that researchers and developers must demonstrate as they usher a candidate biomarker from the bench to the bedside. Analytical validity, Clinical validity, Clinical utility and Ethical, legal and social aspects (ACCE) guidelines describe the characteristics and assessment of diagnostic and prognostic tests of a biological or genomic nature [4, 5]. The foundation of each investigation conducted according to the ACCE guidelines is first and foremost the appropriately documented good operational practices for sample collection, storage and handling. The development (and later the evaluation) process starts with the requirement to demonstrate *analytical validity*, demonstrating the assay is highly specific for the analyte, highly sensitive to changes in the analyte, and sufficiently reproducible to ensure the technical accuracy in capturing the targeted genetic polymorphisms (or genomic or other properties) in almost any setting. To establish the *clinical validity* of the assay, the sensitivity, specificity and reproducibility of the test for accurately indicating the presence of the phenotype of interest must be measured. In the case of a genetic marker, clinical validity is also a function of the penetrance of the marker for the clinical phenotype. Clinical validity can be studied retrospectively, using previously collected samples, as long as the appropriate samples were collected from the vast majority of cohort participants (ideally >95%) with the appropriate informed consent. The third step, establishing the *clinical utility* of the biomarker, is the ultimate test of the expected impact of the biomarker on health outcomes. While during the course of clinical validity testing the value of the diagnostic is projected based on technical performance and estimated medical and commercial value, the "clinical utility" step is designed to test these hypotheses prospectively in the context of healthcare management. An understanding of the clinical utility of a biomarker test is a pivotal component of decision making, whether as part of a drug development pipeline or in medical practice settings. These studies can range in design and goals, but primarily aim to quantify and resolve the predictive value (positive and negative) of the biomarker in the framework of controlled clinical settings. Further studies may then be conducted to assess cost-effectiveness and adoption characteristics in real-world clinical settings. Comparative effectiveness analyses may also be performed in order to weigh the novel, biomarker-integrated treatment paradigm against standard-of-care [6]. Other aspects that need to be considered are the Ethical, Legal and Social Issues (ELSI) that arise when introducing the diagnostic to the market – e.g., who benefits from the test and are other groups disadvantaged by it [7]? Eventually, the success of a clinically validated biomarker will depend on its endorsement by regulators, its incorporation into medical practice guidelines, its influence on physician behaviors, and its adoption into reimbursement policies [8].

In the "omics" field, the validation and qualification of biomarkers involves complex, high dimensional data encompassing sets of large numbers of biological molecules, and potentially including DNA sequences, gene transcripts, proteins or metabolites [9]. With the notable exception of DNA sequence variation, the majority of "omics" biomarkers (e.g. gene transcription, protein expression, and metabolomics) exhibit dynamic profiles across tissues and time-points. High dimensional data are particularly prone to statistical over-fitting, which in turn increases the likelihood of ascertainment bias and reduces the correlation between predicted clinical utility (as assessed through ACCE evaluations) and actual medical value. The Institute of Medicine (prompted by the National Cancer Institute (NCI) in the US) has published recommendations to clarify and improve the pathway from discovery to first use of "omics"-based tests in a clinical trial, and to assess the potential for new "omics"-based tests to benefit patients [9].

The single most important component of the ACCE framework – testing the clinical utility of candidate biomarkers as personalized medicine tools – requires a prospective study design in parallel to a drug efficacy trial. To this end, the US Food and Drug Administration (FDA) issued a concept paper that proposes a framework for the co-development of drugs and diagnostics via integrated regulatory paths [10]. A pioneering example for drug/diagnostic (Rx/Dx) co-development for early intervention in the development of AD is outlined in the following sections [3, 11].

3 Alzheimer's Disease Treatment and Early Intervention – Gaps Between the Current State of Affairs and Growing Medical Needs

AD is a neurological disorder that presents as decline in domains of cognition, memory, and activities of daily living. Typically, the memory loss involves impaired recall of learned or previously known information, indicating medial temporal lobe (and possibly basal forebrain) dysfunction. While overlapping, this pattern is distinct from normal aging where learning efficiency, working memory and psychomotor speed (activities related to frontal lobe function) are reduced [12]. The archetypal clinical disease course includes cognitive impairment that progresses from early stages of MCI and prodromal AD through possible and probable AD. Upon death, a definitive AD diagnosis can be established, contingent on histopathological findings in the brain: (1) extracellular senile plaques, primarily consisting amyloid beta (Aβ) protein depositions; and (2) intracellular neurofibrillary tangles (NFT) comprised of aggregates of hyperphosphorylated tau [13].

The gradual progression of the disease has led researchers to focus their attention on earlier phases of the disease continuum, and particularly MCI, as the optimal point to target for screening programs, prognostic biomarkers, and preventive agents, aimed at arresting or delaying the progression to the more severe AD stages. Targeting early phases of the disease (e.g. MCI), or the preceding pre-symptomatic stage, is further supported by the failure of multiple treatment trials of AD dementia over the past decade. The latter sought to intervene when brain pathology may have been too advanced to allow effective treatment of clinical symptoms or to modify disease progression. The challenge for an early intervention program, however, is to identify subjects with MCI, or those who are at increased risk of developing MCI, and to then predict who will eventually progress to AD. The annual rate of AD diagnosis among subjects with MCI is generally 10–15% and MCI may also progress to other forms of non-AD dementia, such as frontotemporal dementia (FTD), or even remain unchanged without any worsening of impairment [14]. Overall, the late age of onset of AD and its intermediate phenotypes, the heterogeneous presentation patterns and its sporadic manifestation make research, diagnosis and prognosis challenging, and these challenges establish immediate priorities for identifying personalized medicine approaches that utilize reliable risk prediction methodologies.

Currently, more than five million people in the United States, or more than 10% of citizens 71 years of age or older suffer from AD, with 1 new case developing every 7 seconds, or even more frequently in developing countries [15–17]. In the absence of effective therapies for preventing or delaying AD onset, it is projected that >16 million people in the US and >107 million people worldwide will suffer from AD by 2050. The projected benefits of developing therapeutics that delay AD onset by even 1 or 2 years are striking. Such therapies are speculated to decrease the worldwide disease burden in 2050 by 12 million (1 year delay) or 23 million (2 year delay) cases [2]. This would relieve the overwhelming economic burden on tax-payers and patients' caregivers, with the latter estimated to spend greater than $94 billion annually (2004) in the US on AD-related healthcare costs [18]. Finally, given the aging of the large "baby-boom" generation and increase in life expectancy, actuary forecasts indicate a steady rise in disease prevalence and healthcare costs. This gloomy conjecture is augmented by the fact that currently no treatments or lifestyle changes have been proven to substantially delay or prevent AD onset [19].

4 The Discovery of the *TOMM40* rs10524523 AD Genetic Biomarker

The heritability of AD is estimated to be as high as 80% [20] yet to-date only a few genes have been definitively associated with late onset AD (www.alzgene.org). The only consistently replicated gene accounting for an appreciable proportion of the variability in both risk and age of disease onset is *APOE* (encoding apolipoprotein E), a discovery first reported by our group 20 years ago [21–25]. Of the three alleles of *APOE*, ε4 is associated with increased risk and decreased age of onset of disease. Indeed, research conducted by various groups in the last two decades has identified multiple pathoetiological pathways in which the apoE protein is involved [26]. Abnormality in these processes has been shown to associate with the sequence, structure and biochemical distinctions between the apoE4, apoE3 and apoE2 isoforms. The apoE isoforms are the only molecules associated thus far with the entire spectrum of biochemical disturbances characteristic of AD, including Aβ deposition, NFT formation, oxidative stress, lipid homeostasis deregulation, synaptic plasticity loss, and cholinergic dysfunction [26, 27].

The *APOE* gene is located on chromosome 19 in a region with extended linkage disequilibrium (LD) among genes. This genomic region has been repeatedly reported to have strong evidence of association in AD genome-wide association studies (GWAS). In addition to *APOE* the region includes the neighboring genes *TOMM40* (the translocase of the outer mitochondrial membrane 40) and *APOC1* (apolipoprotein C-I). This association signal, reported by one group with a statistical significance of $p < 10^{-299}$ [28], is unprecedented for common, non-Mendelian diseases. In a region with high LD and limited recombination, such as the case of the genomic sequence emcompassing the *TOMM40-APOE-APOC1* genes, this statistical association signal may actually be attributed to any genetic variant that accumulated on the 'risk' haplotype over evolution. In other words, because of LD, combinations of genetic variants that share evolutionary history show the same statistical association with the phenotype of interest. Further, in order to identify and disentangle the risk contribution of these combinations of variants (or haplotypes) it is therefore essential to investigate regions of the genome in the genomic context instead of the conventional single marker by single marker approach. A phylogenetic analysis strategy can be used to model the evolutionary history of the region of interest in the process of identifying distinct haplotypes, since it is assumed that haplotypes that contain disease-related mutations will belong to the same evolutionary pedigree [29]. Another advantage to phylogenetic mapping methodologies is the possibility to incorporate into the analysis polymorphic structural variants (e.g. poly-T tracts, Alu repeats, insertion-deletion polymorphisms, etc.) along with single nucleotide polymorphisms (SNPs) (Table 1).

The *APOE* allele system (ε2, ε3 and ε4) is a three-allele two-SNP haplotype system. The two SNPs that determine *APOE* genotype are not included on commercially available GWAS platforms. However, because of the high LD of the region, researchers, with the exception of Li et al. [39], have reflexively attributed all the genetic association in these GWAS to *APOE*. In depth investigation using phylogenetically-derived haplotypes to disentangle the substructure of the *APOE-TOMM40* LD region and refine its association with clinical phenotypes, revealed that a highly polymorphic poly-T repeat variant (rs10524523, hereafter 523) in an intron of the *TOMM40* gene was associated with mean onset age of AD [44, 45]. *TOMM40* encodes the essential, channel subunit, of the mitochondrial import system [46] and is likely to affect mitochondrial dynamics and function [47, 48]. When tested in prospectively observed epidemiological cohorts of subjects developing incipient AD, for whom accurate ascertainment of age of disease onset has been established, the association between *TOMM40* genotypes and age of onset distributions is replicated (publication in preparation).

Table 1 Case-control GWAS top reported results for late onset AD in Caucasians

Reference	Top non-*APOE* region SNPs (nearest gene)	Genotype (Allele) frequency in controls[a]	Genotype (Allele) frequency in cases	OR (HmZ)	combined p value	Combined sample size (cases, controls)	*TOMM40/ APOE/APOC1/* selected markers	Minor genotype (Allele) frequency in healthy controls[a]	Minor genotype (Allele) frequency in cases	OR (HmZ)	*TOMM40/ APOE/ APOC1/* marker combined p value
[30]						(664; 422)	rs4420638 *APOC1*	0.03 (0.18)		4.01	5.3×10^{-34}
[31]	rs10074258 *EFNA5*	0.13 (0.36)			2.5×10^{-7}	(172; 209)	rs429358 *APOE*	0.02 (0.15)	0.17 (0.41)		2.3×10^{-16}
							rs2075650 *TOMM40*	0.02 (0.14)	0.09 (0.30)		7.48×10^{-7}
[32]	rs11136000 *CLU (APOJ)*	0.16 (0.40)	0.13 (0.36)	0.86	8.5×10^{-10}	(5,964; 10,188)	**rs405509** ***APOE***	**0.27 (0.52)**	**0.18 (0.42)**	**0.7**	$\mathbf{4.9 \times 10^{-37}}$
	rs3851179 *PICALM*	0.14 (0.37)	0.12 (0.34)	0.86	1.3×10^{-9}		**rs2075650** ***TOMM40***	**0.02 (0.15)**	**0.09 (0.30)**	**2.53**	$\mathbf{1.8 \times 10^{-157}}$
[33]	rs11136000 *CLU (APOJ)*	0.14 (0.38)	0.12 (0.35)	0.86	7.5×10^{-9}	(6,010; 8,625)	rs2075650 *TOMM40*	0.01 (0.11)	0.11 (0.33)		$<2 \times 10^{-16}$
	rs6656401 *CR1*	0.04 (0.19)	0.04 (0.21)	1.21	3.5×10^{-9}		rs405509 *APOE*	0.28 (0.53)	0.21 (0.46)		$<2 \times 10^{-16}$
[34]	rs727153 *LRAT*	0.29 (0.54)	0.23 (0.48)		3.4×10^{-6}	(1,082; 2,639)	rs8106922 *TOMM40*	0.36 (0.60)	0.49 (0.70)		4.0×10^{-14}
							rs405509 *APOE*	0.31 (0.56)	0.21 (0.46)		4.8×10^{-6}
[35]	rs690705 *RFC3*	0.06 (0.25)	0.11 (0.33)		6.3×10^{-7}	(331; 368)	rs2075650 *TOMM40*	0.02 (0.15)	0.09 (0.30)		3.22×10^{-11}
[36]	rs5984894 *PCDH11X*	0.21 (0.46)	0.27 (0.52)	1.3 *(1.75 females)*	2.2×10^{-7} *(females 3.9×10^{-12})*	(2,356; 2,384) *(females 1,461; 1,304)*	**rs2075650** ***TOMM40***	**0.02 (0.13)**	**0.11 (0.33)**		$\mathbf{3.7 \times 10^{-120}}$
							rs405509 *APOE*	0.28 (0.53)	0.18 (0.42)		6.2×10^{-27}
[37]						(1,808; 2,062)	rs157581 *TOMM40*	0.04 (0.19)	0.15 (0.39)	2.73	$<1 \times 10^{-8}$

(continued)

Table 1 (continued)

Reference	Top non-*APOE* region SNPs (nearest gene)	Genotype (Allele) frequency in controls[a]	Genotype (Allele) frequency in cases	OR (HmZ)	combined p value	Combined sample size (cases, controls)	*TOMM40/ APOE/APOC1/* selected markers	Minor genotype (Allele) frequency in healthy controls[a]	Minor genotype (Allele) frequency in cases	OR (HmZ)	*TOMM40/ APOE/ APOC1/* marker combined p value
							rs405509 *APOE*	0.27 (0.52)	0.18 (0.43)	1.43	$<1 \times 10^{-8}$
[38]	rs2373115 *GAB2*	0.50 (0.71)	0.85 (0.92)	4.06	9.7×10^{-11}	(861; 550)	E4+ *APOE*	0.04 (0.21)	0.37 (0.61)	6.07	
[39]						(1,171; 985)	**rs4420638** ***APOC1***	**0.03 (0.18)**			$\mathbf{2.3 \times 10^{-44}}$
[40]	rs11610206 *FAM113B*	0.02 (0.10)			2.5×10^{-7}	(730; 718)	E4+ *APOE*	0.02 (0.15)	0.18 (0.42)		
[41]	rs3764650 *ABCA7*	0.01 (0.11)		1.23	5.0×10^{-21}	(17,683, 34,269)					
[28]	rs7561528 *BIN1*	0.12 (0.35)		1.17	5.2×10^{-14}	(10,011, 8,949)	**rs2075650** ***TOMM40***	**0.05 (0.23)**		**2.81**	$\mathbf{1.28 \times 10^{-299}}$
							rs4420638 *APOC1*	0.07 (0.26)		3.64	6.2×10^{-27}
[42]	rs11136000 *CLU*	0.15 (0.39)		0.85	1.6×10^{-16}	(8,371, 26,629)	**rs2075650** ***TOMM40***[b]	**0.02 (0.14)**		**2.53**	$\mathbf{1.04 \times 10^{-295}}$
[43]	rs11754661 *MTHFD1L*	0.005 (0.07)		2.10	1.9×10^{-10}	(931, 1,104)	**rs2075650** ***TOMM40***	**0.04 (0.2)**		**2.94**	$\mathbf{4.87 \times 10^{-36}}$
							rs405509 *APOE*	0.23 (0.48)		0.61	8.13×10^{-14}

Highlighted in bold are *p* values $< 1 \times 10^{(-35)}$

Legend: Top associated SNP markers in GWAS of case-control late-onset Alzheimer's disease Caucasian cohorts published between 2007 and 2011. Presented are the top associations in the *TOMM40-APOE-APOC1* region on the right hand-side and a few top associations (p < 10(−6)) across the remainder of the genome. Allele and genotype frequencies, odds ratios (*OR*) and p values are included if reported in the original paper. P values, sample sizes and ORs presented are for the final combined dataset (i.e. discovery and replication cohort results together). Note that some of the reports utilized partially overlapping cohorts

OR odds ratio, *HmZ* homozygote, *SNP* single nucleotide polymorphism

[a] When missing from the original report, allele and genotype frequencies in controls are adopted from CEU dataset as reported in HapMap

[b] rs2075650 is reported as located in the APOE gene by Seshadri et al. [42]

5 What Is the Clinical Utility of the Novel *TOMM40* Poly-T rs10524523 Discovery?

The initial finding of association between *APOE* ε4 and AD age of onset (Fig. 1) has been validated in cohort and case-control studies over the past two decades [50]; however, only *APOE* ε4 homozygosity (~2% of Caucasians) is useful for individual risk and age of onset predictions. Additionally, *APOE* ε2 has shown evidence of disease protection and association with later age of AD onset. Specific lengths of the *TOMM40* poly-T 523 variant are inherited with specific *APOE* alleles, and the length of the poly-T repeat is associated with age of AD onset [44, 45]. Roses et al. established a convention for categorical classification of the 523 poly-T lengths; Short (S) (≤19 T nucleotide repeats), Long (L, 20–29 Ts), Very Long (VL, ≥30 Ts) [44, 45]. Using these conventions, genotypes of 523 have been shown to be informative of individual age-dependent risk estimates for ~97% of Caucasians, now including carriers of *APOE* ε3 and heterozygous *APOE* ε4 carriers (Fig. 2), thanks to the additional resolution for *APOE* ε3 carriers granted by considering *TOMM40* 523 genotype. This is particularly important as *APOE* ε3 is, by far, the most frequent allele in all world populations.

The previously described age of onset distribution for *APOE* ε4/4 subjects (Fig. 1) is virtually identical to the 523 long-long (L/L) genotype distribution in Caucasians (Fig. 2). However, the genotypes bearing the *APOE* ε3 allele, which has been associated by most investigators with 'average' population risk, can be ascribed to disease onset distribution curves depending on the length of the *TOMM40* 523 allele linked to *APOE* ε3 (Fig. 2). The *APOE* ε3/4 curve is now

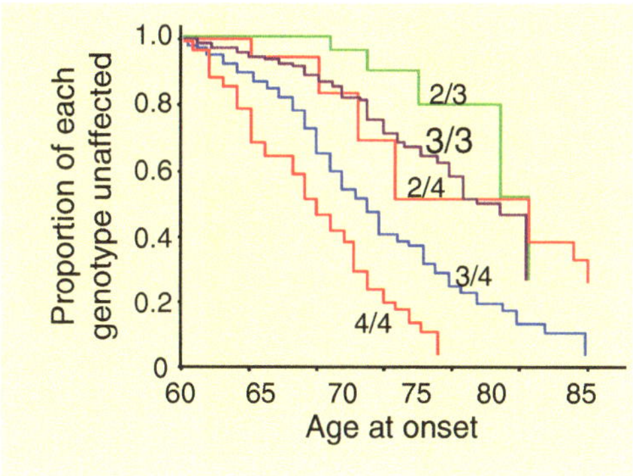

Fig. 1 Alzheimer's disease age of onset curves by *APOE* genotype, based on the knowledge current to 1994 (adopted from [49]). *Legend:* Age of onset for late onset AD, sporadic AD and control subjects as published in 1994 [49]. The age-of-onset is scored as a function of the individual's *APOE* genotype. *Onset curves* were estimated by Kaplan-Meier product limit distributions

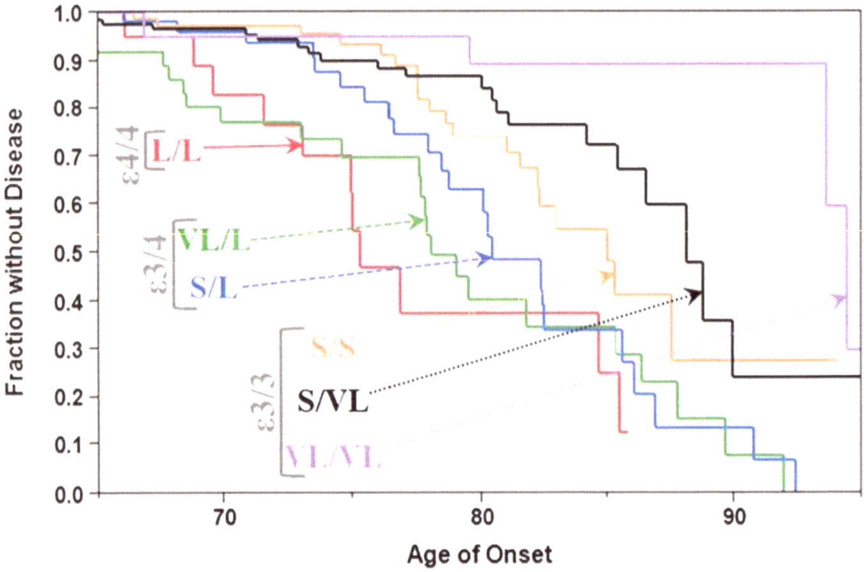

Fig. 2 Alzheimer's disease age of onset curves by *TOMM40-APOE* haplotype, based on the information available today (2012). *Legend:* Kaplan-Meyer survival curves, where the Y axis shows the percent survival without cognitive impairment, and the X axis represents age. Data was obtained from the Duke Bryan ADRC cohort N = 438 subjects: 106 diagnosed with cognitive impairment, 332 cognitively normal. N for each genotype: L,L:23; VL,L:54; S,L:72; S,S:100; S,VL:138; VL,VL:51

replaced by two curves, according to which of the two *523* alleles (S or VL) is connected to *APOE* ε3 (i.e. in the genotypes S/L or VL/VL). The single age of onset distribution that would previously have been described for *APOE* ε3/3 can now be de-convoluted into three age of onset distributions according to the *TOMM40* genotypes S/S, S/VL, or VL/VL. Of *APOE* ε2 allele carriers, in the entire dataset used for Fig. 2 only 6 of 70 (<10%) developed cognitive impairment, all of whom presented after age 75 years. Of those, the *APOE* ε2/4 genotype, like *APOE* ε3/4, was observed on the 523 S/L or VL/L curves, while *APOE* ε2/2 and *APOE* ε2/3 genotypes, like *APOE* ε3/3, were observed on the 523 S/S, S/VL, or VL/VL curves in Fig. 2. The fact that these *APOE* ε2 carrier cases all occurred at later ages, regardless of 523 genotype, confirms earlier observations of a protective effect of the *APOE* ε2 allele. Analyses of the data conducted with and without the *APOE* ε2 carriers showed similar results. To conclude, combining *TOMM40* 523 and *APOE* genotypes provides an effective means to distinguish people at higher and lower age-dependent risk of developing MCI, and subsequently AD.

6 The Early Intervention Clinical Trial: Simultaneous Biomarker Qualification and Treatment Efficacy Evaluation

As a whole, an effective enrichment strategy is paramount to the feasibility of early intervention study design. Without reliably identifying individuals at *high risk* of developing the investigated disease within a pre-specified timeframe, large numbers of individuals must be followed over an extended time period to accrue sufficient number of "events" (i.e. conversions from normal health to diagnosed condition) in order to draw a statistically significant and convincing conclusion. By considering the important risk factors, and enriching for these risks in the studied population, the number of individuals required for a statistically valid study can be reduced and the duration of the study may be abbreviated. Akin to other common methods used for enrichment of clinical trial populations, the biomarker in the exemplar study is being used to identify a pool of subjects with higher risk of developing the event (i.e. MCI due to AD) in order to accumulate during the course of the study the number of events required for statistically acceptable evidence of the intended indication (i.e. delay of onset of MCI due to AD). Without enrichment more time and/or more subjects would be required to achieve the same goal. Moreover, providing treatment only for those who are at risk also minimizes exposure of individuals who are at *low risk* of developing the condition to the active agent. Therefore, an algorithm that takes into account all relevant factors that affect susceptibility and/or age of disease onset (e.g. genetic variants, demographic characteristics, etc.) must be reliably compiled prior to launch of a clinical trial that utilizes it. A risk profile can then be composed to maximize the enrichment features. For instance, in the upcoming AD early intervention study we have created a risk algorithm that is comprised of two genetic loci (*APOE* and *TOMM40* 523) and the age of study participants at study entry (68–83 years inclusive).

The regulatory validation and prospective qualification of the risk biomarker for prognosis of high or low risk of developing MCI due to AD in 5 years in cognitively normal individuals is being planned. Biomarker qualification will be conducted simultaneously with an efficacy clinical trial, where subjects in the high risk group (classified by the biomarker) will be evaluated for the efficacy of a drug to delay the onset of MCI due to AD (as defined by Albert et al. [51]) (Fig. 3). In order to satisfy

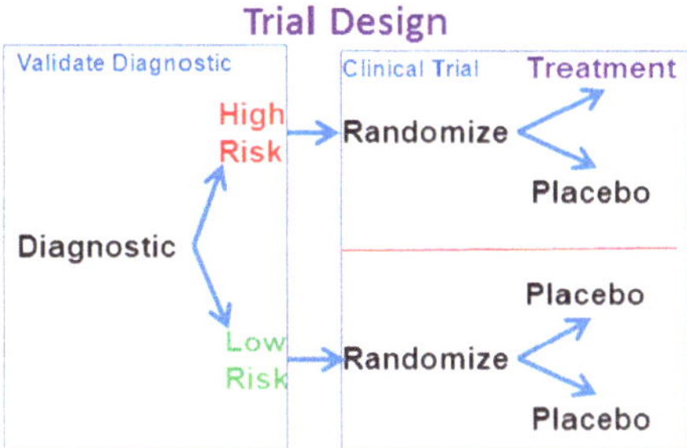

Fig. 3 Design of a co-development plan for treatment and prognostics (Rx/Dx) for early intervention for Alzheimer's disease

both components of this co-development program and to ensure the integrity of the double-blind design, all subjects in the low risk group will be treated with placebo, while subjects in the high risk group will be randomized to active treatment or placebo, at a 1:1 ratio. Zinfandel Pharmaceuticals, Inc. and Takeda Pharmaceutical Company Ltd. have entered into a worldwide licensing agreement regarding Zinfandel's TOMM40 assay as a biomarker for the risk of AD. The companies will attempt to prospectively validate the TOMM40 biomarker as a test of individual risk. The clinical study is planned for launch in 2013. Use of the biomarker for clinical prognosis will not be permitted until the successful qualification of this double-blind, pharmacogenetic-assisted clinical trial is completed.

7 How to Define and Ascertain the Relevant Target Population?

Genetic attributes tend to vary across world populations [52, 53]. Therefore, in order to generalize the applicability of biomarkers, each must be assessed relative to the phenotype of interest in diverse races and ethnicities. There is a paucity of longitudinal, large epidemiological cohorts of elderly individuals that are not Caucasians of European descent. Until such data become available, investigation of allele and genotype frequencies in healthy cohorts of diverse populations may serve to generate hypotheses regarding the possible appropriateness of analyzing Caucasian and non-Caucasian data in aggregate. In order to investigate this aspect in the context of the *TOMM40* 523 locus, we compared 523 allele and genotype (and APOE genotypes in some cases) in samples from different ethnicities from the USA (i.e., non-Hispanic Caucasians, African-Americans and Hispanic Caucasians), the Far East (i.e. Koreans, Japanese and Han Chinese) and West Africa (i.e., Ghanaians) [54]. Unique patterns of distribution of allele frequencies were observed, as well as subtle differences in the poly-T length range within each allele group (i.e. S, L and VL groupings). Non-Hispanic Caucasians of European descent (Whites) and Hispanic Caucasians had similar *TOMM40* and *APOE* allele frequencies, both of which differed from the frequencies in African Americans. Interestingly, the frequency of the *TOMM40* 523 S allele crudely reflected the diaspora of Homo Sapiens out of Africa, as this allele was most common in Ghanaians residing in Africa (71%) and in African Americans (65%), but its frequency decreased through non-Hispanic and Hispanic Caucasians (44–45%) residing in the USA, and its lowest frequency was observed in Far Easterners (20–38%).

The level of co-occurrence of the loci under investigation is another important genomic property that needs to be taken into account when constructing multi-marker biomarkers. This is particularly important for markers that are in LD, as is the case of the *TOMM40* poly-T 523 and the two *APOE* allele-defining SNPs. In Caucasian populations the *TOMM40* 523 L allele almost always co-occurs with the *APOE* ε4 allele (Fig. 4, [44, 45]). In our recent study, we also inferred differences in the co-inheritance pattern between *TOMM40* 523 and *APOE* alleles among

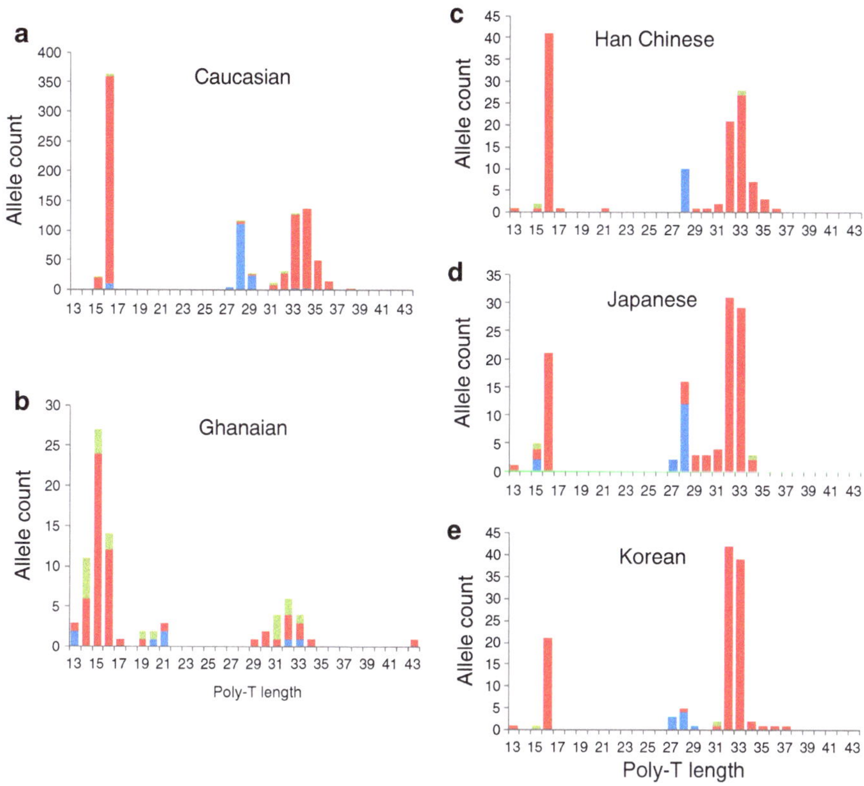

Fig. 4 Allele frequencies of the *TOMM40* 523 poly-T variant and *APOE* in diverse world populations (adopted from [55]). *Legend: TOMM40* 523 poly-T length is presented in number of deoxy-T residues on the X axis, with the linked *APOE* alleles indicated by color: *blue* for ε4, *red* for either ε2 or ε3 and *green* denoting uncertainty in prediction of the linked *APOE* allele. The Caucasian samples (n = 463) in panel (**a**) were obtained from the Wisconsin Registry for Alzheimer's Prevention. The *Ghanaian* samples (**b**, n = 41) were from elderly, non-demented subjects. The Far Eastern samples (**c**–**e**, n = 60 in each case) were from young, healthy subjects. Allele lengths are determined by a genotyping assay (**a**, **d**–**e**) or sequencing (**b**)

the different ethnic groups [54]. In the group of non-Hispanic Caucasians the co-occurrence of genotypes at each locus was consistent with the linkage observed in a different cohort of Caucasians of European descent. Caucasian Hispanic subjects living in the USA also showed high co-occurrence of the two variants, although to a slightly lesser extent. On the other hand, African Americans and Ghanaians exhibited different frequencies of co-occurrence, most notably a significant number of *TOMM40* 523 S/S subjects also carried at least one *APOE* ε4 allele (∼13 and 7% of the S alleles, respectively).

The intricate pattern of genomic variability identified in these population studies emphasizes the need to explore additional ethnic groups in the USA and worldwide prior to generalizing the use of the prognostic biomarker to non-Caucasians.

In particular, within each group the association between *TOMM40* and *APOE* genotypes and age of disease onset must be assessed. At the present time, the vast majority of well-phenotyped cohorts with prospectively followed disease onset data are composed of European descent Caucasians. As a result, the assessment of the primary endpoints in the AD early intervention study introduced in this chapter is powered for non-Hispanic Caucasian subjects; although enrollment is open to all ethnicities in all study arms.

8 How to Convert the Biomarker into a Feasible Tool for Use in High-Throughput Registration Clinical Trials?

Once a genetic (or other) biomarker has been developed and sufficient experimental data have been collected, including multiple retrospective association studies demonstrating the expected utility of the biomarker, it is necessary to confirm its performance characteristics in prospective, well-designed clinical trials. In order to *validate* a biomarker assay, its performance characteristics are being assessed and documented, as are the optimal conditions that will generate the reproducibility and accuracy of the assay [56]. Biomarker *qualification* is a conclusion by regulators that the data generated about the biomarker reliably supports its stated context of use and adequately reflects the biological processes and/or clinical endpoints of interest [57]. Qualification of a biomarker via confirmatory and prospective clinical trials for its intended clinical use adheres to stricter development practices. The recommendations recently published by the Institute of Medicine in the USA for the qualification of all translational "omics" state that at the "lock down" point any test being evaluated must be fully defined and validated (analytical and clinical/biological validation) and submitted to the appropriate Institutional Review Board (IRB). Depending upon the intended use, it may be appropriate to consult with the regulatory authorities [9]. In general, changes to any proposed test after "lock down" require a return to the test validation phase and approval by the IRB and possibly the regulators. It should be noted that all aspects of the test validation process should be applied not only to the assay that generates the data, but also to the fully specified computational procedures used to analyze the data and report the result. In these settings, and particularly when the biomarker serves to prospectively stratify a clinical trial population, there is a need to develop an assay that is simple, lends itself to high throughput and demonstrates quick turn-around. This assay also needs to be standardized prior to study initiation so as to perform within strict parameters in multiple testing laboratories. Finally, for wide-spread use of a test for clinical trials or for later use in clinical practice, the test should be affordable and readily interpreted.

All clinical laboratory tests that support treatment decisions, including "omics"-based tests, should be developed, validated, and placed into clinical use either through review by the appropriate regulatory bodies (In-Vitro Devices, IVD) or through validation and performance in laboratories that meet the Clinical Laboratory Improvement Amendments (CLIA) certification, also called Laboratory Developed Tests (LDT). Although it reserves the right to examine all LDTs, the FDA currently reviews only LDT tests determined to be of high complexity or to be associated with high risk to patients (as is the case for tests employed in the prospective stratification scheme of therapeutic clinical trials). For the latter, a CLIA-certified laboratory should design, optimize, validate, and implement the LDT under current clinical laboratory standards. The process of transforming research level assays into high-throughput, feasible tests in CLIA-certified settings requires robust demonstration of analytical validity, which is a key component of the Investigational Device Exemption (IDE) process (Title 21, Code of Federal Regulations (21 CFR) Part 812) [58], as are the qualification and quality control evaluations [57, 59].

9 How to Conduct Pharmacogenetically Assisted Clinical Trials *Effectively*?

Preparations for the AD early intervention study described above, expected to launch in 2013, started early on. The typical order of operational activities was reversed in order to achieve significant time savings. To accomplish this, Zinfandel Pharmaceuticals, Inc., which developed the risk algorithm, took a number of pro-active steps: (1) Multiple disciplines of the regulatory bodies at the FDA were consulted, prior to identifying the study drug, via the Voluntary Exploratory Data Submission (VXDS) mechanism [60]; (2) Key investigators at appropriate sites around the world were identified; (3) Registries of cognitively normal subjects with interest in participating in an AD early intervention study were initiated; and (4) Qualified personnel were engaged. As a first step, sites located around the world with access to either a sizeable population, >60 years of age, receiving long-term healthcare from a central medical provider, or to ongoing, longitudinal epidemiological studies of older adults were identified.

Institutional support and appropriate infrastructure were important factors in determining suitability of candidate sites. These sites were identified over a period of 1 to 3 years prior to expected study launch in 2013; the countries currently being considered are the USA, Australia, Switzerland, the UK, Russia, and Israel. This process was a key enabler for the expected clinical trial since even trials that use enrichment biomarkers typically enroll thousands of subjects and follow them for multiple years. In the context of delay of onset of AD, some of the published studies were conducted in a handful of clinical sites (e.g., ADAPT [61]), each with hundreds of subjects retained throughout the study. Employing a similar study design, sites in the upcoming registration clinical study have been designed to act with efficiently staffed neuropsychologists and other professionals, trained to conduct standardized neuropsychological tests, thus ensuring continuity of capabilities for subject monitoring, despite the significant staff turnover that is anticipated during multi-year studies. These specialized and focused testing centers, pivotal to each site's clinical development structure, are expected to be autonomous with respect to staffing, patient recruitment and follow-up, and will be centrally monitored for standardization of testing procedures, subject management practices, and accreditation.

While this example specifically applies to early intervention studies along the continuum of cognitive impairment through to AD, the design and operational methodology could be easily applied to any number of other diseases, and certainly to other neurodegenerative conditions.

References

1. National Prevention Council (2011) National prevention strategy. US Department of Health and Human Services, Office of the Surgeon General, Washington, DC
2. Brookmeyer R, Johnson E, Ziegler-Graham K, Arrighi HM (2007) Forecasting the global burden of Alzheimer's disease. Alzheimers Dement 3(3):186–191
3. Grossman I, Lutz MW, Crenshaw DG, Saunders AM, Burns DK, Roses AD (2010) Alzheimer's disease: diagnostics, prognostics and the road to prevention. EPMA J 1(2):293–303. doi:10.1007/s13167-010-0024-3
4. Haddow JE, Palomaki GE (2003) ACCE: a model process for evaluating data on emerging genetic tests. In: Khoury M, Little J, Burke W (eds) Human genome epidemiology: a scientific foundation for using genetic information to improve health and prevent disease. Oxford University Press, Oxford/New York, pp 217–233
5. Matchar DB, Thakur ME, Grossman I, McCrory DC, Orlando LA, Steffens DC, Goldstein DB, Cline KE, Gray RN (2007) Testing for cytochrome P450 polymorphisms in adults with non-psychotic depression treated with selective serotonin reuptake inhibitors (SSRIs). Evid Rep Technol Assess (Full Rep) 146:1–77
6. Hughes B (2009) The comparative effectiveness challenge. Nat Rev Drug Discov 8(4):261–263
7. Surbone A (2011) Social and ethical implications of BRCA testing. Ann Oncol 22(Suppl 1):i60–i66. doi:10.1093/annonc/mdq668
8. Ginsburg GS, Willard HF (2009) Genomic and personalized medicine: foundations and applications. Transl Res 154(6):277–287. doi:S1931-5244(09)00274-6[pii]10.1016/j.trsl.2009.09.005
9. Institute of Medicine IOM (2012) Evolution of translational omics: lessons learned and the path forward. The National Academies Press, Washington, DC
10. US Food and Drug Administration (2005) Drug-diagnostic co-development concept paper (draft). Available at http://www.fda.gov/downloads/Drugs/ScienceResearch/ResearchAreas/Pharmacogenetics/UCM116689.pdf. Accessed 13 Dec 2012
11. Roses AD (2010) An inherited variable poly-T repeat genotype in TOMM40 in Alzheimer disease. Arch Neurol 67(5):536–541. doi:67/5/536 [pii]10.1001/archneurol.2010.88
12. West RL (1996) An application of prefrontal cortex function theory to cognitive aging. Psychol Bull 120(2):272–292
13. McKhann GM, Knopman DS, Chertkow H, Hyman BT, Jack CR Jr, Kawas CH, Klunk WE, Koroshetz WJ, Manly JJ, Mayeux R, Mohs RC, Morris JC, Rossor MN, Scheltens P, Carrillo MC, Thies B, Weintraub S, Phelps CH (2011) The diagnosis of dementia due to Alzheimer's disease: recommendations from the National Institute on Aging-Alzheimer's Association workgroups on diagnostic guidelines for Alzheimer's disease. Alzheimers Dement 7(3):263–269. doi:10.1016/j.jalz.2011.03.005

14. Visser PJ, Verhey FR (2008) Mild cognitive impairment as predictor for Alzheimer's disease in clinical practice: effect of age and diagnostic criteria. Psychol Med 38(1):113–122
15. Hebert LE, Scherr PA, Bienias JL, Bennett DA, Evans DA (2003) Alzheimer disease in the US population: prevalence estimates using the 2000 census. Arch Neurol 60(8):1119–1122
16. National Institutes of Health (2007) Journey to discovery: 2005–2006 progress report on Alzheimer's disease. Available at http://www.nia.nih.gov/sites/default/files/20062007_progress_report_on_alzheimers_disease.pdf. Accessed 13 Dec 2012
17. Ferri CP, Prince M, Brayne C, Brodaty H, Fratiglioni L, Ganguli M, Hall K, Hasegawa K, Hendrie H, Huang Y, Jorm A, Mathers C, Menezes PR, Rimmer E, Scazufca M, Alzheimer's Disease International (2005) Global prevalence of dementia: a Delphi consensus study. Lancet 366(9503):2112–2117. doi:10.1016/S0140-6736(05)67889-0
18. World Health Organization (2004) The global burden of disease (2004 update). WHO Press, Geneva
19. Blennow K, de Leon MJ, Zetterberg H (2006) Alzheimer's disease. Lancet 368(9533):387–403
20. Gatz M, Reynolds CA, Fratiglioni L, Johansson B, Mortimer JA, Berg S, Fiske A, Pedersen NL (2006) Role of genes and environments for explaining Alzheimer disease. Arch Gen Psychiatry 63(2):168–174
21. Bertram L, Lill CM, Tanzi RE (2010) The genetics of Alzheimer disease: back to the future. Neuron 68(2):270–281. doi:10.1016/j.neuron.2010.10.013
22. Strittmatter WJ, Saunders AM, Schmechel D, Pericak-Vance M, Enghild J, Salvesen GS, Roses AD (1993) Apolipoprotein E: high-avidity binding to beta-amyloid and increased frequency of type 4 allele in late-onset familial Alzheimer disease. Proc Natl Acad Sci U S A 90(5):1977–1981
23. Corder EH, Saunders AM, Strittmatter WJ, Schmechel DE, Gaskell PC, Small GW, Roses AD, Haines JL, Pericak-Vance MA (1993) Gene dose of apolipoprotein E type 4 allele and the risk of Alzheimer's disease in late onset families. Science 261(5123):921–923
24. Saunders AM, Schmader K, Breitner JC, Benson MD, Brown WT, Goldfarb L, Goldgaber D, Manwaring MG, Szymanski MH, McCown N et al (1993) Apolipoprotein E epsilon 4 allele distributions in late-onset Alzheimer's disease and in other amyloid-forming diseases. Lancet 342(8873):710–711
25. Saunders AM, Strittmatter WJ, Schmechel D, George-Hyslop PH, Pericak-Vance MA, Joo SH, Rosi BL, Gusella JF, Crapper-MacLachlan DR, Alberts MJ et al (1993) Association of apolipoprotein E allele epsilon 4 with late-onset familial and sporadic Alzheimer's disease. Neurology 43(8):1467–1472
26. Bu G (2009) Apolipoprotein E and its receptors in Alzheimer's disease: pathways, pathogenesis and therapy. Nat Rev Neurosci 10(5):333–344
27. Cedazo-Minguez A (2007) Apolipoprotein E and Alzheimer's disease: molecular mechanisms and therapeutic opportunities. J Cell Mol Med 11(6):1227–1238
28. Naj AC, Jun G, Beecham GW et al (2011) Common variants at MS4A4/MS4A6E, CD2AP, CD33 and EPHA1 are associated with late-onset Alzheimer's disease. Nat Genet 43(5):436–441. doi:10.1038/ng.801
29. Templeton AR (2010) The diverse applications of cladistic analysis of molecular evolution, with special reference to nested clade analysis. Int J Mol Sci 11(1):124–139. doi:10.3390/ijms11010124
30. Coon KD, Myers AJ, Craig DW, Webster JA, Pearson JV, Lince DH, Zismann VL, Beach TG, Leung D, Bryden L, Halperin RF, Marlowe L, Kaleem M, Walker DG, Ravid R, Heward CB, Rogers J, Papassotiropoulos A, Reiman EM, Hardy J, Stephan DA (2007) A high-density whole-genome association study reveals that APOE is the major susceptibility gene for sporadic late-onset Alzheimer's disease. J Clin Psychiatry 68(4):613–618
31. Potkin SG, Guffanti G, Lakatos A, Turner JA, Kruggel F, Fallon JH, Saykin AJ, Orro A, Lupoli S, Salvi E, Weiner M, Macciardi F, Alzheimer's Disease Neuroimaging Initiative (2009) Hippocampal atrophy as a quantitative trait in a genome-wide association study identifying novel susceptibility genes for Alzheimer's disease. PLoS One 4(8):e6501. doi:10.1371/journal.pone.0006501

32. Harold D, Abraham R, Hollingworth P et al (2009) Genome-wide association study identifies variants at CLU and PICALM associated with Alzheimer's disease. Nat Genet 41(10):1088–1093
33. Lambert JC, Heath S, Even G et al (2009) Genome-wide association study identifies variants at CLU and CR1 associated with Alzheimer's disease. Nat Genet 41(10):1094–1099
34. Abraham R, Moskvina V, Sims R, Hollingworth P, Morgan A, Georgieva L, Dowzell K, Cichon S, Hillmer AM, O'Donovan MC, Williams J, Owen MJ, Kirov G (2008) A genome-wide association study for late-onset Alzheimer's disease using DNA pooling. BMC Med Genomics 1:44
35. Heinzen EL, Need AC, Hayden KM, Chiba-Falek O, Roses AD, Strittmatter WJ, Burke JR, Hulette CM, Welsh-Bohmer KA, Goldstein DB (2010) Genome-wide scan of copy number variation in late-onset Alzheimer's disease. J Alzheimers Dis 19(1):69–77. doi:V7N8Q30275184532 [pii]10.3233/JAD-2010-1212
36. Carrasquillo MM, Zou F, Pankratz VS, Wilcox SL, Ma L, Walker LP, Younkin SG, Younkin CS, Younkin LH, Bisceglio GD, Ertekin-Taner N, Crook JE, Dickson DW, Petersen RC, Graff-Radford NR, Younkin SG (2009) Genetic variation in PCDH11X is associated with susceptibility to late-onset Alzheimer's disease. Nat Genet 41(2):192–198
37. Grupe A, Abraham R, Li Y, Rowland C, Hollingworth P, Morgan A, Jehu L, Segurado R, Stone D, Schadt E, Karnoub M, Nowotny P, Tacey K, Catanese J, Sninsky J, Brayne C, Rubinsztein D, Gill M, Lawlor B, Lovestone S, Holmans P, O'Donovan M, Morris JC, Thal L, Goate A, Owen MJ, Williams J (2007) Evidence for novel susceptibility genes for late-onset Alzheimer's disease from a genome-wide association study of putative functional variants. Hum Mol Genet 16(8):865–873
38. Reiman EM, Webster JA, Myers AJ, Hardy J, Dunckley T, Zismann VL, Joshipura KD, Pearson JV, Hu-Lince D, Huentelman MJ, Craig DW, Coon KD, Liang WS, Herbert RH, Beach T, Rohrer KC, Zhao AS, Leung D, Bryden L, Marlowe L, Kaleem M, Mastroeni D, Grover A, Heward CB, Ravid R, Rogers J, Hutton ML, Melquist S, Petersen RC, Alexander GE, Caselli RJ, Kukull W, Papassotiropoulos A, Stephan DA (2007) GAB2 alleles modify Alzheimer's risk in APOE epsilon4 carriers. Neuron 54(5):713–720. doi:S0896-6273(07)00379-0 [pii]10.1016/j.neuron.2007.05.022
39. Li H, Wetten S, Li L, Wang LS, Buros J, Gallins PJ, Buxbaum JD, Ertekin-Taner N, Fallin MD, Friedland R, Inzelberg R, Kramer P, Rogaeva E, St George-Hyslop P, Alzheimer's Disease Genetics Consortium, Cantwell LB, Dombroski BA, Saykin AJ, Reiman EM, Bennett DA, Morris JC, Lunetta KL, Martin ER, Montine TJ, Goate AM, Blacker D, Tsuang DW, Beekly D, Cupples LA, Hakonarson H, Kukull W, Foroud TM, Haines J, Mayeux R, Farrer LA, Pericak-Vance MA, Schellenberg GD (2008) Candidate single-nucleotide polymorphisms from a genomewide association study of Alzheimer disease. Arch Neurol 65(1):45–53. doi:10.1001/archneurol.2007.3
40. Beecham GW, Martin ER, Li YJ, Slifer MA, Gilbert JR, Haines JL, Pericak-Vance MA (2009) Genome-wide association study implicates a chromosome 12 risk locus for late-onset Alzheimer disease. Am J Hum Genet 84(1):35–43
41. Hollingworth P, Harold D, Sims R et al (2011) Common variants at ABCA7, MS4A6A/MS4A4E, EPHA1, CD33 and CD2AP are associated with Alzheimer's disease. Nat Genet 43(5):429–435. doi:10.1038/ng.803
42. Seshadri S, Fitzpatrick AL, Ikram MA et al (2010) Genome-wide analysis of genetic loci associated with Alzheimer disease. JAMA 303(18):1832–1840. doi:10.1001/jama.2010.574
43. Naj AC, Beecham GW, Martin ER, Gallins PJ, Powell EH, Konidari I, Whitehead PL, Cai G, Haroutunian V, Scott WK, Vance JM, Slifer MA, Gwirtsman HE, Gilbert JR, Haines JL, Buxbaum JD, Pericak-Vance MA (2010) Dementia revealed: novel chromosome 6 locus for late-onset Alzheimer disease provides genetic evidence for folate-pathway abnormalities. PLoS Genet 6(9):e1001130
44. Lutz MW, Crenshaw DG, Saunders AM, Roses AD (2010) Genetic variation at a single locus and age of onset for Alzheimer's disease. Alzheimers Dement 6(2):125–131. doi:S1552-5260(10)00016-6 [pii]10.1016/j.jalz.2010.01.011

45. Roses AD, Lutz MW, Amrine-Madsen H, Saunders AM, Crenshaw DG, Sundseth SS, Huentelman MJ, Welsh-Bohmer KA, Reiman EM (2010) A TOMM40 variable-length polymorphism predicts the age of late-onset Alzheimer's disease. Pharmacogenomics J 10(5):375–384. doi:tpj200969 [pii]10.1038/tpj.2009.69
46. Humphries AD, Streimann IC, Stojanovski D, Johnston AJ, Yano M, Hoogenraad NJ, Ryan MT (2005) Dissection of the mitochondrial import and assembly pathway for human Tom40. J Biol Chem 280(12):11535–11543. doi:M413816200 [pii]10.1074/jbc.M413816200
47. Manczak M, Calkins MJ, Reddy PH (2011) Impaired mitochondrial dynamics and abnormal interaction of amyloid beta with mitochondrial protein Drp1 in neurons from patients with Alzheimer's disease: implications for neuronal damage. Hum Mol Genet 20(13):2495–2509. doi:10.1093/hmg/ddr139
48. Santel A, Fuller MT (2001) Control of mitochondrial morphology by a human mitofusin. J Cell Sci 114(Pt 5):867–874
49. Roses AD (1994) Apolipoprotein E affects the rate of Alzheimer disease expression: beta-amyloid burden is a secondary consequence dependent on APOE genotype and duration of disease. J Neuropathol Exp Neurol 53(5):429–437
50. Williams JW, Plassman BL, Burke J, Holsinger T, Benjamin S. Preventing Alzheimer's disease and cognitive decline. Evidence report/technology assessment no. 193. (Prepared by the Duke Evidence-based Practice Center under Contract No. HHSA 290-2007-10066-I.) AHRQ Publication No. 10-E005, Agency for Healthcare Research and Quality, Rockville. Source: http://www.ahrq.gov/downloads/pub/evidence/pdf/alzheimers/alzcog.pdf. Accessed 13 Dec 2012
51. Albert MS, DeKosky ST, Dickson D, Dubois B, Feldman HH, Fox NC, Gamst A, Holtzman DM, Jagust WJ, Petersen RC, Snyder PJ, Carrillo MC, Thies B, Phelps CH (2011) The diagnosis of mild cognitive impairment due to Alzheimer's disease: recommendations from the National Institute on Aging-Alzheimer's Association workgroups on diagnostic guidelines for Alzheimer's disease. Alzheimers Dement 7(3):270–279. doi:10.1016/j.jalz.2011.03.008
52. Nelson MR, Bryc K, King KS, Indap A, Boyko AR, Novembre J, Briley LP, Maruyama Y, Waterworth DM, Waeber G, Vollenweider P, Oksenberg JR, Hauser SL, Stirnadel HA, Kooner JS, Chambers JC, Jones B, Mooser V, Bustamante CD, Roses AD, Burns DK, Ehm MG, Lai EH (2008) The population reference sample, POPRES: a resource for population, disease, and pharmacological genetics research. Am J Hum Genet 83(3):347–358. doi:10.1016/j.ajhg.2008.08.005
53. Bustamante CD, Burchard EG, De la Vega FM (2011) Genomics for the world. Nature 475(7355):163–165. doi:10.1038/475163a
54. Linnertz C, Saunders AM, Lutz MW et al (2012) Characterization of the poly-T variant in the TOMM40 gene in diverse populations. PLoS One 7(2):e30994. doi:10.1371/journal.pone.0030994
55. Lutz MW, Saunders AM, Crenshaw DG et al (2010) Frequencies of the Alzheimer's disease associated TOMM40 poly-T allele in different ethnic groups. Poster presented at the 60th annual meeting of The American Society of Human Genetics, Washington D.C., USA. Abstract available at http://www.ashg.org/2010meeting/abstracts/fulltext/f22260.htm. Accessed 13 Dec 2012
56. Wagner JA (2002) Overview of biomarkers and surrogate endpoints in drug development. Dis Markers 18(2):41–46
57. European Medicines Agency (2010) ICH guideline E16 Genomic biomarkers related to drug response: context, structure, and format of qualification submissions, step 4. Available at http://www.ema.europa.eu/docs/en_GB/document_library/Scientific_guideline/2010/09/WC500097060.pdf. Accessed 13 Dec 2012
58. US Food and Drug Administration (1998) Guidance on IDE policies and procedures. Available at http://www.fda.gov/downloads/MedicalDevices/DeviceRegulationandGuidance/GuidanceDocuments/ucm080203.pdf. Accessed 13 Dec 2012

59. US Food and Drug Administration (2010) Guidance for industry qualification process for drug development tools (draft). Available at http://www.fda.gov/downloads/Drugs/GuidanceComplianceRegulatoryInformation/Guidances/UCM230597.pdf. Accessed 13 Dec 2012
60. Goodsaid FM, Amur S, Aubrecht J, Burczynski ME, Carl K, Catalano J, Charlab R, Close S, Cornu-Artis C, Essioux L, Fornace AJ Jr, Hinman L, Hong H, Hunt I, Jacobson-Kram D, Jawaid A, Laurie D, Lesko L, Li HH, Lindpaintner K, Mayne J, Morrow P, Papaluca-Amati M, Robison TW, Roth J, Schuppe-Koistinen I, Shi L, Spleiss O, Tong W, Truter SL, Vonderscher J, Westelinck A, Zhang L, Zineh I (2010) Voluntary exploratory data submissions to the US FDA and the EMA: experience and impact. Nat Rev Drug Discov 9(6):435–445. doi:10.1038/nrd3116
61. Lyketsos CG, Breitner JC, Green RC, Martin BK, Meinert C, Piantadosi S, Sabbagh M (2007) Naproxen and celecoxib do not prevent AD in early results from a randomized controlled trial. Neurology 68(21):1800–1808. doi:10.1212/01.wnl.0000260269.93245.d2

An Update on CSF Biomarkers of Parkinson's Disease

Peter A. LeWitt, Bertrand R. Huber, and Jing Zhang

Abstract Disease-specific markers have the potential to advance understanding of Parkinson's disease (PD) pathogenesis as well as its treatments. Although its clinical features can be unmistakable, PD has not been easy to measure by laboratory methods. Over five decades of research, there have been many attempts to tease out neurochemical indicators of its diagnosis and progression. The most logical starting point of this search has been cerebrospinal fluid (CSF), a biospecimen contiguous with the sites of major pathological changes in this disorder. Guiding the search for biomarkers is the understanding that neurons undergoing degeneration do so in concert with abnormal protein aggregation and a decline in neurotransmitter output. Therefore, investigations for PD biomarkers have sought clues in CSF proteins and small molecular-weight biochemicals that might mirror disease-specific changes in the brain. For this purpose, dozens of CSF constituents have been screened in the search PD biomarkers that might directly and indirectly reflecting the impact of the disease. This chapter will review the highlights of recent developments as well as the limitations of these findings. The ultimate hope of the authors is that biomarkers discovered in CSF might also appear in specimens more easily accessible (such as

P.A. LeWitt (✉)
Departments of Neurology, Wayne State University School of Medicine and Henry Ford Hospital, Detroit, MI, USA

Parkinson's Disease and Movement Disorders Program, Henry Ford West Bloomfield Hospital, 6777 West Maple Road, West Bloomfield, MI 48034, USA
e-mail: plewitt1@hfhs.org

B.R. Huber
Mental Illness Research, Education and Clinical Center, Veterans Administration – Puget Sound Health Care System, S-116-6E S. Columbian Way, Seattle, WA 98108, USA

J. Zhang
Department of Pathology, University of Washington School of Medicine, 359635, RM 302B R&T, 325 9th Avenue, Seattle, WA 98104-2499, USA
e-mail: zhangj@uw.edu

S. Mandel (ed.), *Neurodegenerative Diseases: Integrative PPPM Approach as the Medicine of the Future*, Advances in Predictive, Preventive and Personalised Medicine 2, DOI 10.1007/978-94-007-5866-7_8,

from a blood test). If so, screening of populations at risk for PD or monitoring a future neuroprotective treatment might be practical applications. A PD biomarker might also provide new insights into the disease process.

Keywords Parkinson's disease • Molecular mechanism • Person at risk • Population screening • Predictive diagnostics • Personalized medicine • Disease specific biomerker • Neuroprotecton • Disease monitoring • Tailored treatment

1 Introduction

Finding biomarkers for Parkinson's disease (PD) has become one of the most challenging directions in current research for this disorder [1–5]. A robust disease-specific indicator would be extremely valuable for better understanding of PD's pathophysiology. In addition, a PD biomarker might offer a practical means for the early detection of PD or monitoring of its progression. The critical need for PD biomarkers has even attracted the attention of a venture capitalist, who has commented that they could become "paradigm changers" in therapeutics research, helping "... to make drug discovery and development more cost-efficient, predictable, and truly innovative" [6]. The past quarter-century has seen an evolution in understanding what constitutes a PD biomarker, especially since there has been a major revision as to the identity of PD itself. What was once regarded simply as syndrome of dopaminergic depletion is now recognized to have a more complex story involving non-motor prodromal features, altered protein chemistry and mitochondrial disturbance throughout the body, disrupted signaling pathways beyond deficient dopamine neurotransmission, and several other changes. There appears to be some genetic input to the disorder (at least in some patients). Even in sporadic forms of PD, there appears to be considerable heterogeneity in clinical features and outcomes. As a result of a multi-dimensional expansion in the identity of PD, options for uncovering its biomarkers also have increased.

The notion of a biomarker deserves some special considerations in the context of PD. Ideally, it should be useful for sorting out PD from other neurological disorders that can be challenging to distinguish, such as early stages of progressive supranuclear palsy (PSP) or multiple system atrophy (MSA). An effective biomarker should also permit PD to be distinguished from changes found in normal aging. In the elderly brain, small numbers of Lewy bodies (the histological hallmark of PD, when present in large number and in association with nigrostriatal neuronal loss) can be found in otherwise healthy individuals. If a particular biomarker was derived specifically from this cellular inclusion, for example, it would need to recognize the difference between incidental Lewy bodies and the greater abundance of them as found in the neurodegenerative disorder (that is, if there actually is a difference!). Conceptually, biomarkers for PD can be categorized in two groups that, while distinct for their applications, are not mutually exclusive (that is, a particular biomarker might be found to share both identities). The first concept, that of *trait marker*, refers to a disease-specific change. A trait marker can be present even before clinical expression of the disorder and is not necessarily modified as a function of disease duration or severity. A trait marker might also tie into a particular disease clinical subtype or might be nothing more than a risk factor for acquiring PD. In contrast, the concept of a *state marker* is for more than an indicator of disease presence; it should offer a gradation as to disease severity, duration, or rate of change.

Among various biospecimen options, cerebrospinal fluid (CSF) might seem to be an ideal medium for the detection of disease-specific biochemicals and proteins. Since it arises and circulates on the inside compartment of the blood-brain barrier, CSF has been sampled extensively for the study of CNS disease. Although CSF is not identical with the extra-cellular fluid (ECF) of the brain, both fluids appear to be contiguous and because of mixing, most of their constituents are similar. Nevertheless, the origin of CSF differs from ECF in that it arises from secretion by several structures found within the cerebral ventricles – the choroid plexi – unlike extracellular fluid, which diffuses out of the cellular elements of the brain. Furthermore, CSF does not percolate through the brain mass but, after secretion in the ventricle, circulates caudally towards the lumbar space or outwards around the cortical surface of the brain, where it can be absorbed through Pacchionian granulations. Its complex circulation pattern and continued synthesis – its volume is reconstituted several times in the course of a day – means that factors other than its relative abundance in brain tissue govern the concentration of a substance measured in CSF.

Over the time that CSF travels from the brain region to the lumbar space, the concentration of some constituents will change. One reason for this is that extraction of water occurs as CSF passes caudally in the spinal canal. This means that proteins and other biochemicals will become more concentrated as the fluid column moves further down the spinal canal. Other substances that arise in the brain and have mechanisms in brain or spinal cord tissue that can extract them into the circulation also will decline in concentration in a rostral-caudal gradient. Passive diffusion might also be operative for certain biochemicals or proteins. Clinical considerations require that specimens of CSF used in most research does not come from the lateral cerebral ventricles or the cisterna magna; specimens from the lumbar space need to serve for sampling the CSF compartment even though the fluid there has been secreted many hours earlier in the brain. The practical message to be gained from knowledge of CSF physiology is that one should not assume that *concentration* of a CSF constituent will necessarily represent its synthesis rate or concentration in the brain. This point is well exemplified by markers of dopamine metabolism, as discussed below.

The sections that follow will review available information as to monoamine metabolites, neuropeptides, metal ions, other biochemicals, and proteins that have been studied in the context of CSF biomarkers in PD. Though not an exhaustive list of all research on this topic, this chapter will cover the highlights of biomarker studies in the past two decades. A thorough review of findings up to 1990 can be found in an earlier review [7]. The latter chapter discusses findings pertaining to several studies not reviewed in the current chapter, including measurements of immunoreactivity for several neuropeptides, among them, somatostatin, cholecystokinin, enkephalin, endorphin, arginine-vasopressin, and delta-sleep-inducing peptide.

2 CSF Biochemicals as Biomarkers of Parkinson's Disease

2.1 Monoamine Neurotransmitters

A logical starting point for CSF biomarkers of PD is investigation of the monoamine neurotransmitters and their metabolites, especially those related to dopamine. When first recognized, the marked loss of dopamine synthesis by neurons of the substantia nigra pars compacta (SNpc) led to a revolution in understanding about PD [8]. A marked depletion in content of dopamine and its major metabolite, homovanillic acid (HVA), is evident in brain tissue from the SNpc and the regions to which these neurons project (the putamen and caudate nuclei, which are termed collectively as the striatum). On this basis, there has been a longstanding expectation that the diminished neuronal output of dopamine would be reflected also in biochemical profiling of CSF. One caveat countering for this assumption is that the nigrostriatal pathway is only one of several brain regions involved in dopamine synthesis and its turnover. Hence, even a marked change in dopamine produced within the nigrostriatal pathway may be dwarfed by the absence of change in other dopaminergic pathways that comprise much larger brain structures.

Many (but not all) of dozens of PD CSF investigations found that dopamine and HVA concentrations were diminished [7]. Measurement of dopamine concentrations presented a technical challenge for most laboratories due to its low concentrations. In contrast, HVA concentration is 2 orders of magnitude greater and at levels more conducive to measurement by conventional high-performance liquid chromatography. In the past, HVA has been regarded as a more likely candidate for reflecting dopamine turnover since it accumulates to reach a steady-state concentration in CSF. This view has been challenged by Goldstein and colleagues [9], who have pointed out that HVA is actually a more indirect marker of dopamine turnover than another dopamine metabolite, dihdroxyphenylacetic acid (DOPAC).

One of the most definitive studies of CSF HVA in PD investigated unmedicated patients whose CSF specimens were collected in a standardized manner. Its mean concentration in 99 subjects was 30.0 ± 7.3 ng/ml. HVA was measured by gas chromatography-mass spectrometry using pre-loaded internal standards and a rigorous collection technique that included overnight bed rest [10]. This study also recognized that the measured HVA concentration depended upon where in a caudal-rostral gradient the specimen was collected, as this value almost doubled from the first to the 23rd ml of CSF during collection by lumbar puncture. Hence, interpretation of HVA content in CSF, like that of many other substances, needs to consider the change in this compound's concentration from the lumbar region to the more rostral source of CSF secretion and brain metabolism. In the latter study, the sum of molar concentrations of DA and its metabolites in CSF was approximately 12.6% that of HVA. The next two most abundant metabolites were 3-O-methyldopa (mean concentration: 10.95 ± 4.96 nM) and dopamine-3-sulfate (2.35 ± 1.30 nM). Other compounds measured in smaller quantities included DOPAC, 3-methoxytyramine, levodopa, and homovanillol [10]. The CSF specimens used in this study were a subset of a larger group whose CSF specimens were collected as part of a clinical trial of neuroprotective strategies for PD (the DATATOP study). Measurements of CSF HVA for the larger group of DATATOP participants revealed a wide range of concentrations, reported to be a mean (± S.D.) of 34.7 ± 17.0 ng/ml [11] (a greater value than reported in the 99-subject DATATOP subset discussed above). Over up to 2 years of follow-up, time, there was a relatively minor reduction in CSF HVA concentration. However, the study concluded that the measured HVA concentrations were quite close to values reported in normal individuals and lacked correlation with either the duration of PD symptomatology or the severity of ratings for Parkinsonian features. These conclusions were in contrast to claims from several other clinical reports that found HVA a good correlate of PD's clinical features [7], although many of these studies were carried out with small sample sets and suboptimal study methods (such as controlling for the CSF rostro-caudal gradient effect or even the possible confounding effect from administration of levodopa or other PD medications).

Another challenge to interpreting CSF HVA concentration in PD is that it is not specific to this disorder. Other neurodegenerative conditions such as PSP, MSA, and corticobasal degeneration (CBD) also involve neuronal dropout from the SNpc and so CSF HVA concentrations have been reported to be diminished [12–14]. Among numerous studies reporting CSF HVA concentrations in PD with comparisons to healthy and age-matched controls (e.g., among many others, [15–17]), the values have ranged widely. In one study, CSF HVA concentration was compared to the findings of a radioisotope-measured dopamine reuptake in the striatum [18]. With these two approaches for gauging nigrostriatal dopaminergic innervation in the living PD patient, a strong correlation was found.

CSF monoamine metabolism in PD was the topic of a recent report that the highlighted the value of DOPAC as a differentiating factor between PD and pure autonomic failure (another disorder that shares with PD marked alpha-synuclein pathology in affected neurons) [9]. The authors make the point that HVA found in CSF, while more abundant than other dopamine metabolites, is subject to several intervening metabolic processes carried out in non-dopaminergic cells. In contrast, DOPAC offers the best index of dopamine synthesis capability in PD, as its formation in neurons by monoamine oxidase type-A reflects neuronal dopamine stores [19]. This study compared unmedicated PD subjects with age-matched healthy controls, finding that there was a significantly reduced CSF concentration of endogenous levodopa but not dopamine (which was measured close to its limits of detection). In these studies, the DOPAC concentration of 0.87 ± 0.09 nmol/l in PD CSF was significantly below the control value of 2.15 ± 0.18 nmol/l. For another monoamine neurotransmitter of importance in PD, norepinephrine, its concentrations and that of its major metabolite dihydroxyphenylglycol measured showed a story similar to that of dopamine. Only the metabolite was decreased in PD as compared to controls (8.82 ± 0.44 versus 11.02 ± 0.62 nmol/l). This study and the subsequent study by this research group [9] also provides a wealth of information with respect to clinical correlations between CSF and plasma dopamine metabolites, and comparisons of PD findings to those in other neurodegenerative disorders with synucleinopathy.

Although measuring the concentration of CSF HVA was, by itself, uninformative as an indicator of PD, the possibility of using HVA as an index for the measurement of other CSF substances presented another opportunity for investigation of a biomarker. Animal studies had shown that pharmacological manipulations of striatal dopamine metabolism led to alterations purine metabolism [20] and in PD patients, CSF concentration of uric acid levels are inversely correlated to the rate of PD progression [21]. On this basis, CSF concentrations of the purine metabolite xanthine and HVA were evaluated as a ratio, since their concentrations had a correlation coefficient of approximately 0.7 for both PD and controls ($p < 0.001$). The [HVA]/[xanthine] ratio revealed that, in contrast to measurements of HVA or xanthine alone, their quotient provided strong differentiation between PD and control specimens. In addition, this ratio increased for those PD specimens collected up to 24 months later [22].

Another monoamine metabolite of importance in PD is serotonin. Measurements of its major metabolite 5-hydroxyindoleacetic acid (5-HIAA) in PD have been reported in some studies to be decreased as compared to controls [10, 23]. One investigation involving 16 controls and 16 PD subjects did not find a statistically significant difference although the mean value was decreased by 15% [24]. The latter study also measured total serotonin and found it to be decreased by 17% ($p < 0.01$). Since depression is associated with decreased CSF serotonin and its metabolites, this can be a confounding factor if PD patients in a study also have concomitant depression. An alternative interpretation might be that the risk for depression is greater in PD, possibly because of the loss or (down-regulation) of serotonin synthesis as compared to non-Parkinsonian individuals.

Pertinent to measurements of dopamine and serotonin neurotransmitter biomarkers are data pertaining to CSF tetrahydrobiopterin (THB). This compound is the obligatory cofactor involved with both tyrosine and tryptophan hydroxylases and so exerts an important rate-limiting influence on their synthesis. THB concentrations in unmedicated PD subjects are decreased by one-third as compared to age matched controls [25]. Other studies have also confirmed a substantial reduction of this cofactor in PD [10].

2.2 *Markers of Oxidative Stress*

One of the leading hypotheses for the pathogenesis of PD is that it may be the consequence of inadequate defense against reactive oxidative species, especially for dopaminergic neurons in the SNpc projecting to the striatum. Various mechanisms have been linked to oxidative stress, including increased tissue content of iron (a catalyst for hydroxyl radical formation), glycation end-products, decreased generation of glutathione, and free radical generation by the auto-oxidation or catabolism of the dopamine molecule. Several studies have sought biomarker evidence for oxidative stress in PD. In one study of CSF specimens, formation of 8-hydroxyguanosine (a stable oxidation product formed in response to RNA free radical exposure) was measured. Although serum specimens showed no difference between PD and control subjects, the PD subjects showed an approximately 3-fold increase in this compound's CSF concentrations [26]. Other CSF investigations have also targeted indicators of oxidative stress. Lipid peroxidation products in 22 untreated PD patients were reported to be approximately 40% greater ($p = 0.043$) than for 41 control specimens [27]. In addition, the sum of nitrite and nitrate concentrations (as indicators of nitric oxide generation) were approximately double ($p = 0.012$) in the PD specimens. Another clue for oxidative stress has been the measurement of glutathione, which is a major intracellular anti-oxidant and capable of recycling. The available studies pertaining to glutathione have shown a diversity of results. The first report on the topic investigated 84 PD subjects and 21 age-matched controls, finding no significant difference in total glutathione concentration [28]. A recent study have found a 40% decrease in the concentration of the oxidized form of glutathione ($p < 0.005$) for post-mortem CSF specimens in PD versus controls [29], while in another report using specimens from living subjects, reduced (but not oxidized) glutathione concentration was significantly decreased [30].

Concentration of the antioxidant alpha-tocopherol was measured in a study of 11 healthy controls and 11 medicated PD patients; this value was decreased by about half ($p < 0.005$) [31]. This study also investigated CSF lipoprotein oxidation and found it to be increased for the PD CSF specimens. Another clue as to oxidative stress has come from studies of CSF 5-cysteinyldopamine [32]. This compound, a product of dopamine auto-oxidation, can be found in PD brain as well as in CSF. Although its concentration was not significantly increased in unmedicated PD patients versus controls, the ratio of the 5-S-Cys-DA to HVA was significantly increased. In this study, HVA concentration was also diminished, suggesting a relationship between dopamine depletion and a biomarker of oxidative stress.

2.3 *Amino Acids*

The wide variety of metabolic, neuro-signaling, and structural roles for amino acids has raised interest to this class of biochemicals for possible biomarker properties. Amino acids in PD CSF have been the topic of investigation going back to the 1970s. Not surprisingly, results have varied greatly among the various publications that have appeared on the topic. The most prominent change in CSF amino acids has been a decrease in CSF concentration of γ-aminobutyric acid (GABA) in CSF from unmedicated PD patients [33–36]. These findings are of particular interest in that dopaminergic pathways and the SN in PD exhibit decreased activity of glutamic acid decarboxylase, the rate-limiting enzyme of GABA synthesis. A diversity of findings has come from screenings of CSF amino acids in PD. Other than decreased GABA concentration, one study found no other amino acid changes [37].

One study, which examined cisternal CSF from ten unmedicated Parkinsonian subjects and ten controls, found no difference in concentrations for most of the amino acids. The exception was an approximately two-thirds decrease in glutamate concentration in the PD specimens. Smaller decreases were found for four others: lysine, alanine, isoleucine, and glutamine [38]. The latter amino acid is of particular interest since glutamine and glutamate can be interconverted. Interpreting amino acid concentration in CSF is subject to a number of influences besides disease states [39] and so the results of this and other studies are potentially artefactual (or at least not directly representative of what is actually going on in brain tissue).

2.4 Metal Ions

The chemical milieu has been searched for clues in many of the metallic elements that can be found in CSF. In an older study, median CSF copper concentration in PD was 2.5 times the control value ($p < 0.001$) [40]. Free copper ions are highly reactive in catalyzing the formation of free radicals, and so unbound copper in the fluid environment poses a threat to lipid membranes, nucleic acids, and other cellular structures. However, among other studies, results have differed widely. For example, one study that investigated 20 PD patients and 15 patients with "unspecified neurological diseases" found that copper was increased by 84% while zinc elevation was almost three times control values (both $p < 0.01$), and manganese was increased by 74% ($p < 0.05$) [41]). In this study, iron and magnesium did not differ in PD from controls. Another report involved specimens from 42 PD patients and 20 age-matched controls who were living in the same region [42]. For most of the 26 metal ions measured, the results indicated substantial overlap between PD and control specimens. However, in the PD subjects, the mean values for six of the elemental metals were much lower. Among these were chromium, iron lead, silicon, cobalt, and strontium. The other metals (including copper) were not altered in concentration.

Somewhat different results were found in a study of 18 controls and 91 PD subjects investigated by the same research group as the previous study: in this report, the differentiating findings were lower concentrations of iron and silicon in PD, but not the other metals mentioned above [43]. The disparity between these studies suggests the possibility of methodological, dietary, or other factors unrelated to disease state, and hypotheses based on mechanisms linked to these metals need to be tempered as to the uncertainties associated with this data.

2.5 Miscellaneous Biochemicals

Among other CSF biomarker findings in PD, significantly increased concentration of the long-chain omega-3 fatty acid, eicosapentaenoic acid has been described as increased as compared to controls [44]. The same research group also reported that the pattern of polyamine compounds also was altered in PD specimens versus controls. Although the measurements were subject to considerable variability between subjects, the group of seven polyamines in total was increased by 69% ($p < 0.001$) for the 9 PD subjects versus 24 controls [45]. In addition, several of the polyamines individually differed greatly from the control values. A strong differentiation of PD from controls was observed in data in a scatter plot of the ratio between putrescine and spermidine ($p < 0.001$).

Another CSF constituent of interest has been *N*-methyl(*R*)salsalinol. This compound is derived from a minor metabolic pathway acting on dopamine and is toxic to dopaminergic neurons. For 16 unmedicated PD subjects as compared to 29 controls, the mean concentration value was almost double ($p < 0.0001$) [46]. A subsequent followup of nine of the PD subjects at close to 2 years later found that the concentration had decreased somewhat but still was greater than control values [47].

A study comparing CSF β-phenylethylamine concentration found lower values in 23 PD subjects than in 12 controls [48]. Since the PD subjects were medicated and the controls had some manner of neurological disease, the significance of this finding for biomarker purposes is uncertain.

2.6 Metabolomic Analysis of CSF

Assay methods for characterizing small-molecule biochemicals now permit hundreds of compounds to be measured simultaneously by liquid chromatography and mass spectrometry techniques. These advances have provided the opportunity for untargeted analysis of biospecimens, much like the proteonomic techniques (to be described below) for profiling of proteins. With these *metabolomic* techniques, biomarkers can be characterized from large datasets, focusing on either individual compounds or from groupings of them. The various 'omics' approaches have broadened the concept of a biomarker to include patterns of change in a metabolic sequence or in other ways that compounds can be related [49].

To date, the only metabolomic analysis of PD CSF was carried out with post mortem specimens removed from the lateral ventricles at autopsy. In this study of CSF from 48 histologically-confirmed PD and 57 control brains, the measurement of small-molecule constituents was carried out with ultra-high liquid chromatography linked to mass spectrometry [29]. More than 250 compounds were identified in this manner. Excluded from the subsequent analysis were dietary compounds and dopamine metabolites (since, in this study, the PD subjects had received levodopa before death). Of 192 compounds analyzed for those whose concentrations differentiated between PD and control specimens at $p < 0.02$, three of the compounds had increased PD:control ratios (1.15–1.57-fold) and 16 had decreased ratios (0.52–0.87-fold) after a multiple-comparison adjustment at a 20% false-discovery rate level. One of the few compounds previously found to differentiate PD from controls was glutathione [30], which in this study was also diminished (see above). Most of the other compounds with statistically significant alterations in their PD:control ratios were not known to be linked to neurodegenerative processes (or in some instances, even had known roles in CNS metabolism). These findings underscore the value of untargeted discovery methods for discovery of PD biomarkers.

3 CSF Proteins as Biomarkers of Parkinson's Disease

4 Background

The protein composition present in cells and tissues is a snapshot of the functional state of a biological system. Protein expression levels are tightly regulated as post-translational modifications (PTMs) that modulate protein function [50]. As CSF is contiguous with the extracellular fluid that bathes the cells of the central nervous system, many proteins that are synthesized in the brain can be found in CSF either as full length proteins or as peptide fragments. The composition of proteins in CSF changes as a consequence of alterations in brain function, physiologically or pathologically, or changes related to crossover between CSF and blood. One important caveat is that CSF proteins also undergo constant change due to replacement of CSF, which occurs several times each day [51]. In other words, the concentration of a CSF should be thought of as a "snapshot" of the protein at the given time but subject to variability if measured at a different time of day. Nonetheless, protein biomarker discovery for PD has been extremely productive and has advanced in two major directions, (1) targeted searches based on genetic mutation leading to familial PD or important disease processes, and (2) unbiased profiling. In the next sections, advances in both fields are summarized.

4.1 Proteins Altered in Parkinson's Disease CSF That Are Related to Genetic Searches

A number of proteins are known to be involved in familial cases of PD. These include α-synuclein, LRRK2, parkin, DJ-1 and PINK1. However, fewer than 10% of PD cases can be attributed to these genes, leaving the cause of the vast majority of PD cases unknown. Nonetheless, most proteins coded by these genes, e.g. α-synuclein and DJ-1 that have been studied extensively as biomarkers in CSF, are also features found in idiopathic PD. Genetic mutation of SNCA (the gene coding α-synuclein) results in autosomal dominant familial PD [52], and α-synuclein is also the principal component of Lewy bodies, the hallmark pathological finding in idiopathic PD (iPD) [53, 54]. Although physiological roles of α-synuclein remains to be defined, dysregulation of the protein is clearly important to many critical cellular processes, including mitochondrial dysfunction, increased oxidative stress and overt neuroinflammation, all of which are thought to be key components of the disease process in iPD [55–59]. DJ-1, whose mutation leads to autosomal-recessive familial PD, behaves as an antioxidant and may have other interactions with the control of oxidative stress in PD [60, 61]. Both DJ-1 and α-synuclein interact with several of the same proteins [62].

4.2 α-Synuclein as a Biomarker of Parkinson's Disease

α-Synuclein has been the major focus of study in PD, both as a biomarker for diagnosis and as an objective biomarker of disease progression or endpoint of a potential targeted therapy. Several groups have examined α-synuclein concentration in CSF using ELISA and Luminex assays. The common finding amongst the groups with large cohorts of patients has been that α-synuclein levels are reduced in patients with PD and that α-synuclein levels offer reasonable sensitivity and specificity in distinguishing PD from controls [63, 64]. Unfortunately, α-synuclein levels alone are not a good predictor of either severity or duration of the disease in most of these studies. It also fails to provide sufficient differential diagnostic sensitivity and specificity from other overlapping forms of Parkinsonism. It should be noted that some of the smaller-scale studies of CSF α-synuclein have either confirmed the results of the larger studies [64–66] or failed to find a significant difference between PD and control CSF samples [67, 68]. The discrepancies between groups are most likely attributable to differences in sampling protocols, control for blood contamination and differences in detection/capture antibodies [69]. Recently, oligomeric α-synuclein has been shown to be increased in PD CSF; when this measurement is used in combination with total α-synuclein, sensitivity and specificity are improved substantially [70]. More recently, Wang et al. identified a phosphorylated α-synuclein (ps129) in human CSF. This work revealed that ps129, alone or in combination with total α-synuclein, not only improved diagnostic accuracy, but also showed a correlation with PD severity [71]. These research claims are preliminary results and require further validation by independent investigations.

4.3 DJ-1 as a CSF Biomarker of Parkinson's Disease

An abundant cellular protein, DJ-1 is secreted into the CSF and plasma both in cases of PD and in individuals without the disease. The initial investigation of DJ-1 CSF concentration showed no significant differences between PD and control groups, nor was there any correlation with age [72]. Waragai et al. demonstrated that DJ-1 levels are increased in the CSF of PD patients as compared to control patients [73]. However, a more recent study with a relatively large cohort of subjects contradicted this result and reached a different conclusion; using a Luminex assay, these investigators found that DJ-1 levels were significantly decreased in PD patient CSF. Another principal difference in this research project (in addition to a larger study cohort) was that the investigators rigorously controlled for blood contamination in the CSF specimens [63]. The discrepancy of assay findings amongst different investigators, as is the case for α-synuclein, can be attributed to many issues. However, for DJ-1 measurements, blood contamination from hemolysis is likely the most significant factor responsible for the disparate results. Future studies of DJ-1 may focus on oxidized forms of DJ-1, which may show greater sensitivity to the presence of oxidative stress [74–76].

4.4 Alterations of Tau and $A\beta_{1-42}$ in Parkinson's Disease CSF

Another targeted search has looked for tau and related proteins. This research direction was greatly facilitated on the basis of recent discovery by several genome-wide association studies (GWAS; [77–81]) that MAPT (a gene coding for tau) is the only other gene (besides SNCA) consistently associated with sporadic PD. Tau, along with amyloid beta (Aβ), has been previously considered to be an Alzheimer's disease (AD) marker, and for this reason, measurements of both are typically carried out [82]. In CSF from patients with AD, the concentrations of both t-tau and p-tau (a phosphorylated form) are significantly higher than in control CSF, while $A\beta_{1-42}$ is decreased in AD specimens relative to controls [82]. Interestingly, decreased levels of t-tau and p-tau are observed in PD as well as in multisystem atrophy (MSA), a disorder that is also an α-synucleinopathy. Both MSA and PD manifest the aggregation of α-synuclein in the form of Lewy bodies. The primary difference between these disorders is the distribution of involved brain structures and the presence of glial cytoplasmic inclusions, which are only seen in MSA. These diseases are particularly difficult to differentiate on clinical grounds, particularly in their early stages and even later because of overlap in their signs and symptoms [83, 84]. The CSF findings demonstrate that decreased levels of tau are a shared feature of α-synucleinopathies that can be used to differentiate them from AD; however, measurement of tau concentration is not, by itself, a useful way for differentiating among the various α-synucleinopathies. Alterations in CSF Aβ in PD are similar to AD, especially in those patients presenting cognitive impairment [85]. Remarkably, tau, p-tau and Aβ have been shown recently to be decreased even in patients with the LRRK2 mutation but without PD symptomatology (LRRK2 mutation being another cause of autosomal dominant familial PD) cases. From these findings, it seems that alterations in the pathways of these proteins might be pathologically important during early stages of PD [86]. From a biomarker point of view, however, the changes are probably too subtle to be diagnostically useful. Consequently, tau and Aβ levels are likely used in the context of disease profiling using a broader panel of protein biomarkers.

4.5 CSF Inflammatory Biomarkers Associated with Parkinson's Disease

The immune system plays an essential role in PD pathogenesis, leading some researchers to look for inflammatory markers that might be specifically associated with PD. Several studies have found an increase of IL-1β, IL-2, IL-4, IL-6 and TNF-α in CSF of PD patients. However, although none of these inflammatory protein changes has been demonstrated sufficient sensitivity or specificity for practical application as diagnostic biomarkers [87–90]. More recently, in an investigation involving screening of 42 cytokines and chemokines, 33 were detected in CSF (unpublished results from Shi et al.). Two of these chemokines, Flt-3 ligand and fractalkine, were found to be diagnostically useful. Flt-3 ligand is a cytokine that has both neurotrophic and anti-apoptotic properties. In both neurons and glia, it can induce proliferation, differentiation and promote survival. Flt-3 ligand was found to be decreased in the CSF of patients with MSA [16]. Interestingly, concentration of Flt-3 ligand was not significantly altered in PD, AD, or healthy controls. Quite remarkably, Flt-3 ligand was shown to be an excellent marker for differentiating PD from MSA, with 99% sensitivity and 95% specificity [66].

Another important area in biomarker discovery is predicting or correlating severity and progression of PD, which has limited success. In addition to ps129 [71] discussed above, another marker that has demonstrated utility in predicting severity and progression is fractalkine (CX3CL1). Fractalkine is a cytokine released from neurons modulating microglial activation. In animal models of Parkinsonism, fractalkine has been shown to be neuroprotective [91]. The ratio of CSF fractalkine to $A\beta_{1-42}$ was shown to be significantly increased both with PD severity (in a cross-sectional cohort) and progression (in a longitudinal cohort), a correlation that was independent of hemoglobin content derived from hemolyzed blood [66]. This result also demonstrates the utility of multianalyte panels in differentiating very subtle biochemical differences in disease states and severity. To this end, it should be stressed that it is the independent markers, not the proteins closely linked together (e.g. a-synuclein and DJ-1), that could enhance the performance of one another when analyzed in a grouped panel of multiple markers.

4.6 CSF Proteomic Analysis in Parkinson's Disease

Proteins involved in PD pathogenesis are likely to be more than just α-synuclein, DJ-1 and a few inflammatory biomarkers that can be detected in CSF. Additional evidence has accumulated that CSF has a greater complexity to it than previously suspected. CSF is not only a carrier of nutrients and wastes but is also involved in cellular signaling via volume transmission or synaptic exocytosis [92, 93]. Consequently, structures and mechanisms involved in the transport, protection from degradation, targeting, and transduction of the signal are in place, and all of which could influence proteins unique to PD (whether or not such proteins have important roles in the pathophysiology of PD or just bystanders to the disease process).

Mass spectrometry (MS)-based proteomics is a promising method for unbiased discovery of a broad spectrum of proteins related to PD pathological process. The output from this technique can be adapted to other analytical methods such as Luminex or ELISA. One of the principal advantages of proteomics over transcriptomics (the quantification of mRNA transcripts) is its capacity to identify and quantify PTMs that may be associated with a specific disease process, possibly helping to predict its severity or progression.

4.7 Methodologies and Principles of Proteomics

The technologies involved in proteomic studies generally involve several components integrated together to address the specific challenges presented by particular tissue or fluid of interest. Large scale non-biased approaches typically involve separation [94], MS [95], and data processing steps. One of the principal challenges in proteomics is sufficiently separating protein/peptide mixtures to identify low abundance species that would be masked in unfractionated samples. Separation technologies applied to proteomics include liquid chromatography, electrophoresis, solid phase extraction and immunoaffinity methods. These separation strategies are often performed sequentially to achieve multidimensional separation based on different bimolecular properties. With this methodology, the analytical dynamic range and detection sensitivity can be enhanced by reducing the signal from other proteins present in mixed samples.

Once the samples are fractionated, the proteins/peptides within the fractions need to be identified. This is most often accomplished with mass spectrometers, which can utilize a variety of strategies for identification and quantification. High-resolution instruments include Fourier transform–ion cyclotron resonance, Orbitrap, quadrupole time-of-flight (TOF) and TOF/TOF [96–98]. All of these different methods add specific benefits, but also can present tradeoffs in the form of added technical complexity or longer analysis time. All of these instruments require ionization of sample proteins with either electrospray ionization or matrix-assisted laser desorption ionization (MALDI) ion source [99, 100]. Once a peptide is ionized, additional steps have been developed that allow more sophisticated analysis of the sample. These methods include collision-induced dissociation for ion fragmentation and soft collision techniques, such as electron transfer dissociation [101], which can aid in the identification of PTMs, such as phosphorylation and glycosylation. Once the data is gathered it must be interpreted through database development, search algorithms for peptide identification, PTM identification and quantitative analysis. These functions are generally combined into a data analysis pipeline that involves data conversion, database search and peptide/protein identification.

The most common approach to quantitative proteomics involves two-dimensional gel electrophoresis that quantifies proteins before protein identification. Another method involves labeling proteins from different sources with labels that are chemically identical but contain different isotopes. Since the labeled peptides do not differ in their chemical properties, they ionize at the same rates, but subsequently can be identified based on changes in their mass. These strategies allow both relative quantification between different specimens as well as absolute quantification when a labeled standard of known concentration is employed. Labeling methods include isobaric tag for relative abundance and quantification (iTRAQ), isotope-coded affinity tag (ICAT), stable isotope labeling with amino acids in cell culture, and ^{18}O labeling [102–106]. There are also label-free methods involving spectra count, which have been utilized for quantitative study [107–110]. Another recently-introduced technique that has gained popularity is candidate-based quantitative proteomics. This method, often called MRM, employs peptide standards to quantify known proteins of interest [111].

4.8 Proteomics of CSF Proteins

Protein profiling of CSF has revealed the presence of more than 3,000 proteins. Among these are many proteins involved in structural or functional roles in the CNS as well as many others that mediate immune processes [112, 113]. Subcellular structures derived from the constituents of synaptic vesicle and exosomes have been described by Harrington et al. [114]. These spherical structures range in size from 30 to 200 nm and carry a wide assortment of proteins including polyunsaturated fatty acids (PUFAs), synaptobrevin, synaptotagmin, and syntaxin. The presence of these synaptic vesicles suggests that signaling transduction is enacted by the contents of CSF. Exosomes also appear to be involved in the endosomal pathway and removal of potentially detrimental proteins by fusion with the plasma membrane and release of contents into the extracellular space [115, 116]. Exosomes also can be released intact into the extracellular space where they might be able to interact with neighboring cells. Studies of neurodegenerative conditions such as AD and prion disease, have demonstrated that potentially damaging proteins are sequestered into exosomes and transported to neighboring cells [117, 118]. In PD, this process hypothetically allows α-synuclein to enter into the extracellular space, where it can then be taken up by other neurons and damage them [119, 120]. This process is not fully understood at present; however, in a recent report, Rab11a, an endosomal protein was shown to be involve in the secretion of α-synuclein [120].

The identification of biomarkers by proteomic approaches is still at relatively early stage. A recent MS study using specimens of human CSF from AD, PD, and DLB (dementia with Lewy bodies) patients and healthy controls detected more than 1,500 proteins. Among the proteins identified, 136, 72, and 101 proteins, respectively, appeared to be associated with AD, PD, and DLB [97]. Some of the proteins were found to be associated uniquely with PD; these included apolipoprotein H/β-2-glycoprotein 1 (Apo-H), ceruloplasmin, and chromogranin B (secretogranin I). Other proteins were uniquely associated with DLB including ApoC1 and T-cadherin. In the above study, eight of the markers (BDNF, IL-8, VDBP, β2-microglobulin, ApoAII, ApoE, plus tau and $A\beta_{1\text{-}42}$) were evaluated in a subsequent Luminex assay and were able to classify PD patients from controls with 95% sensitivity and 95% specificity [121]. Independent validation of these results will be required, however.

Despite all of the recent advances and refinements of proteomics, the routine analysis of clinical specimens presents many challenges. The complexity of analysis for CSF samples and the nonlinear dynamic range in protein abundance requires careful attention to methodology in sample preparation and sophisticated data analysis to account for the many factors that can cause variability among samples, such as caudo-rostral gradients of CSF contents. Many of the current technical limitations of proteomics may prove to be transient as the field progresses to solve current methodological challenges. For example, new technologies are being developed enhance the dynamic range of assays [122] by further reducing the complexity of protein mixtures. Some of the upcoming methods offer ways to select a panel of markers from a large candidate pool of markers interrogated by discovery proteomics and which generally do not have specific commercially-available antibodies [101, 123]. At some point, refinement of proteomic methods may lead to analytical platforms that can be employed in routine clinical settings.

4.9 *Concluding Remarks on CSF Proteins in Parkinson's Disease*

In recent research studies, targeted and proteomic analysis of CSF have provided a number of biomarkers that allow excellent differentiation of PD from other neurodegenerative diseases. Among these results are biomarkers that correlate with severity and duration of PD. Despite these achievements, there are still major hurdles ahead to refine and improve current sample purification, separation, MS and data processing method. Moving forward, several important issues need to be addressed to facilitate PD biomarker discovery and validation. These include: (1) standardization of protocols for sample collection and storage, (2) repeated (longitudinal) collection of samples with thorough clinical databases to guide interpretations of CSF results, and (3) application of newer proteomic technologies to well-characterized disease and control cohorts. With collaborative effort of both academic and industry, the day may come where healthy individuals will have the opportunity to be tested for PD and other neurodegenerative diseases before any clinical signs or symptoms are manifested. Such testing capabilities are critical needs if neuroprotection therapies are developed that would have to be started at the earliest stages of neurodegeneration.

5 Conclusions

The PD world awaits one or more biomarkers that could serve in clinical and research applications. Reliable state and trait markers would help to advance the search for better therapeutics and, possibly, the underlying cause(s) of PD. The research findings to date, as described above, have offered a promising start in this direction. Though the CSF biomarker search has taken many directions, it still is uncertain whether proteins or smaller biochemicals will yield the best indicators of the disease [124]. Perhaps the compilation of multiple measurements into a composite biomarker will lead to the most successful outcome. It is also possible that biomarkers need to be stratified based on clinical phenotypes, e.g. tremor dominant PD.

The methodological challenges facing biomarker research still need refinement. Many of the reports reviewed above and others in the medical literature represent suboptimal science in one or more ways. First, the diagnosis of PD and normal controls needs to meet a high standard of certainty, and this information should be clearly provided in reports. Clinical criteria that have achieved wide acceptance for the diagnosis of PD include the United Kingdom PD Society Brain Bank criteria [125], but even with these, researchers should recognize that a certain diagnosis of PD cannot be made on clinical grounds alone. In some of the studies reported, drugs with CNS effects, including dopaminergic drugs were in use by the PD subjects. The pharmacological effects of such drugs have been shown to confound interpretation of dopamine metabolites and other CSF constituents, and so definitive studies should be carried out only in drug-free subjects. The effect of a rostrocaudal gradient, mentioned earlier in the context of HVA studies, may be relevant to all substances measured in CSF (either as concentrations of individual substances or even as ratios between concentration measurements). Age, sex, body size, specimen storage techniques, presence of preservatives or antioxidants, and even time of day for CSF collection are all factors that might influence findings in CSF. One of the most common flaws of many of the published studies is sample sets of insufficient size for firm conclusions to be drawn. Reports in which the measurement data shows relatively large standard error, or sample sets of fewer than 15 per group can be subject to the bias of small numbers. The tremendous variability of CSF measurement data is exemplified by the diversity of findings reported for CSF HVA and amino acid concentrations. Finally, longitudinal cohorts followed for years will also be an important direction for future biomarker studies to take.

Acknowledgements Support was provided by the Michael J. Fox Foundation for Parkinson's Research (to P.A.L. and J.Z.), and by the National Institute of Neurological and Communicative Disorders, National Institutes of Health (RO1-NS27892, to P.A.L), and RO1-AG033398, P42-ES004696 (sub-award 5897), P30-ES007033 (sub-award 6364), RO1-ES016873, R01-ES019277, 5-R01-NS057567, and P50-NS062684 (sub-award 6221 to J.Z). Office of Academic Affiliations, Advanced Fellowship Program in Mental Illness Research and Treatment, Department of Veterans Affairs (B.R.H.).

References

1. Graeber MB (2009) Biomarkers for Parkinson's disease. Exp Neurol 216:249–253
2. Antoniades CA, Barker RA (2008) The search for biomarkers in Parkinson's disease: a critical review. Expert Rev Neurother 8:1841–1852
3. Sherer TB (2011) Biomarkers for Parkinson's disease. Sci Transl Med 3:79ps14
4. Chahine LM, Stern MB (2011) Diagnostic markers for Parkinson's disease. Curr Opin Neurol 24:309–317
5. Gerlach M, Maetzler W, Broich K, Hampel H, Rems L, Reum T, Riederer P, Stöffler A, Streffer J, Berg D (2012) Biomarker candidates of neurodegeneration in Parkinson's disease for the evaluation of disease-modifying therapeutics. J Neural Transm 119:39–52
6. Eckstein JW (2010) Biomarkers in Parkinson's disease: a venture capitalists's perspective. Biomark Med 4:731–735
7. LeWitt PA, Galloway MP (1990) Neurochemical markers of Parkinson's disease. In: Koller WC, Paulson G (eds) Therapy of Parkinson's disease. Marcel Dekker, New York, pp 63–93
8. Hornykiewicz O (1998) Biochemical aspects of Parkinson's disease. Neurology 51(suppl 2): S2–S9
9. Goldstein DS, Holmes C, Sharabi Y (2012) Cerebrospinal fluid biomarkers of central catecholamine deficiency in Parkinson's disease and other synucleinopathies. Brain 135: 1900–1913
10. LeWitt PA, Galloway MP, Matson W, Milbury P, McDermott M, Srivastava DK, Oakes D (1992) Markers of dopamine metabolism in Parkinson's disease. Neurology 42:2111–2117
11. Parkinson Study Group (1995) CSF homovanillic acid in the DATATOP study on Parkinson's disease. Arch Neurol 52:237–245
12. Ruberg M, Javoy-Agid F, Hirsch E, Scatton B, LHeureux R, Hauw JJ, Duyckaerts C, Gray F, Morel-Maroger A, Rascol A, Serdaru M, Agid Y (1985) Dopaminergic and cholinergic lesions in progressive supranuclear palsy. Ann Neurol 18:523–529
13. Abdo WF, de Jong D, Hendriks JC, Horstink MW, Kremer BP, Bloem BR, Verbeek MM (2004) Cerebrospinal fluid analysis differentiates multiple system atrophy from Parkinson's disease. Mov Disord 19:571–579
14. Kanemaru K, Mitani K, Yamanouchi H (1998) Cerebrospinal fluid homovanillic acid levels are not reduced in early corticobasal degeneration. Neurosci Lett 245:121–122
15. Hildebrand J, Bourgeois F, Buyse M, Przedborski S, Goldman S (1990) Reproducibility of monoamine metabolite measurements in human cerebrospinal fluid. Acta Neurol Scand 81:427–430
16. Toghi H, Abe T, Takahashi S, Ueno M, Nozaki Y (1990) Cerebrospinal fluid dopamine, norepinephrine, and epinephrine concentrations correlated with clinical symptoms. Adv Neurol 53:277–282
17. Poseta JS, Parsons L, Engelland S, Kripke DF (2009) Circadian rhythm of CSF monoamines and hypocretin-1 in restless leg syndrome and Parkinson's disease. Sleep Med 10:129–133
18. Ishibashi K, Kanemaru K, Saito Y, Murayama S, Oda K, Ishiwata K, Mizusawa H, Ishii K (2010) Cerebropinal fluid metabolite and nigrostriatal dopaminergic function in Parkinson's disease. Acta Neurol Scand 122:46–51
19. Goldstein DS, Holmes C, Kopin IJ, Sharabi Y (2011) Intra-neuronal vesicular uptake of catecholamines is decreased in patients with Lewy body diseases. J Clin Invest 121: 3320–3330
20. Loeffler DA, LeWitt PA, Juneau PL, Camp DM, DeMaggio AJ, Milbury P, Matson WR, Rathbone MP (1998) Altered guanosine and guanine concentrations in rabbit striatum following increased dopamine turnover. Brain Res Bull 45:297–299
21. Ascherio A, LeWitt PA, Xu K, Eberly S, Watts A, Matson WR, Marras C, Kieburtz K, Rudolph A, Bogdanov MB, Schwid SR, Tennis M, Tanner CM, Beal MF, Lang AE, Oakes D, Fahn S, Shoulson I, Schwarzschild MA; Parkinson Study Group DATATOP Investigators (2009) Urate as a predictor of the rate of clinical decline in Parkinson disease. Arch Neurol 66:1460–1468

22. LeWitt P, Schultz L, Auinger P, Lu M; Parkinson Study Group DATATOP Investigators (2011) CSF xanthine, homovanillic acid, and their ratio as biomarkers of Parkinson's disease. Brain Res 1408:88–97
23. Mayeux R, Stern Y, Cote L, Williams JB (1984) Altered serotonin metabolism in depressed patients with Parkinson's disease. Neurology 34:642–646
24. Toghi H, Abe T, Takahashi S, Takahashi J, Hamato H (1993) Concentrations of serotonin and its related substances in the cerebrospinal fluid of Parkinsonian patients and their relations to the severity of symptoms. Neurosci Lett 150:71–74
25. LeWitt PA, Miller LP, Newman RP, Burns RS, Insel T, Levine RA, Lovenberg W, Calne DB (1984) Tyrosine hydroxylase cofactor (tetrahydrobiopterin) in Parkinsonism. Adv Neurol 40:459–462
26. Abe T, Isobe C, Murata T, Sato C, Toghi H (2003) Alteration of 8-hydroxyguanosine concentrations in the cerebrospinal fluid and serum from patients with Parkinson's disease. Neurosci Lett 336:105–108
27. Boll M-C, Alcaraz-Zubeldia M, Montes S, Rios C (2008) Free copper, ferroxidase and SOD1 activities, lipid peroxidation and NO_x content in the CSF, a different marker profile in four neurodegenerative diseases. Neurochem Res 33:1717–1723
28. Konings CH, Kuiper MA, Teerlink T, Mulder C, Scheltens P, Wolters EC (1999) Normal cerebrospinal fluid glutathione concentrations in Parkinson's disease, Alzheimer's disease, and multiple system atrophy. J Neurol Sci 168:112–115
29. LeWitt PA, Lu M, Poisson L, Beach T, Adler C, Guo L (2011) PD biomarker search: metabolomic analysis of post-mortem CSF. Mov Disord 26(suppl 2):S193
30. Maetzler W, Schmid SP, Wurster I, Liepelt I, Gaenslen A, Gasser T, Berg D (2011) Reduced but not oxidized cerebrospinal fluid glutathione levels are lowered in Lewy body diseases. Mov Disord 26:176–181
31. Buhmann C, Arlt S, Kontush A, Möller-Bertram T, Sperber S, Oechsner M, Stuerenburg HJ, Beisiegel U (2004) Plasma and CSF markers of oxidative stress are increased in Parkinson's disease and influenced by antiparkinsonian medication. Neurobiol Dis 15:160–170
32. Cheng F-C, Kuo J-S, Chia L-G, Dryhurst G (1996) Elevated 5-S-cysteinyldopamine/homovanillic acid ratio and reduced homovanillic acid in cerebrospinal fluid: possible markers for and potential insights into the pathophysiology of Parkinson's disease. J Neural Transm 103:433–446
33. Gjessing LR, Gjesdahl P, Dietrichson P, Presthuis J (1974) Free amino acids in the cerebrospinal fluid in old age and in Parkinson's disease. Eur Neurol 12:33–37
34. Araki K, Takino T, Ida S, Kuriyama K (1986) Alteration of amino acids in cerebrospinal fluid from patients with Parkinson's disease and spinocerebellar degeneration. Acta Neurol Scand 71:105–110
35. Manyam BV, Ferraro TN, Hare TA (1988) Cerebrospinal fluid amino compounds in Parkinson's disease: alterations due to carbidopa/levodopa. Arch Neurol 45:48–50
36. Molina JA, Jiménez-Jiménez FJ, Gomez P, Vargas C, Navarro JA, Ortí-Pareja M, Gasalla T, Benito-León J, Bermejo F, Arenas J (1997) Decreased cerebrospinal fluid levels of neutral and basic amino acids in patients with Parkinson's disease. J Neurol Sci 150:123–127
37. Hare TA, Beasley BL, Chambers RA, Boehme DH, Vogel WH (1973) GABA and amino acid levels in plasma and cerebrospinal fluid of patients with Parkinson's disease before and during treatment with L-DOPA. Clin Chim Acta 45:273–280
38. Mally J, Szalai G, Stone TW (1997) Changes in the concentration of amino acids in serum and cerebrospinal fluid of patients with Parkinson's disease. J Neurol Sci 151:159–162
39. Kornhuber J, Kornhuber ME, Hartmann GM, Kornhuber AW (1988) In vivo influences on CSF amino acids. Neurochem Int 12:25–31
40. Pall HS, Williams AC, Blake DR, Lunec J, Gutteridge JM, Hall M, Taylor A (1987) Raised cerebrospinal fluid copper concentration in Parkinson's disease. Lancet 2:238–241
41. Hozumi I, Hasegawa T, Honda A, Ozawa K, Hayashi Y, Hashimoto K, Yamada M, Koumura A, Sakurai T, Kimura A, Tanaka Y, Satoh M, Inuzuka T (2011) Patterns of levels of biological metals in CSF differ among neurodegenerative diseases. J Neurol Sci 2011(303):95–99

42. Alimonti A, Bocca B, Pino A, Ruggieri F, Forte G, Sancesario G (2007) Elemental profile of cerebrospinal fluid in patients with Parkinson's disease. J Trace Elem Med Biol 21:234–241
43. Bocca B, Alimonti A, Senofonte O, Pino A, Violante N, Petrucci F, Sancesario G, Forte G (2006) Metal changes in CSF and peripheral compartments of parkinsonian patients. J Neurol Sci 248:23–30
44. Lee PH, Lee G, Paik M-J (2008) Polyunsaturated fatty acid concentrations polyamine patterns in the cerebrospinal fluid of patients with Parkinson's disease and multiple system atrophy. Mov Disord 23:309–310
45. Paik M-J, Ahn Y-H, Lee PH, Kang H, Park CB, Choi S, Lee G (2010) Polyamine patterns in the cerebrospinal fluid of patients with Parkinson's disease and multiple system atrophy. Clin Chim Acta 411:1532–1535
46. Maruyama W, Abe T, Toghi H, Dostert P, Naoi M (1996) A dopaminergic neurotoxin, *N*-methyl(*R*)salsalinol, increases in parkinsonian CSF. Ann Neurol 40:119–122
47. Maruyama W, Abe T, Toghi H, Naoi M (1999) An endogenous MPTP-like dopaminergic neurotoxin, *N*-methyl(*R*)salsalinol, in the cerebrospinal fluid decreases with progression of Parkinson's disease. Neurosci Lett 262:13–16
48. Zhou G, Shoji H, Yamada S, Matsuishi T (1997) Decreased beta-phenylethylamine in CSF in Parkinson's disease. J Neurol Neurosurg Psychiatry 63:754–758
49. Caudle WM, Bammler TK, Lin Y, Pan S, Zhang J (2010) Using 'omics' to define pathogenesis and biomarkers of Parkinson's disease. Expert Rev Neurother 10:925–942
50. Cohen P (2000) The regulation of protein function by multisite phosphorylation – a 25 year update. Trends Biochem Sci 25:596–601
51. Silverberg GD, Mayo M, Saul T, Rubenstein E, McGuire D (2003) Alzheimer's disease, normal-pressure hydrocephalus, and senescent changes in CSF circulatory physiology: a hypothesis. Lancet Neurol 2:506–511
52. Polymeropoulos MH, Lavedan C, Leroy E, Ide SE, Dehejia A, Dutra A, Pike B, Root H, Rubenstein J, Boyer R, Stenroos ES, Chandrasekharappa S, Athanassiadou A, Papapetropoulos T, Johnson WG, Lazzarini AM, Duvoisin RC, Di Iorio G, Golbe LI, Nussbaum RL (1997) Mutation in the α-synuclein gene Identified in families with Parkinson's disease. Science 276:2045–2047
53. Braak H, Braak E, Yilmazer D, Schultz C, de Vos RA, Jansen EN (1995) Nigral and extranigral pathology in Parkinson's disease. J Neural Transm Suppl 46:15–31
54. Forno LS (1996) Neuropathology of Parkinson's disease. J Neuropathol Exp Neurol 55: 259–272
55. Schapira AH, Cooper JM, Dexter D, Clark JB, Jenner P, Marsden CD (1990) Mitochondrial complex I deficiency in Parkinson's disease. J Neurochem 54:823–827
56. Schapira AH, Cooper JM, Dexter D, Jenner P, Clark JB, Marsden CD (1989) Mitochondrial complex I deficiency in Parkinson's disease. Lancet 333:1269
57. Devi L, Raghavendran V, Prabhu BM, Avadhani NG, Anandatheerthavarada HK (2008) Mitochondrial import and accumulation of α-synuclein impair complex I in human dopaminergic neuronal cultures and Parkinson disease brain. J Biol Chem 283:9089–9100
58. Su X, Maguire-Zeiss KA, Giuliano R, Prifti L, Venkatesh K, Federoff HJ (2008) Synuclein activates microglia in a model of Parkinson's disease. Neurobiol Aging 29:1690–1701
59. Zhang W, Wang T, Pei Z, Miller D, Wu X, Block M, Wilson B, Zhang W, Zhou Y, Hong J, Zhang J (2005) Aggregated alpha-synuclein activates microglia: a process leading to disease progression in Parkinson's disease. FASEB J 19:533–542
60. Aleyasin H, Rousseaux MW, Marcogliese PC, Hewitt SJ, Irrcher I, Joselin AP, Parsanejad M, Kim RH, Rizzu P, Callaghan SM, Slack RS, Mak TW, Park DS (2010) DJ-1 protects the nigrostriatal axis from the neurotoxin MPTP by modulation of the AKT pathway. Proc Natl Acad Sci U S A 107:3186–3191
61. Duan X, Kelsen SG, Merali S (2008) Proteomic analysis of oxidative stress-responsive proteins in human pneumocytes: insight into the regulation of DJ-1 expression. J Proteome Res 7:4955–4961

62. Jin J, Li GJ, Davis J, Zhu D, Wang Y, Pan C, Zhang J (2007) Identification of novel proteins associated with both alpha-synuclein and DJ-1. Mol Cell Proteomics 6:845–859
63. Hong Z, Shi M, Chung KA, Quinn JF, Peskind ER, Galasko D, Jankovic J, Zabetian CP, Leverenz JB, Baird G, Montine TJ, Hancock AM, Hwang H, Pan C, Bradner J, Kang UJ, Jensen PH, Zhang J (2010) DJ-1 and alpha-synuclein in human cerebrospinal fluid as biomarkers of Parkinson's disease. Brain 133:713–726
64. Mollenhauer B, Locascio JJ, Schulz-Schaeffer W, Sixel-Döring F, Trenkwalder C, Schlossmacher MG (2011) α-Synuclein and tau concentrations in cerebrospinal fluid of patients presenting with parkinsonism: a cohort study. Lancet Neurol 10:230–240
65. Mollenhauer B, Cullen V, Kahn I, Krastins B, Outeiro TF, Pepivani I, Ng J, Schulz-Schaeffer W, Kretzschmar HA, McLean PJ, Trenkwalder C, Sarracino DA, Vonsattel JP, Locascio JJ, El-Agnaf OM, Schlossmacher MG (2008) Direct quantification of CSF alpha-synuclein by ELISA and first cross-sectional study in patients with neurodegeneration. Exp Neurol 213:315–325
66. Shi M, Bradner J, Hancock AM, Chung KA, Quinn JF, Peskind ER, Galasko D, Jankovic J, Zabetian CP, Kim HM, Leverenz JB, Montine TJ, Ginghina C, Kang UJ, Cain KC, Wang Y, Aasly J, Goldstein D, Zhang J (2011) Cerebrospinal fluid biomarkers for Parkinson disease diagnosis and progression. Ann Neurol 69:570–580
67. Ohrfelt A, Grognet P, Andreasen N, Wallin A, Vanmechelen E, Blennow K, Zetterberg H (2009) Cerebrospinal fluid alpha-synuclein in neurodegenerative disorders – a marker of synapse loss? Neurosci Lett 450:332–335
68. Borghi R, Marchese R, Negro A, Marinelli L, Forloni G, Zaccheo D, Abbruzzese G, Tabaton M (2000) Full length alpha-synuclein is present in cerebrospinal fluid from Parkinson's disease and normal subjects. Neurosci Lett 287:65–67
69. Mollenhauer B, El-Agnaf OM, Marcus K, Trenkwalder C, Schlossmacher MG (2010) Quantification of α-synuclein in cerebrospinal fluid as a biomarker candidate: review of the literature and considerations for future studies. Biomark Med 4:683–699
70. Park MJ, Cheon SM, Bae HR, Kim SH, Kim JW (2011) Elevated levels of α-synuclein oligomer in the cerebrospinal fluid of drug-naive patients with Parkinson's disease. J Clin Neurol 7:215–222
71. Wang Y, Shi M, Chung KA, Zabetian CP, Leverenz JB, Berg D, Srulijes K, Trojanowski JQ, Lee VM-Y, Siderowf AD, Hurtig H, Litvan I, Schiess MC, Peskind ER, Masuda M, Hasegawa M, Lin X, Pan C, Galasko D, Goldstein DS, Jensen PH, Yang H, Cain KC, Zhang J (2012) Phosphorylated α-synuclein in Parkinson's disease. Sci Transl Med 4(121):121ra120
72. Maita C, Tsuji S, Yabe I, Hamada S, Ogata A, Maita H, Iguchi-Ariga SM, Sasaki H, Ariga H (2008) Secretion of DJ-1 into the serum of patients with Parkinson's disease. Neurosci Lett 431:86–89
73. Waragai M, Wei J, Fujita M, Nakai M, Ho GJ, Masliah E, Akatsu H, Yamada T, Hashimoto M (2006) Increased level of DJ-1 in the cerebrospinal fluids of sporadic Parkinson's disease. Biochem Biophys Res Commun 345:967–972
74. Inden M, Kitamura Y, Takahashi K, Takata K, Ito N, Niwa R, Funayama R, Nishimura K, Taniguchi T, Honda T, Taira T, Ariga H (2011) Protection against dopaminergic neurodegeneration in Parkinson's disease-model animals by a modulator of the oxidized form of DJ-1, a wild-type of familial Parkinson's disease-linked PARK7. J Pharmacol Sci 2011(117):189–203
75. Saito Y, Hamakubo T, Yoshida Y, Ogawa Y, Hara Y, Fujimura H, Imai Y, Iwanari H, Mochizuki Y, Shichiri M, Nishio K, Kinumi T, Noguchi N, Kodama T, Niki E (2009) Preparation and application of monoclonal antibodies against oxidized DJ-1. Significant elevation of oxidized DJ-1 in erythrocytes of early-stage Parkinson disease patients. Neurosci Lett 465:1–5
76. Ren H, Fu K, Wang D, Mu C, Wang G (2011) Oxidized DJ-1 interacts with the mitochondrial protein BCL-XL. J Biol Chem 286:35308–35317

77. Edwards TL, Scott WK, Almonte C, Burt A, Powell EH, Beecham GW, Wang L, Züchner S, Konidari I, Wang G, Singer C, Nahab F, Scott B, Stajich JM, Pericak-Vance M, Haines J, Vance JM, Martin ER (2010) Genome-wide association study confirms SNPs in SNCA and the MAPT region as common risk factors for Parkinson disease. Ann Hum Genet 74:97–109
78. Hamza TH, Zabetian CP, Tenesa A, Laederach A, Montimurro J, Yearout D, Kay DM, Doheny KF, Paschall J, Pugh E, Kusel VI, Collura R, Roberts J, Griffith A, Samii A, Scott WK, Nutt J, Factor SA, Payami H (2010) Common genetic variation in the HLA region is associated with late-onset sporadic Parkinson's disease. Nat Genet 42:781–785
79. Pankratz N, Wilk JB, Latourelle JC, DeStefano AL, Halter C, Pugh EW, Doheny KF, Gusella JF, Nichols WC, Foroud T, Myers RH; PSG-PROGENI and GenePD Investigators, Coordinators and Molecular Genetic Laboratories (2009) Genomewide association study for susceptibility genes contributing to familial Parkinson disease. Hum Genet 124:593–605
80. Satake W, Nakabayashi Y, Mizuta I, Hirota Y, Ito C, Kubo M, Kawaguchi T, Tsunoda T, Watanabe M, Takeda A, Tomiyama H, Nakashima K, Hasegawa K, Obata F, Yoshikawa T, Kawakami H, Sakoda S, Yamamoto M, Hattori N, Murata M, Nakamura Y, Toda T (2009) Genome-wide association study identifies common variants at four loci as genetic risk factors for Parkinson's disease. Nat Genet 41:1303–1307
81. Simon-Sanchez J, Schulte C, Bras JM, Sharma M, Gibbs JR, Berg D, Paisan-Ruiz C, Lichtner P, Scholz SW, Hernandez DG, Krüger R, Federoff M, Klein C, Goate A, Perlmutter J, Bonin M, Nalls MA, Illig T, Gieger C, Houlden H, Steffens M, Okun MS, Racette BA, Cookson MR, Foote KD, Fernandez HH, Traynor BJ, Schreiber S, Arepalli S, Zonozi R, Gwinn K, van der Brug M, Lopez G, Chanock SJ, Schatzkin A, Park Y, Hollenbeck A, Gao J, Huang X, Wood NW, Lorenz D, Deuschl G, Chen H, Riess O, Hardy JA, Singleton AB, Gasser T (2009) Genome-wide association study reveals genetic risk underlying Parkinson's disease. Nat Genet 41:1308–1312
82. Hampel H, Bürger K, Teipel SJ, Bokde AL, Zetterberg H, Blennow K (2008) Core candidate neurochemical and imaging biomarkers of Alzheimer's disease. Alzheimers Dement 4:38–48
83. Jellinger KA (2003) Neuropathological spectrum of synucleinopathies. Mov Disord 18(Suppl 6):S2–S12
84. Litvan I, Halliday G, Hallett M, Goetz CG, Rocca W, Duyckaerts C, Ben-Shlomo Y, Dickson DW, Lang AE, Chesselet MF, Langston WJ, Di Monte DA, Gasser T, Hagg T, Hardy J, Jenner P, Melamed E, Myers RH, Parker D Jr, Price DL (2007) The etiopathogenesis of Parkinson disease and suggestions for future research. Part I. J Neuropathol Exp Neurol 66:251–257
85. Montine TJ, Shi M, Quinn JF, Peskind ER, Craft S, Ginghina C, Chung KA, Kim H, Galasko DR, Jankovic J, Zabetian CP, Leverenz JB, Zhang J (2010) CSF Aβ42 and tau in Parkinson's disease with cognitive impairment. Mov Disord 25:2682–2685
86. Aasly JO, Shi M, Sossi V, Stewart T, Johansen KK, Wszolek ZK, Uitti RJ, Hasegawa K, Yokoyama T, Zabetian CP, Kim HM, Leverenz JB, Ginghina C, Armaly J, Edwards KL, Snapinn KW, Stoessl AJ, Zhang J (2012) Cerebrospinal fluid amyloid β and tau in LRRK2 mutation carriers. Neurology 78:55–61
87. Mogi M, Harada M, Kondo T, Riederer P, Inagaki H, Minami M, Nagatsu T (1994) Interleukin-1β, interleukin-6, epidermal growth factor and transforming growth factor-α are elevated in the brain from parkinsonian patients. Neurosci Lett 180:147–150
88. Mogi M, Harada M, Narabayashi H, Inagaki H, Minami M, Nagatsu T (1996) Interleukin (IL)-1β, IL-2, IL-4, IL-6 and transforming growth factor-α levels are elevated in ventricular cerebrospinal fluid in juvenile parkinsonism and Parkinson's disease. Neurosci Lett 211: 13–16
89. Mogi M, Harada M, Riederer P, Narabayashi H, Fujita K, Nagatsu T (1994) Tumor necrosis factor-α (TNF-α) increases both in the brain and in the cerebrospinal fluid from parkinsonian patients. Neurosci Lett 165:208–210
90. Nagatsu T, Mogi M, Ichinose H, Togari A (2000) Cytokines in Parkinson's disease. In: Mizuno Y, Calne DB, Horowski R, Poewe W, Riederer P, Youdim MBH (eds) Advances in Research on Neurodegeneration. Springer, Vienna, pp 143–151

91. Pabon MM, Bachstetter AD, Hudson CE, Gemma C, Bickford PC (2011) CX3CL1 reduces neurotoxicity and microglial activation in a rat model of Parkinson's disease. J Neuroinflammation 8:9
92. Nilsson C, Lindvall-Axelsson M, Owman C (1992) Neuroendocrine regulatory mechanisms in the choroid-plexus cerebrospinal-fluid system. Brain Res Rev 17:109–138
93. Vigh B, Marzano e Silva MJ, Frank CL, Vincze C, Czirok SJ, Szabo A, Lukats A, Szel A (2004) The system of cerebrospinal fluid-contacting neurons: its supposed role in the nonsynaptic signal transmission of the brain. Histol Histopathol 19:607–628
94. Ly L, Wasinger VC (2011) Protein and peptide fractionation, enrichment and depletion: tools for the complex proteome. Proteomics 11:513–534
95. Domon B, Aebersold R (2006) Mass spectrometry and protein analysis. Science 312:212–217
96. Senko MW, Hendrickson CL, Emmett MR, Shi SDH, Marshall AG (1997) External accumulation of ions for enhanced electrospray ionization Fourier transform ion cyclotron resonance mass spectrometry. J Am Soc Mass Spectrom 8:970–976
97. Hardman M, Makarov AA (2003) Interfacing the Orbitrap mass analyzer to an electrospray ion source. Anal Chem 75:1699–1705
98. Medzihradszky KF, Campbell JM, Baldwin MA, Falick AM, Juhasz P, Vestal ML, Burlingame AL (2000) The characteristics of peptide collision-induced dissociation using a high-performance MALDI-TOF/TOF tandem mass spectrometer. Anal Chem 72:552–558
99. Fenn JB, Mann M, Meng CK, Wong SF, Whitehouse CM (1989) Electrospray ionization for mass spectrometry of large biomolecules. Science 246:64–71
100. Karas M, Hillenkamp F (1988) Laser desorption ionization of proteins with molecular masses exceeding 10,000 daltons. Anal Chem 60:2299–2301
101. Syka JE, Coon JJ, Schroeder MJ, Shabanowitz J, Hunt DF (2004) Peptide and protein sequence analysis by electron transfer dissociation mass spectrometry. Proc Nat Acad Sci U S A 101:9528–9533
102. Boersema PJ, Raijmakers R, Lemeer S, Mohammed S, Heck AJ (2009) Multiplex peptide stable isotope dimethyl labeling for quantitative proteomics. Nat Protoc 4:484–494
103. Gygi SP, Rist B, Gerber SA, Turecek F, Gelb MH, Aebersold R (1999) Quantitative analysis of complex protein mixtures using isotope-coded affinity tags. Nat Biotechnol 17:994–999
104. Hsu JL, Huang SY, Chen SH (2006) Dimethyl multiplexed labeling combined with microcolumn separation and MS analysis for time course study in proteomics. Electrophoresis 27:3652–3660
105. Ross PL, Huang YN, Marchese JN, Williamson B, Parker K, Hattan S, Khainovski N, Pillai S, Dey S, Daniels S, Purkayastha S, Juhasz P, Martin S, Bartlet-Jones M, He F, Jacobson A, Pappin DJ (2004) Multiplexed protein quantitation in Saccharomyces cerevisiae using amine-reactive isobaric tagging reagents. Mol Cell Proteomics 3:1154–1169
106. Yao X, Afonso C, Fenselau C (2003) Dissection of proteolytic 18O labeling: endoprotease-catalyzed 16O-to-18O exchange of truncated peptide substrates. J Proteome Res 2:147–152
107. May D, Fitzgibbon M, Liu Y, Holzman T, Eng J, Kemp CJ, Whiteaker J, Paulovich A, McIntosh M (2007) A platform for accurate mass and time analyses of mass spectrometry data. J Proteome Res 6:2685–2694
108. Mueller LN, Rinner O, Schmidt A, Letarte S, Bodenmiller B, Brusniak MY, Vitek O, Aebersold R, Muller M (2007) SuperHirn – a novel tool for high resolution LC-MS-based peptide/protein profiling. Proteomics 7:3470–3480
109. Liu H, Sadygov RG, Yates JR 3rd (2004) A model for random sampling and estimation of relative protein abundance in shotgun proteomics. Anal Chem 76:4193–4201
110. Zybailov BL, Florens L, Washburn MP (2007) Quantitative shotgun proteomics using a protease with broad specificity and normalized spectral abundance factors. Mol Biosyst 3:354–360
111. Pan S, Rush J, Peskind ER, Galasko D, Chung K, Quinn J, Jankovic J, Leverenz JB, Zabetian C, Pan C, Wang Y, Oh JH, Gao J, Zhang J, Montine T, Zhang J (2008) Application of targeted quantitative proteomics analysis in human cerebrospinal fluid using a liquid chromatography matrix-assisted laser desorption/ionization time-of-flight tandem mass spectrometer (LC MALDI TOF/TOF) platform. J Proteome Res 7:720–730

112. Abdi F, Quinn JF, Jankovic J, McIntosh M, Leverenz JB, Peskind E, Nixon R, Nutt J, Chung K, Zabetian C, Samii A, Lin M, Hattan S, Pan C, Wang Y, Jin J, Zhu D, Li GJ, Liu Y, Waichunas D, Montine TJ, Zhang J (2006) Detection of biomarkers with a multiplex quantitative proteomic platform in cerebrospinal fluid of patients with neurodegenerative disorders. J Alzheimers Dis 9:293–348
113. Pan S, Zhu D, Quinn JF, Peskind ER, Montine TJ, Lin B, Goodlett DR, Taylor G, Eng J, Zhang J (2007) A combined dataset of human cerebrospinal fluid proteins identified by multi-dimensional chromatography and tandem mass spectrometry. Proteomics 7:469–473
114. Harrington M, Fonteh A, Oborina E, Liao P, Cowan R, McComb G, Chavez J, Rush J, Biringer R, Huhmer A (2009) The morphology and biochemistry of nanostructures provide evidence for synthesis and signaling functions in human cerebrospinal fluid. Cerebrospinal Fluid Res 6:10
115. Johnstone RM, Adam M, Hammond JR, Orr L, Turbide C (1987) Vesicle formation during reticulocyte maturation – association of plasma-membrane activities with released vesicles (exosomes). J Biol Chem 262:9412–9420
116. Thery C, Zitvogel L, Amigorena S (2002) Exosomes: composition, biogenesis and function. Nat Rev Immunol 2:569–579
117. Ecroyd H, Sarradin P, Dacheux JL, Gatti JL (2004) Compartmentalization of prion isoforms within the reproductive tract of the ram. Biol Reprod 71:993–1001
118. Vingtdeux V, Hamdane M, Loyens A, Gele P, Drobeck H, Begard S, Galas MC, Delacourte A, Beauvillain JC, Buee L, Sergeant N (2007) Alkalizing drugs induce accumulation of amyloid precursor protein by-products in luminal vesicles of multivesicular bodies. J Biol Chem 282(25):18197–18205
119. El-Agnaf OM, Salem SA, Paleologou KE, Cooper LJ, Fullwood NJ, Gibson MJ, Curran MD, Court JA, Mann DMA, Ikeda S, Cookson MR, Hardy J, Allsop D (2003) Alpha-synuclein implicated in Parkinson's disease is present in extracellular biological fluids, including human plasma. FASEB J 17:1945–1947
120. Liu J, Zhang JP, Shi M, Quinn T, Bradner J, Beyer R, Chen S, Zhang J (2009) Rab11a and HSP90 regulate recycling of extracellular alpha-synuclein. J Neurosci 29:1480–1485
121. Zhang J, Sokal I, Peskind ER, Quinn JF, Jankovic J, Kenney C, Chung KA, Millard SP, Nutt JG, Montine TJ (2008) CSF multianalyte profile distinguishes Alzheimer and Parkinson diseases. Am J Clin Pathol 129:526–529
122. Selvaraju S, El Rassi ZE (2012) Liquid-phase-based separation systems for depletion, prefractionation and enrichment of proteins in biological fluids and matrices for in-depth proteomics analysis – an update covering the period 2008–2011. Electrophoresis 33:74–88
123. Lehnert S, Jesse S, Rist W, Steinacker P, Soininen H, Herukka S-K, Tumani H, Lenter M, Oeckl P, Ferger B, Hengerer B, Otto M (2012) iTRAQ and multiple reaction monitoring as proteomic tools for biomarker search in cerebrospinal fluid of patients with Parkinson's disease dementia. Exp Neurol 234:499–505
124. Breen DP, Michell AW, Barker RA (2011) Parkinson's disease – the continuing search for biomarkers. Clin Chem Lab Med 49:393–401
125. Hughes AJ, Daniel SE, Ben-Shlomo Y, Lees AJ (2002) The accuracy of diagnosis of parkinsonian syndromes in a specialist movement disorder service. Brain 125:861–870

Part IV

Glaucomatous Optic Neuropathy: Risk Assessment and Potential Targets for Effective Prevention and Treatments Tailored to the Patient

Olga Golubnitschaja, Kristina Yeghiazaryan, and Josef Flammer

Abstract **Prevalence of glaucoma:** Neurodegenerative eye disease glaucoma is the second leading cause of blindness with an estimated 67 million of patients worldwide.

Potential prediction and targeted prevention: Molecular pathomechanisms of glaucoma demonstrate remarkable particularities compared to other neurodegenerative disorders. Identification of pathology-specific biomarker-patterns is essential for risk assessment, early diagnosis and targeted prevention.

Risk assessment as the reliable platform for glaucoma prediction and treatments tailored to the patient: Molecular pathways affected in glaucoma pathology provide information for the patient-profiling. The key pathways are the following: stress response, apoptosis and DNA-repair, adhesion, blood–brain-barrier-breakdown, tissue remodelling, transcription regulation, multidrug resistance and energy metabolism. Individualised profiling allows for the cost-effective treatments tailored to the patient.

Keywords Personalised medicine • Glaucoma • Blood biomarker-patterns • Patient-profiles • Early/predictive diagnosis • Targeted preventive measures

---

O. Golubnitschaja (✉) • K. Yeghiazaryan
Division of Molecular/Experimental Radiology, Department of Radiology, Rheinische Friedrich-Wilhelms-University of Bonn, Sigmund-Freud-Str. 25, D-53105 Bonn, Germany
e-mail: Olga.Golubnitschaja@ukb.uni-bonn.de

J. Flammer
Department of Ophthalmology, University Hospital Basel, Basel, Switzerland

S. Mandel (ed.), *Neurodegenerative Diseases: Integrative PPPM Approach as the Medicine of the Future*, Advances in Predictive, Preventive and Personalised Medicine 2, DOI 10.1007/978-94-007-5866-7_9,

1 Glaucoma Pathology in the Context of Neurodegenerative Diseases

Glaucoma is a complex neurodegenerative disease the molecular pathomechanisms of which demonstrate both a considerable overlap and remarkable particularities compared to some other neurodegenerative disorders such as Alzheimer's disease. As an example, *neuronal thread protein* (*NTP*) demonstrates enhanced expression levels in glaucoma patients against controls [1], the same observation occurring also for Down Syndrome, Alzheimer's and some other diseases [2], indicating axonal lesions as the common feature but giving no information about the corresponding aetiology. Simultaneous monitoring of expressional rates of further *TAU*-protein enables a distinction between the pathologies: in contrast to the accumulation of *TAU*-protein characteristic for Alzheimer's disease and other tauopathies, glaucoma patients demonstrate no increase in the target protein expression *versus* controls [3]. Therefore, the selection of molecular targets specifically affected by glaucoma pathology is of high value for its practical application by advanced diagnostic approaches, targeted prevention and cost-effective treatments tailored to the patient [4].

2 Characteristic Attributes of the Glaucomatous Optic Neuropathy

Glaucoma is a neurodegenerative eye disease and the second leading cause of permanent vision loss with an estimated 67 million patients worldwide [5]. Glaucomatous optic neuropathy (GON) is characterised by a combination of a loss of retinal ganglion cells and their axons with disease typical optic nerve head (ONH) atrophy (see Fig. 1).

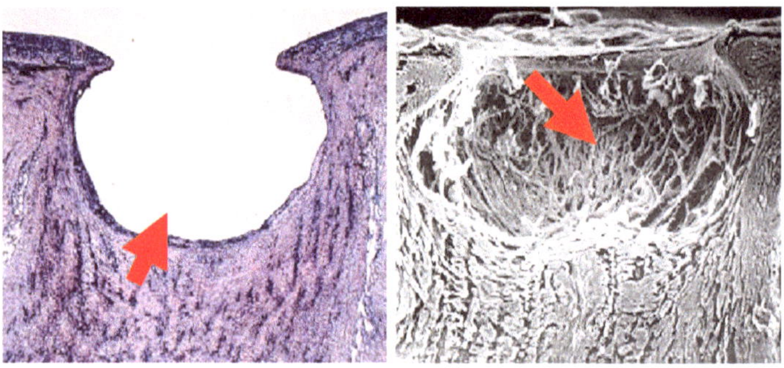

Fig. 1 Morphologic examinations result in the discovery of degenerative tissue remodelling (marked with the *red arrow*) characteristic for the glaucomatous damage [6]. An extensive tissue remodelling accompanying GON is a very complex process that involves several molecular pathways affected by GON

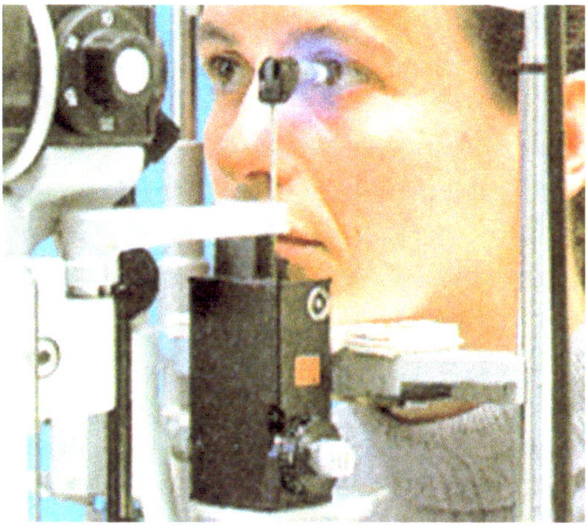

Fig. 2 Measurements of the intraocular pressure (IOP) are essential for the diagnosis as one of the major risk factors in glaucoma [6]

3 Risk Factors of Glaucoma

The aetiology of glaucoma is only partially known. The risk factors known are highly complex:

- intraocular pressure (IOP) (see Fig. 2),
- race,
- female gender,
- genetics,
- myopia,
- corneal thickness,
- optic disc haemorrhages,
- haemodynamics,
- others [7, 8].

4 Heterogeneity of Glaucoma Patient Cohort

Intraocular pressure (IOP) is considered the most important and only modifiable risk factor in glaucoma. Although elevated IOP has been shown to be the major risk factor, there is a cohort of patients even at younger ages with normal IOP developing normal-tension glaucoma (NTG) (Fig. 3).

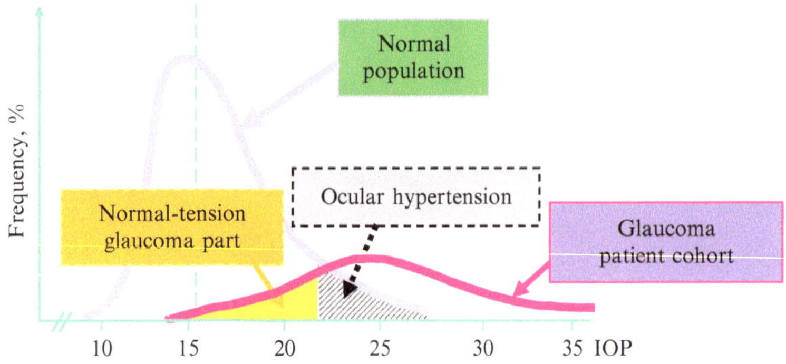

Fig. 3 Although increased intraocular pressure (IOP) is a well-known risk factor, the IOP values of the part of the glaucoma patient cohort, namely normal-tension glaucoma patients are within the statistical norm

The frequency and incidence of different glaucoma forms vary from country to country, depending on race, sex and age [9–11].

5 A Genetic Component in the Development of Glaucoma Pathology

A family history of the disease is common among glaucoma patients, suggesting impacts of a genetic element for this disorder. The existence of a genetic linkage associated with particular forms of glaucoma is well documented in the literature [11–18]. The growing body of evidence suggests the idea that genetic-based approaches in early diagnosis of glaucoma and targeted treatment might yield better results than medical, surgical and rehabilitation interventions alone [18]. Although mutation research in glaucoma detected some attractive potential targets, the mutated genes found – like *TIGR* – play a limited role in the pathogenesis of glaucoma and do not explain the usual clinical picture [10, 19–23]. Therefore, genomic and pharmacogenetic approaches alone may provide only minor success in the prediction of the pathology, its targeted prevention and treatments.

Are there any diagnostic approaches developed to identify glaucoma-predisposed individuals, to distinguish between glaucoma forms and to predict a disease progression? We should state a current lack of clinical approaches for predictive diagnostics of glaucoma. However, the large effort made and recent progress achieved in the identification of disease characteristic molecular patterns should be clearly acknowledged – the issue that is reviewed in this chapter.

6 Impacts of Cardiovascular Component in Glaucoma Pathology and Corresponding Relevance of Molecular Patterns in Blood

A wealth of literature points to the potential importance of haemodynamics and cardiovascular component in glaucoma pathology [24]. Healthy individuals with vasospastic syndrome demonstrate molecular profiles in blood similar to those of glaucoma patients [25]. Particular predisposition of vasospastic individuals to glaucoma has been described by the authors earlier [26] and is also discussed in the separate chapter of this book [27].

Although animal models for glaucoma provide useful information about affected molecular pathways and potential targets in the eye, they are barely applicable for non-invasive diagnostics on man. Since blood is easily accessible, a molecular analysis of just a few millilitres may offer a practical application in population screening for high-risk subjects. How relevant are gene expression patterns in blood serum and white blood cells for the eye? Glaucoma patients frequently exhibit abnormal T-cell subsets and increased titres of serum antibodies to retina and optic nerve proteins; these alterations in the cellular subset indicate that the immune system plays an important role in the initiation and/or progression of glaucomatous optic neuropathy [28]. Therefore, serum antibodies to retina and optic nerve proteins might be considered as potential indicators for the diagnosis of glaucoma pathology [29]. In addition, it has been demonstrated that gene expression patterns of both trabecular meshwork and Schlemm's canal are similar to those of circulating leukocytes (CL) [30]. Moreover, blood serum is the biggest reservoir of signalling molecules in human beings: many metabolites are simultaneously secreted from different types of cells in blood that represents a universal way of communication between cells. Although this communication is extremely complex, one of the known natural sensors affected by the tremendous number of metabolites present in blood serum is CL. The spectrum of the molecules affecting expression patterns in CL is very broad; their regulation is extremely complex and not yet completely understood. However, it is known that the resulting regulation of CL is triggered by altered gene expression patterns on the levels of transcription and translation. The resulting shift in a gene expression pattern referred to a corresponding physiologic/pathologic condition can be measured on both levels.

Since both vascular and immune components play a role in the pathomechanisms of glaucoma [31], the pathology-specific molecular patterns in blood are highly relevant for diagnostic purposes. Taking the above into consideration, the search for disease specific gene expression patterns in blood of glaucoma is of great value for a better understanding of the molecular pathomechanisms and concomitant creation of non-invasive approaches in predictive glaucoma diagnostics.

7 Development of Innovative Diagnostic Tools to Predict Glaucoma in Predisposed Individuals

Molecular pathways affected by glaucoma pathology have been identified and described by the authors in journals and books specifically dedicated to predictive diagnostics of neurodegenerative diseases and personalised medicine [26, 32–34].

The majority of molecular patterns specifically altered in blood and tissues of glaucoma patients belong to well-defined functional groups of gene-products which, altogether, represent molecular pathways affected by the pathology (see Fig. 4).

8 Molecular Targets for Patient-Profiling

8.1 Stress Response, Apoptosis and DNA-Repair

An increased oxidative stress has been well documented under glaucoma pathology. Several proteins modified by oxidative stress have been demonstrated in retinal protein lysates isolated from ocular hypertensive eyes of a rat-model of glaucoma [36]. Consequently, altered gene expression patterns, particularly for stress response factors, can be expected. Thus, the alteration of *heat shock 27 kDa protein 1* expression was stable in retinas of rat hypertensive eyes [37]. "Gene hunting" technology of "subtractive hybridisation" followed by quantification of target transcripts (see Fig. 5) performed in CL of glaucoma patients revealed an up-regulation of both the pro-apoptotic factor *p53* and proteolytic enzyme *20S proteasome subunit XAPC7* [1], the activity of which is usually increased during reperfusion [39]. In agreement with this, both an apoptotic inhibitor *survivin* and DNA-repair *Xeroderma pigmentosum gene C* have been found to be down-regulated in CL of glaucoma patients [1]. Moreover, *ex vivo* subcellular imaging by "Comet Assay"-analysis demonstrates significantly increased (un-repaired) DNA-damage in circulating leukocytes of both normal- and high-tension glaucoma patients [32, 40].

8.2 Adhesion, Blood–Brain-Barrier (BBB)-Breakdown and Tissue Remodelling

Gene expression profiles characteristic for an activated adherent function have been identified in CL of glaucoma patients: up-regulated transcripts of *lymphocyte-IgE-receptor* (*Fc-epsilon-RII/CD23*), *T cell-specific tyrosine kinase* (*ITK*), *thromboxan-A2-receptor*, and *alkaline-phosphatase* have been detected by "subtractive hybridisation" [41]. Adherent circulating leukocytes (CL) could be an important contributor

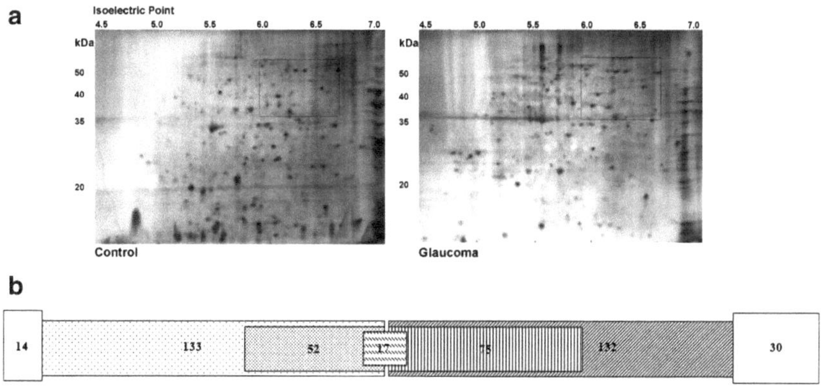

c

SSP number	Control group	Glaucoma group	NCBI /Acc. No	Protein identified	Peptide Number Measured/Matched	Sequence coverage (%)	Potential function
2304	*=1*	***1.6↑***	HSP70_HUMAN AAB06397	*HSP70*	47/18	35	Stress protein
3302	*=1*	***1.6↓***	-	-	-	-	-
4401	*=1*	***2.0↑***	AP-2 beta CAA64990	*AP-2β*	22/19	55	Basic transcription regulator
4502	=1	1	K1C10_HUMAN P13645	*Cytokeratin 10*	31/21	69	Cytoskeletal protein
4703	*=1*	***1.6↓***	-	-	-	-	-
4803	*=1*	***3.4↓***	ATPB_HUMAN P06576	*ATP synthase beta chain, mitochondrial precursor*	55/14	27	Energy metabolism
5503	=1	1	ACTB_HUMAN P60709	*Beta-actin*	34/19	59	Cytoskeletal protein
5703	*=1*	***1.3↓***	-	-	-	-	-
6404	*=1*	***1.4↓***	-	-	-	-	-
6406	*=1*	***2.7↑***	GSTP1-HUMAN P09211	*Glutathione S-transferase P*	11/8	46	Red-ox control
6601	*=1*	***1.6↑***	PDIA1_HUMAN P07237	*Protein disulfide-isomerase precursor*	18/6	21	Stress-response protein
6603	*=1*	***1.8↓***	-	-	-	-	-
6704	*=1*	***2.4↓***	-	-	-	-	-
6705	=1	1	PROF1_HUMAN P07737	*Profilin 1*	13/5	24	Cytoskeletal protein
6706	*=1*	***2.0↓***	-	-	-	-	-
7601	=1	1	-	-	-	-	-
7803	=1	1	ACTG_HUMAN P63261	*Gamma-actin*	31/16	38	Cytoskeletal protein

Fig. 4 Protein expression patterns characteristic for circulating leukocytes of glaucoma patients; data taken from [35]. (**a**) 2D-PAGE images of differential gene expression patterns in circulating leukocytes of glaucoma patients *versus* controls with particular changes marked. (**b**) The scheme represents summarised results from the evaluation of comparative 2D-PAGE images. 132 protein spots were constantly present in all control images, whereby 75 spots were expressed at the same level. 133 protein spots were constantly present in patient images, whereby 52 spots were expressed at the same level. 17 proteins represent the overlap between 75 and 52 protein spots. Thirty spots were present only in the control group whereas 14 spots were present only in the patient group; all spots were detectable only in single images and, therefore, were not included in the above 133 and 132 pools. (**c**) Quantitative analysis of 2D-PAGE images: among 75 and 52 conserved protein spots in the control and patient groups, respectively, 17 overlapping protein spots were found to be conserved for members of the same group. Corresponding expression levels in the control group are taken as standard (=1). Glaucoma characteristic shifts in the expression levels of 12 proteins are marked with *bold italics*; "-" means non-identified protein spots, since the overall protein sequence coverage was lower than 15%

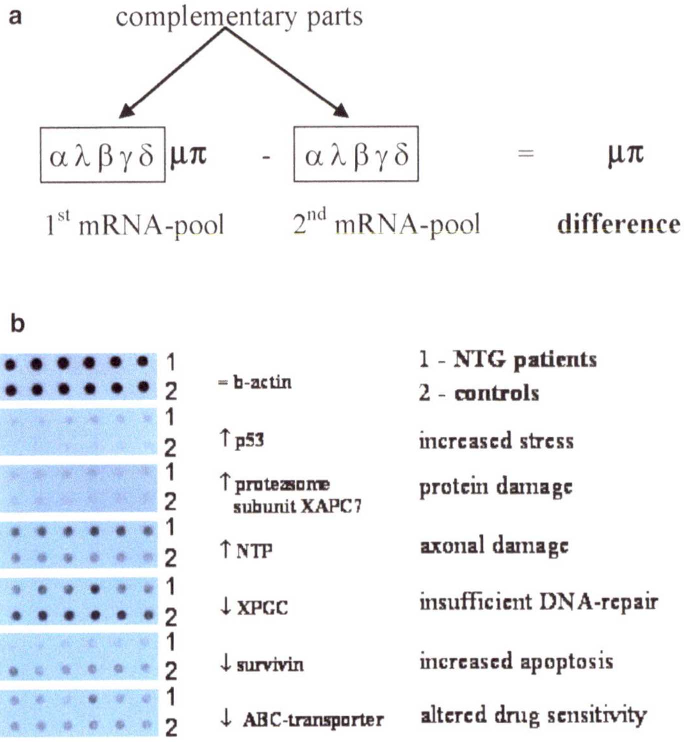

Fig. 5 (**a**) Principle of "gene hunting" by the "subtractive hybridisation" [38]. Each mRNA-pool consists of individual gene transcripts that undergo a comparison by hybridisation of identical (complementary) parts. The subtraction results in identification of non-complementary transcripts differentially expressed in the two pools of comparison; the technology enables a consideration of both qualitative (present or absent) and quantitative (amount of copies per unit weight of mRNA in the corresponding pool) differences. (**b**) Consequent quantification of the target transcripts (mRNA "Dot-Blot"-analysis), further provided a strong indication for increased stress, apoptosis, protein damage and axonal damage as well as insufficient DNA-repair and altered drug-sensitivity in patients with glaucoma compared to the control group

to BBB-breakdown observed in glaucoma pathology [42]. The subtraction of metalloproteinase *MMP-9* and *MT1-MMP* transcripts highly increased in CL of glaucoma patients provides further evidence for this functional link and indicates pathways involved in the extensive tissue remodelling observed in glaucoma [43]. Significantly increased protein expression rates of both latent and active forms of *MMP-9* and *MT1-MMP* in CL correlated well with the enhanced levels of transcription and with glaucoma [43]. Once activated, both hydrolases necessarily contribute to remodelling or even degeneration of the tissue wherein they are secreted by CL. This up-regulation might be a consequence of repeated mild ischaemia/reperfusion postulated for glaucoma patients. However, the question as to whether or not there is a correlation between an increased *MMPs* activity and glaucoma severity should be further clarified. Recently performed comparative gene transcription profiling in TM isolated from post-mortem glaucomatous eyes revealed a significant up-regulation of several cell adhesion molecules including *platelet endothelial cells adhesion molecule-1* and *P-selectin* [44]. An enhanced deposition of the cell aggregation protein *cochlin* in TM and SC of glaucomatous eye supports the crucial role of cellular adhesion in glaucoma pathology and favours cell adhesion molecules as potential disease biomarkers [45].

8.3 Transcription Regulation

Dramatically altered transcription patterns in glaucoma were reported by several research groups [1, 35, 37, 43]. A reason for this extensive shift in transcription regulation was unclear for a very long time till a significant up-regulation of the basic transcriptional regulator *AP-2ß* was attributed to glaucoma pathology: *AP-2ß* has been identified in CL using the smart technology of disease proteomics [35]. Protein mapping is demonstrated in Fig. 4a. *AP-2* proteins play a decisive role, particularly in eye morphogenesis. While the activation of *AP-1* leads to an increased stromelysin (metalloproteinase-3) production in the trabecular meshwork *in vitro* [46, 47], the expression and activity of *AP-2* controls the activity of the *gelatinase B* (*MMP-9*) [48]. In consensus, *MMP-9* was demonstrated to be significantly up-regulated in leukocytes of glaucoma patients and to play an important role in tissue remodelling as a part of glaucomatous degeneration [43]. Therefore, the concerted up-regulation of *AP-2ß* and *MMP-9* in leukocytes of glaucoma patients has been proposed to be an important part of the molecular mechanisms in glaucoma pathology and a reliable biomarker for diagnostic purposes.

8.4 Multidrug Resistance

An extensive deregulation of *ABC-transporters* has been demonstrated in glaucoma pathology [1, 49, 50]. *ABC-transporters* (*ATP-binding cassette transporter*) usually translocate a wide variety of structurally unrelated lipophilic compounds being responsible for drug efflux, and therefore, for multidrug resistance. *ABC 1* has been shown to be stably up-regulated in CL of glaucoma patients [50]. Joyce et al. demonstrated a crucial role for *ABC 1* in protection against atherosclerosis [51]. An activity of *ABC 1* has been shown to have a regulating effect on endothelial function and stimulate nitric oxide bioactivity in arterial walls [52]. The up-regulation of *ABC1* in CL of glaucoma patients might indicate the involvement of this gene in chronic vascular deregulation frequently observed in glaucoma pathology, and has been suggested as a potential marker for early diagnostics of glaucoma [50].

8.5 *Energy Metabolism*

“Gene hunting” in CL of glaucoma patients revealed down-regulated transcripts of *Na^+/K^+-ATPase* [41]. An identification of the subtracted transcripts is demonstrated in Fig. 5. Further, abnormal sodium handling has been proposed to be associated with ocular hypertension and to contribute to a progression of optic nerve damage in both normal-tension and high-tension glaucoma [53, 54]. *Na^+/K^+-ATPase* is known to be down-regulated in lymphocytes of patients with acute myocardial infarction [55], and may be one of the reasons for ventricular arrhythmias and coronary artery spasms. A decrease in intracellular potassium concentration and an increase in intracellular calcium concentration may play a major role in the pathomechanism of coronary artery spasms. Inhibition of *Na^+/K^+-ATPase* is one of functional consequences of oxidative membrane damage caused by an increased oxidative stress [56], which is triggered e.g. under ischaemia/reperfusion. This affects central thermogenic mechanisms, since *Na^+/K^+-ATPase* has been shown to play a key role in cellular energy balance and thermogenesis [57, 58]. A decreased activity of *Na^+/K^+-ATPase* triggers vasospasm [59]. Being an important co-factor of cellular *ATPases*, magnesium can reverse delayed vasospasm and reduce the extent of acute ischaemic lesions [60]. Magnesium therapy demonstrates beneficial effects in vasospastic syndrome and glaucoma [10].

9 Side-effects of Glaucoma Medication on the Predictive Power of Blood Biomarkers

Evaluation of glaucoma biomarkers under treatment conditions is a particular issue in glaucoma research. Indeed, can we use the markers proposed for diagnostics of untreated glaucoma also under glaucoma medication? Are there any reliable markers which can predict the outcome of glaucoma treatment? There are some studies clearly demonstrating that glaucoma medication may definitely mask alterations in crucial biomarkers such as those of the nitric oxide pathway [61]. Additionally, toxic side effects of anti-glaucoma drugs on conjunctiva have been reported: histological studies of conjunctival tissues in patients who underwent a long-term glaucoma treatment demonstrated an abnormal infiltration of inflammatory cells. Consequently, an abnormal expression of inflammatory markers under various anti-glaucomatous treatments has been shown [62].

Much progress has been achieved in the development of markers that can potentially predict the outcome of glaucoma surgery. In the research of Souchier et al., surgery success was defined as reduced IOP $\leq$15 mmHg without any IOP-lowering drugs at 6 months. Conjunctival expression levels of *trefoil factor family 1 (TFF1)* and *MUC5AC* as well as *HLA-DR* in leukocytes have been measured [63] and an increased expression of all three molecular markers has been proposed as

Table 1 Types of glaucoma-specific molecular alterations that can be potentially used for development of advanced tools for early and predictive diagnosis [65]

Type of molecule	Possible type of alteration	Detection technology
Chromosomal DNA, mitochondrial DNA	1. (Oxidative) damage[a]	– Comet assay
	2. Mutations[a]	– Disease genomics
	3. Polymorphism[a]	– PCR, Restriction analysis, etc.
	4. Methylation status of CpG islands[b]	– Methylation-specific PCR
mRNA	1. Multiple alterations in expression patterns[a]	– Disease transcriptomics: Subtractive hybridisation, Expression array, Reverse-Transcriptase-PCR, Real-Time-PCR, etc.
	2. Reduced mRNA editing	– Reverse-Transcriptase-PCR
Proteins	1. Multiple alterations in expression patterns[a]	– Disease Proteomics: 2D-PAGE, MALDI-TOF, Western-blot, etc.
	2. Posttranslational modification[a]	– Western-blot, Activity tests (e.g. Zymography for gelatinases)
	3. Phosphorylation status[a]	– Activity tests
	4. Protein misfolding[a]	– Activity tests
Metabolites (Signalling molecules, amino acids, plasma hormones, etc.)	Altered profiles[a]	Disease Metabolomics: Comparative blood plasma metabolites profiling, HPLC, Activity tests, etc.

[a]Types of molecular alterations reported for glaucoma pathology
[b]Data indicating this kind of alteration collected in our laboratory, which, however, have not been published until now

a potential prediction factor for successful glaucoma surgery. However, the issue of glaucoma medication markers is a multifarious one and requires more research effort.

10 Expert Recommendations and Outlook

High-accuracy proteome maps of human body fluids open new perspectives for early molecular diagnostics of chronic disorders. Abundant scientific evidence from disease proteomics suggests that disease specific protein expression patterns in blood are indicative of many – if not most – human disorders, and might be a powerful tool for early molecular diagnostics of glaucomatous damage. A non-invasive molecular diagnostic approach based on disease specific gene expression patterns in CL has been suggested for glaucoma [64]. The test foresees a precise expression profiling of selected genes in CL isolated from fresh blood samples. These genes (gene products) have been proposed to play a role in glaucoma pathology and belong to following pathways: an apoptosis, stress response, DNA-repair, transcription regulation, drug resistance, tissue remodelling and degeneration. Currently a clinical application of the test is under consideration, and a nanotechnology which should provide a possibly easy and cheap routine application of the test is under development. Possible types of molecular alterations that can be considered for predictive glaucoma diagnosis and corresponding detection technologies are summarised in Table 1.

References

1. Golubnitschaja-Labudova O, Liu R, Decker C, Zhu P, Haefliger IO, Flammer J (2000) Altered gene expression in lymphocytes of patients with normal-tension glaucoma. Curr Eye Res 21:867–876
2. de la Monte SM, Xu YY, Hutchins GM, Wands JR (1996) Developmental patterns of neuronal thread protein gene expression in Down Syndrome. J Neurol Sci 135:118–125
3. Tatton W, Chen D, Chalmers-Redman R, Wheeler L, Nixon R, Tatton N (2003) Hypothesis for a common basis for neuroprotection in glaucoma and Alzheimer's disease: anti-apoptosis by alpha-2-adrenergic receptor activation. Surv Ophthalmol 48:S25–S37
4. Golubnitschaja O (2010) Neurodegeneration: accelerated aging or inadequate healthcare. EPMA J 1:211–215
5. Quigley HA (1996) Number of people with glaucoma worldwide. Br J Ophthalmol 80:389–393
6. Flammer J (2003) Glaucoma. A guide for patients. An introduction for care-providers. A quick reference. Hogrefe Publishing, Bern
7. Gugleta K, Orgül S, Flammer J (1999) Asymmetry in intraocular pressure and retinal nerve fiber layer thickness in normal-tension glaucoma. Ophthalmologica 213:219–223
8. Hayreh SS, Podhajsky P, Zimmerman MB (1999) Role of nocturnal arterial hypotension in optic nerve head ischemic disorders. Ophthalmologica 213:76–96
9. Reddy AB, Panicker SG, Mandal AK, Hasnain SE, Balasubramanian D (2003) Identification of R368H as a predominant CYP1B1 allele causing primary congenital glaucoma in Indian patients. Invest Ophthalmol Vis Sci 44:4200–4203
10. Kanagavalli J, Krishnadas SR, Pandaranayaka E, Krishnaswamy S, Sundaresan P (2003) Evaluation and understanding of myocilin mutations in Indian primary open angle glaucoma patients. Mol Vis 14:606–614
11. Gong G, Kosoko-Lasaki O, Haynatzki GR, Wilson MR (2004) Genetic dissection of myocilin glaucoma. Hum Mol Genet 3:R91–R102
12. Mardin CY, Velten I, Ozbey S, Rautenstrauss B, Michels-Rautenstrauss K (1999) A GLC1A gene Gln368Stop mutation in a patient with normal-tension open-angle glaucoma. J Glaucoma 8:154–156
13. Colomb E, Kaplan J, Garchon HJ (2003) Novel cytochrome P450 1B1 (CYP1B1) mutations in patients with primary congenital glaucoma in France. Hum Mutat 22:496
14. Howell N (2003) LHON and other optic nerve atrophies: the mitochondrial connection. Dev Ophthalmol 37:94–108
15. Leung YF, Fan BJ, Lam DS, Lee WS, Tam PO, Chua JK, Tham CC, Lai JS, Fan DS, Pang CP (2003) Different optineurin mutation pattern in primary open-angle glaucoma. Invest Ophthalmol Vis Sci 44:3880–3884
16. Mackey DA, Craig JE (2003) Predictive DNA testing for glaucoma: reality in 2003. Ophthalmol Clin North Am 16:639–645
17. Polansky JR (2003) Current perspectives on the TIGR/MYOC gene (myocilin) and glaucoma. Ophthalmol Clin North Am 16:515–527, v-vi

18. Panicker SG, Mandal AK, Reddy AB, Gothwal VK, Hasnain SE (2004) Correlations of genotype with phenotype in Indian patients with primary congenital glaucoma. Invest Ophthalmol Vis Sci 45:1149–1156
19. Herndon LW, Challa P, Allingham RR (2003) Glaucoma in Ghana, West Africa: clinical features and the role of mutations in myocilin. Ophthalmol Clin North Am 16:631–637
20. Bruttini M, Longo I, Frezzotti P, Ciappetta R, Randazzo A, Orzalesi N, Fumagalli E, Caporossi A, Frezzotti R, Renieri A (2003) Mutations in the myocilin gene in families with primary open-angle glaucoma and juvenile open-angle glaucoma. Arch Ophthalmol 121:1034–1038
21. Melki R, Idhajji A, Driouiche S, Hassani M, Boukabboucha A, Akhayat O, Garchon H, Belmouden A (2003) Mutational analysis of the myocilin gene in patients with primary open-angle glaucoma in Morocco. Ophthalmic Genet 24:153–160
22. Wiggs JL, Auguste J, Allingham RR, Flor JD, Pericak-Vance MA, Rogers K, LaRocque KR, Graham FL, Broomer B, Del Bono E, Haines JL, Hauser M (2003) Lack of association of mutations in optineurin with disease in patients with adult-onset primary open-angle glaucoma. Arch Ophthalmol 121:1181–1183
23. Sripriya S, Uthra S, Sangeetha R, George RJ, Hemamalini A, Paul PG, Amali J, Vijaya L, Kumaramanickavel G (2004) Low frequency of myocilin mutations in Indian primary open-angle glaucoma patients. Clin Genet 65:333–337
24. Graham SL, Butlin M, Lee M, Avolio AP (2013) Central blood pressure, arterial waveform analysis, and vascular risk factors in glaucoma. J Glaucoma 22:98–103
25. Yeghiazaryan K, Flammer J, Orgül S, Wunderlich K, Golubnitschaja O (2009) Vasospastic individuals demonstrate significant similarity to glaucoma patients as revealed by gene expression profiling in circulating leukocytes. Mol Vis 15:2339–2348
26. Yeghiazaryan K, Flammer J, Golubnitschaja O (2010) Predictive molecular profiling in blood of healthy vasospastic individuals: clue to targeted prevention as personalised medicine to effective costs. EPMA J 1:263–272
27. Yeghiazaryan K, Flammer J, Golubnitschaja O (2013) Individual predispositions in healthy vasospastic individuals: patient profiling for targeted prevention of "down-stream" pathologies as cost-effective personalised medicine. In: Mandel S (ed) Neurodegenerative diseases: integrative PPPM approach as the medicine of the future. Book series: Advances in PPPM. Golubnitschaja O (ed). Springer, Dordrecht, pp. 13–29
28. Yang J, Patil RV, Yu H, Gordon M, Wax MB (2001) T cell subsets and sIL-2R/IL-2 levels in patients with glaucoma. Am J Ophthalmol 131:421–426
29. Grus FH, Joachim SC, Bruns K, Lackner KJ, Pfeiffer N, Wax MB (2006) Serum autoantibodies to alpha-fodrin are present in glaucoma patients from Germany and the United States. Invest Ophthalmol Vis Sci 47:968–976
30. Tomarev SI, Wistow G, Raymond V, Dubois S, Malyukova I (2003) Gene expression profile of the human trabecular meshwork: NEIBank sequence tag analysis. Invest Ophthalmol Vis Sci 44:2588–2596
31. Flammer J, Mozaffarieh M (2007) What is the present pathogenetic concept of glaucoma? Surv Ophthalmol 52(Suppl 2):S162–S173
32. Golubnitschaja O, Yeghiazaryan K, Orgül S, Flammer J (2009) What are the biomarkers for glaucoma? In: Golubnitshaja O (ed) Predictive diagnostics and personalized treatment: dream or reality. Nova Science Publishers, New York
33. Golubnitschaja O (2010) Prädiktive, präventive und personalisierte medizin – eine einführung. In: Niederlag W, Lemke HU, Golubnitschaja O, Rienhoff O (eds) Personalisierte medizin. Health Academy, Dresden
34. Golubnitschaja O, Yeghiazaryan K, Flammer J (2010) Key molecular pathways affected by glaucoma pathology: is predictive diagnosis possible? EPMA J 1:237–244
35. Golubnitschaja O, Yeghiazaryan K, Wunderlich K, Schild HH, Flammer J (2007) Disease proteomics reveals altered basic gene expression regulation in leukocytes of glaucoma patients. Proteomics Clin Appl 1:1316–1323

36. Tezel G, Yang X, Cai J (2005) Proteomic identification of oxidatively modified retinal proteins in a chronic pressure-induced rat model of glaucoma. Invest Ophthalmol Vis Sci 46:3177–3187
37. Ahmed F, Brown KM, Stephan DA, Morrison JC, Johnson EC, Tomarev SI (2004) Microarray analysis of changes in mRNA levels in the rat retina after experimental elevation of intraocular pressure. Invest Ophthalmol Vis Sci 45:1247–1258
38. Golubnitschaja O, Fountoulakis M (2007) Gene hunting by substractive hybridisation in Down Syndrome. Correlation with proteomics analysis. In: Mandel S (ed) Handbook of neurochemistry and molecular neurobiology, Vol. 25: Genomics, proteomics and the nervous system, 3rd edn. Springer, Berlin-Heidelberg
39. Wunderlich K, Golubnitschaja O, Pache M, Eberle AN, Flammer J (2002) Increased plasma levels of 20S proteasome alpha-subunit in glaucoma patients: an observational pilot study. Mol Vis 8:431–435
40. Moenkemann H, Flammer J, Wunderlich K, Breipohl W, Schild HH, Golubnitschaja O (2005) Increased DNA breaks and up-regulation of both G(1) and G(2) checkpoint genes p21(WAF1/CIP1) and 14-3-3 sigma in circulating leukocytes of glaucoma patients and vasospastic individuals. Am Acid 28:199–205
41. Golubnitschaja O, Wunderlich K, Decker C, Mönkemann H, Schild HH, Flammer J (2002) Molecular imaging of perfusion disturbances in glaucoma. Am Acid 23:293–299
42. Grieshaber MC, Flammer J (2007) Does the blood–brain-barrier play a role in glaucoma? Surv Ophthalmol 52(Suppl 2):S144–S154
43. Golubnitschaja O, Yeghiazaryan K, Liu R, Mönkemann H, Leppert D, Schild H, Haefliger IO, Flammer J (2004) Increased expression of matrix metalloproteinases in mononuclear blood cells of normal-tension glaucoma patients. J Glaucoma 13:66–72
44. Diskin S, Kumar J, Cao Z, Schuman JS, Gilmartin T, Head SR, Panjwani N (2006) Detection of differentially expressed glycogenes in trabecular meshwork of eyes with primary open-angle glaucoma. Invest Ophthalmol Vis Sci 47:1491–1499
45. Bhattacharya SK, Rockwood EJ, Smith SD, Bonilha VL, Crabb JS, Kuchtey RW, Robertson NG, Peachey NS, Morton CC, Crabb JW (2005) Proteomics reveal cochlin deposits associated with glaucomatous trabecular meshwork. J Biol Chem 280:6080–6084
46. Fleenor DL, Pang IH, Clark AF (2003) Involvement of AP-1 in interleukin-1alpha-stimulated MMP-3 expression in human trabecular meshwork cells. Invest Ophthalmol Vis Sci 44: 3494–3501
47. Pang IH, Fleenor DL, Hellberg PE, Stropki K, McCartney MD, Clark AF (2003) Aqueous outflow-enhancing effect of tert-butylhydroquinone: involvement of AP-1 activation and MMP-3 expression. Invest Ophthalmol Vis Sci 44:3502–3510
48. Mohan R, Rinehart WB, Bargagna-Mohan P, Fini ME (1998) Gelatinase B/lacZ transgenic mice, a model for mapping gelatinase B expression during developmental and injury-related tissue remodeling. J Biol Chem 273:25903–25914
49. Esser J, Esser P, Mietz H, Hueber A, Kociok N, Schraermeyer U, Krieglstein GK (2000) Multidrug resistance-associated proteins in glaucoma surgery. Graefes Arch Clin Exp Ophthalmol 238:727–732
50. Yeghiazaryan K, Flammer J, Wunderlich K, Schild HH, Orgul S, Golubnitschaja O (2005) An enhanced expression of ABC 1 transporter in circulating leukocytes as a potential molecular marker for the diagnostics of glaucoma. Amino Acid 28:207–211
51. Joyce CW, Amar MJ, Lambert G, Vaisman BL, Paigen B, Najib-Fruchart J, Hoyt RF Jr, Neufeld ED, Remaley AT, Fredrickson DS, Brewer HB Jr, Santamarina-Fojo S (2002) The ATP binding cassette transporter A1 (ABCA1) modulates the development of aortic atherosclerosis in C57BL/6 and apoE-knockout mice. Proc Natl Acad Sci U S A 99:407–412
52. Bisoendial RJ, Hovingh GK, Levels JH, Lerch PG, Andresen I, Hayden MR, Kastelein JJ, Stroes ES (2003) Restoration of endothelial function by increasing high-density lipoprotein in subjects with isolated low high-density lipoprotein. Circulation 107:2944–2948
53. Pechere-Bertschi A, Sunaric-Megevand G, Haefliger IO, Panarello F, Maillard M, Burnier M (2007) Renal sodium handling in patients with normal pressure glaucoma. Clin Sci (Lond) 112:337–344

54. Schwartz B, Abrahamson R, Takamoto T, Wysocki A (2000) Association of ocular pressure and optic disc cup volume with red blood cell sodium-potassium ATPase inhibition. Curr Eye Res 21:897–905
55. Salomon P, Mysiak A, Halawa B (1998) Sodium efflux through lymphocytic cell membranes in patients with acute myocardial infarction. Pol Arch Med Wewn 100:543–550
56. Stark G (2005) Functional consequences of oxidative membrane damage. J Membr Biol 205:1–16
57. Silva JE (1995) Thyroid hormone control of thermogenesis and energy balance. Thyroid 5:481–492
58. Valdemarsson S, Ikomi-Kumm J, Monti M (1992) Increased lymphocyte thermogenesis in hyperthyroid patients. Role of Na/K pump function. Evaluation of aerobic/anaerobic metabolism. Acta Endocrinol (Copenh) 126:291–295
59. Bagrov AY, Dmitrieva RI, Fedorova OV, Kazakov GP, Roukoyatkina NI, Shpen VM (1996) Endogenous marinobufagenin-like immunoreactive substance. A possible endogenous Na, K-ATPase inhibitor with vasoconstrictor activity. Am J Hypertens 9:982–990
60. van den Bergh WM, Dijkhuizen RM, Rinkel GJ (2004) Potentials of magnesium treatment in subarachnoid haemorrhage. Magnes Res 17:301–313
61. Kotikoski H, Moilanen E, Vapaatalo H, Aine E (2002) Biochemical markers of the L-arginine-nitric oxide pathway in the aqueous humour in glaucoma patients. Acta Ophthalmol Scand 80:191–195
62. Baudouin C, Garcher C, Haouat N, Bron A, Gastaud P (1994) Expression of inflammatory membrane markers by conjunctival cells in chronically treated patients with glaucoma. Ophthalmology 101:454–460
63. Souchier M, Buron N, Lafontaine PO, Bron AM, Baudouin C, Creuzot-Garcher C (2006) Trefoil factor family 1, MUC5AC and human leucocyte antigen-DR expression by conjunctival cells in patients with glaucoma treated with chronic drugs: could these markers predict the success of glaucoma surgery? Br J Ophthalmol 90:1366–1369
64. Golubnitschaja O, Flammer J. International Patent No. IB02/00648
65. Golubnitschaja O, Flammer J (2007) What are the biomarkers for glaucoma? Surv Ophthalmol 52:S155–S161

Part V

Toward a Personalized Approach in Amyotrophic Lateral Sclerosis: New Developments in Diagnosis, Genetics, Pathogenesis and Therapies

Beatrice Nefussy and Vivian E. Drory

Abstract Amyotrophic lateral sclerosis (ALS) is a rare neurodegenerative disease that affects upper and lower motor neurons in the brain and spinal cord, with progressive weakness and atrophy of most muscles in the body and is almost always fatal within 3–5 years. A small proportion of cases are familial, and remarkable achievements have been made during the last years in understanding the genetics of the disease. In spite of this, the basic pathogenic mechanisms underlying the sporadic disease are still poorly understood. There is an urgent need for better understanding of the pathogenic processes in order to be able to develop effective treatments.

The present review will focus on recent knowledge gained in diagnosis, genetics, pathogenesis and therapies in ALS. Future development of diagnostic technologies integrating genetic, environmental and individual information will enable us to predict a population at risk for ALS. New treatments actually in development will help improve the medical management of ALS patients, taking into consideration individual traits, as genetic background, and pave a way for a more effective personalized diagnostic and treatment approach.

Keywords Amyotrophic lateral sclerosis • Motor neuron disease • Neurodegeneration • Genetics • Epidemiology • Treatment

B. Nefussy (✉) • V.E. Drory
Neuromuscular Service, Department of Neurology, Tel-Aviv Sourasky Medical Center, 6, Weizmann street, 64239 Tel-Aviv, Israel

Sackler Faculty of Medicine, Tel-Aviv University, Tel-Aviv, Israel
e-mail: beatrisef@tlvmc.gov.il

S. Mandel (ed.), *Neurodegenerative Diseases: Integrative PPPM Approach as the Medicine of the Future*, Advances in Predictive, Preventive and Personalised Medicine 2, DOI 10.1007/978-94-007-5866-7_10,

1 Background

Amyotrophic lateral sclerosis (ALS), also known as Lou Gehrig's disease or Maladie de Charcot, is a progressive, fatal neurodegenerative disease, which causes premature selective death of motor neurons in the primary motor cortex, corticospinal tracts, brainstem and spinal cord. It affects lower motor neurons (LMN) that reside in the anterior horns of the spinal cord and in the brain stem; corticospinal upper motor neurons (UMN) that reside in the precentral gyrus of the brain cortex; and, frequently, prefrontal motor neurons that are involved in planning or orchestrating the work of both UMN and LMN [1]. It progresses rapidly to cause denervation and ultimately weakness of most muscles of the body. The disease is almost always fatal and results in death due to paralysis of the respiratory muscles within 3–5 years of onset [2].

2 Epidemiology

The incidence of ALS is around 1.8–2.5 per 100,000 persons/year and is evenly distributed across Europe and North America [3]. An unusually high incidence was reported around the Pacific ring, on Guam [4], and the Kii peninsula in Japan [5]. Men are slightly more affected than women, with a male: female ratio about 1.5:1, although more recent data suggest that this ratio is smaller, approaching one [6, 7]. The age of onset of sporadic ALS is very variable, from the third to the ninth decade, with a mean of 63 years [8].

3 Clinical Features

The clinical features of ALS are a direct consequence of the progressive loss of upper and lower motor neurons, with secondary denervation and subsequent reinnervation of muscles. As long as reinnervation can compensate for denervation, clinical weakness may not be detectable. However, as the motor units grow larger and their number begins to decrease, the earliest consequence is that affected muscles may fatigue faster than muscles with normal motor units; consequently, one of the first symptoms of ALS may be increased fatigue. As the number of motor units innervating a muscle decreases further, reinnervation is no longer able to compensate for denervation, permanent weakness develops and progresses and the affected muscles gradually undergo atrophy.

The parts of the body affected by early symptoms in ALS depend on which motor neurons are damaged first. In about 75% of patients the disease starts with weakness of lower or upper limb muscles, and patients experience tripping or stumbling, difficulty with tasks such as buttoning a shirt or writing. Affected muscles may

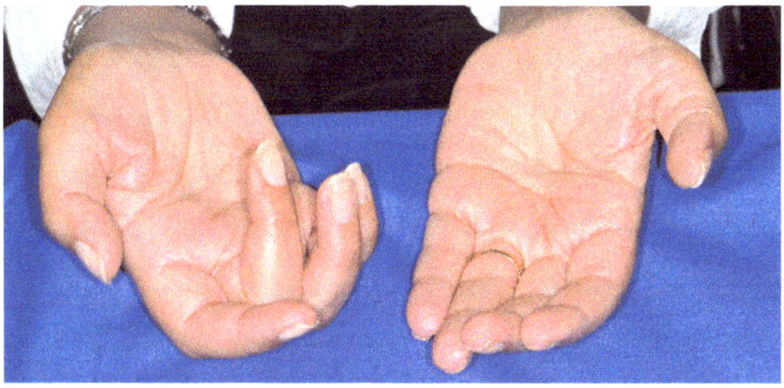

Fig. 1 Atrophy of thenar eminence and weakness of finger extensors, more prominent on the right, in a 70-years-old woman with symptoms of ALS for 6 months

develop twitching, cramping, or stiffness (limb form). In the other 25% of cases the site of onset is the motor nuclei in the brainstem and/or their corticobulbar neurons. These patients experience first slurred and nasal speech, followed by difficulty speaking clearly, difficulty swallowing and loss of tongue mobility (bulbar form).

Regardless of the region of the body first affected, muscle weakness and atrophy spread to other regions of the body as the disease progresses. Patients experience increasing difficulty moving, swallowing (dysphagia), and speaking or forming words (dysarthria). The ineffective swallowing of saliva leads to excessive drooling (sialorrhea). Weakness of breathing muscles leads to respiratory insufficiency.

At neurological examination, a combination of UMN and LMN signs can be found: symptoms of UMN involvement include increased muscle tone (spasticity), exaggerated reflexes (hyperreflexia), pathological pyramidal signs such as extensor plantar responses, loss of dexterity in the presence of normal strength. Symptoms of LMN degeneration include muscle atrophy (Fig. 1) and twitching (fasciculations), diminished to absent tendon reflexes. Fasciculations are not specific for ALS, but can be found in other peripheral nervous system diseases and also at times in healthy individuals, although a finding of large and fast-firing fasciculations at many sites is very indicative for the disease.

Around 15–45% of patients experience uncontrollable emotional symptoms, as inappropriate laughter, crying or smiling, a feature associated with speech and swallowing disturbances in the complex of pseudobulbar syndrome, characteristic of bilateral corticobulbar tracts involvement.

Although the sequence of emerging symptoms and the rate of disease progression are variable, most patients will develop with time major disability in all motor tasks, will not be able to stand or walk, or use their hands and arms. Difficulty swallowing and chewing impair the patient's ability to eat and increase the risk of choking and pulmonary infections. Loss of weight is a consequence of impaired nutrition and muscle wasting. Because the disease usually does not affect cognitive abilities, patients are aware of their progressive loss of function and may become anxious and depressed. On the other hand, the preserved cognition enables patients to take decisions, including those related to their own end-of-life, by themselves, even after they reach a totally immobile state (locked-in).

Up to 50% of patients experience behavioral disturbances, difficulties with word generation, attention, or decision making. 5–10% develop overt frontotemporal dementia, characterized by personality changes and impairment of executive functions and language; this is more common among those with a family history of dementia [9].

As the diaphragm and intercostal muscles weaken, forced vital capacity and inspiratory pressure diminish. Most patients with ALS die of respiratory failure or pneumonia [10]. Death usually occurs within 2–5 years of diagnosis (median 27 months). Less than 5% of patients survive more than 10 years. 6% have an arrested form of disease. In general, survival is better for younger patients and those with limb onset [11].

Interestingly, there are few motor functions that are usually not affected in ALS: control of eye muscles is the most preserved function, although some patients with an extremely long duration of disease may lose eye movement control too. Bladder and bowel control are usually preserved as well, although as a result of immobility and diet changes, intestinal problems such as constipation may require intensive management.

4 Diagnosis

Diagnostic criteria for ALS were developed by expert consensus (El Escorial criteria [12]). By El Escorial criteria the diagnosis of ALS requires all of the following: evidence of UMN degeneration by clinical examination, evidence of LMN degeneration by clinical, electrophysiological or neuropathological examination, evidence of progressive spread of symptoms or signs within a region or to other regions, in the absence of electrophysiological, neuroimaging and pathological evidence of other disease processes that might explain the clinical picture.

The El Escorial criteria define categories of clinical diagnostic certainty. For this purpose four body regions were defined: cranial, including tongue and neck muscles, cervical, thoracic and lumbar. The diagnosis of ALS should be regarded as definite when UMN and LMN signs are found in three body regions. Probable ALS is defined as having either UMN and LMN signs in two regions with at least some UMN signs rostral to LMN signs, or UMN signs in one or more regions and LMN signs by electromyography (EMG) in at least two regions. Possible ALS is the least certain degree and is defined as UMN and LMN signs in a single body region, or UMN signs in two body regions, or UMN and LMN signs in two regions with no UMN signs rostral to LMN signs.

It has to be taken into consideration that the El Escorial criteria were developed for research and may be too restrictive for clinical use. In one series, 44% of patients did not meet criteria for at least probable ALS at diagnosis, and 10% did not meet criteria even at death [13].

The most important diagnostic tool in the diagnosis of ALS is the EMG examination. EMG is able to detect signs of a LMN lesion in muscles that are apparently unaffected, therefore increases the degree of evidence of the diagnosis, mainly in patients with a predominantly UMN picture or a clinically restricted syndrome. The required EMG findings by the El Escorial criteria cited above are signs of denervation (fibrillations) aside fasciculations and signs of re-innervation. Another set of electrodiagnostic criteria, termed "Awaji criteria", was proposed recently in order to facilitate earlier diagnostic characterization for clinical purposes. This set of criteria broadens the electrophysiologic characterization of ALS to include signs of chronic denervation as equivalent to clinical signs of LMN involvement. In addition, the same criteria propose that in the context of suspected ALS, fasciculation potentials should be considered equivalent to signs of active denervation [14]. This set of criteria still remains to be validated in prospective clinical studies.

Neuroimaging studies and other laboratory tests are used to exclude alternative diagnoses. At times, a brain MRI may show bilateral hyperintensities on fluid-attenuated inversion recovery sequence (FLAIR), corresponding to the degeneration of the corticospinal tracts, but the finding is neither specific, nor frequent enough in order to be of use in the diagnostic confirmatory process [15]. Developments of more advanced neuroimaging techniques for confirmation of subclinical UMN involvement are currently underway, as functional MRI (fMRI) [16] and diffusion-tensor imaging (DTI) [17–19].

The differential diagnosis includes other diseases of the UMN, as multiple sclerosis, primary lateral sclerosis, degenerative processes of the cervical spinal cord, syringomyelia, and other diseases that lead to muscle weakness and wasting, as inclusion body myositis [20], post-polio-syndrome, multifocal motor neuropathy with conduction block, paraneoplastic LMN syndromes [21]. Rare motor neuron diseases, as spinobulbar muscular atrophy (Kennedy disease) [22], adult GM2 gangliosidosis (hexosaminidase A deficiency) [23], juvenile muscular atrophy of distal upper extremity (Hirayama disease) [24] have to be ruled out in specific patients.

An incorrect diagnosis of ALS can have devastating emotional and social consequences for patients and their families, and it can delay appropriate treatment of an alternative disease process; therefore, it is important to consider diseases that can mimic ALS or that can contribute to the patient's symptoms.

5 Genetics

5.1 Familial ALS (FALS)

The classical view is that approximately 5–10% of ALS cases are familial [25], whereas most patients diagnosed with the disease are classified as sporadic, as they appear to occur randomly throughout the population. The most commonly reported patterns are dominant inheritance with complete penetrance, dominant inheritance with incomplete penetrance and recessive inheritance. During the last few years it is recognized that the proportion of genetically determined disease might be much larger, as gene defects with low penetrance were described and a large list of genes was found that may possibly not cause the disease directly, but contribute to morbidity in patients without a family history of ALS [26]. A meta-analysis of three twin studies calculated an estimate of the heritability of ALS in twins as high as 61% [27]. The risk to develop ALS in a population-based study in Sweden that made no distinction between FALS and SALS was reported to be 17-fold for siblings and 9-fold for children of patients with ALS compared with a reference group [28]. A clinic-based study reported that over a 16-year period 0.5% of siblings and 1% of offspring of patients with apparently sporadic ALS subsequently developed ALS [29]. The estimate of the risk to a first-degree relative may seem high, but should be put in context. The lifetime risk of not developing ALS by age 85 would drop from the population background of about 99.7–97.6%, which means that the likelihood that a first-degree relative with no other family history will die of causes other than ALS is still almost absolute.

Identification of the genetic causes of ALS has proven difficult even with modern technologies, including genome-wide association studies and exome capture sequencing studies. Many factors lead to underestimation of FALS, among them are the combination of inadequate recording of family history, small family size with resulting smaller chances for other members to be affected, late-onset presentation, phenotypic variability and incomplete disease penetrance (reviewed in [30]).

Linkage studies in ALS have been restricted to few large pedigrees with Mendelian inheritance and high penetrance. As mentioned above, approximately 5–10% of ALS cases have a clear family history of disease in first degree relatives, with a Mendelian pattern of inheritance. The disease is transmitted in an autosomal dominant fashion with complete or incomplete penetrance in most cases. Mean age at onset is 10–20 years younger in patients with FALS than in patients with sporadic disease, and variability between families is greater than variability within families. Most cases of FALS are indistinguishable from sporadic disease; others have unique phenotypes [31, 32].

The first described genetic abnormality is a mutation in the copper/zinc superoxide dismutase (*SOD1*) gene, on chromosome 21q, found in 10–20% of familial cases and 2% of patients with sporadic ALS (SALS) [33]. More than 160 allelic variations have been described in the SOD1 mutants, most of them are missense mutations, leading to single base substitutions with a subsequent single amino acid exchange. The rest are nonsense and deletion mutations that result in a change in length of the SOD1 polypeptide. It is not yet known how mutant SOD1 causes ALS, but mutations in the gene are thought to cause disease through a toxic gain-of-function rather than causing impairment of the antioxidant function of the SOD1 enzyme [34]. The disease age at onset and duration are reported to be closely linked to the type of SOD1 mutation. Certain SOD1 mutations are associated with rapidly progressive disease and short survival, as A4V, while others present with a relatively benign course [35].

The discovery that TDP-43 protein, encoded by *TARDBP* gene on chromosome 1p (encoding the TAR-DNA binding protein TDP-43), is a constituent of ubiquitinated inclusions in the cytoplasm of motor neurons of patients with SALS, as well as in patients with non-SOD1 FALS and in patients with frontotemporal dementia (FTD) [36, 37], constituted an important development in ALS research. TDP-43 is an RNA processing protein that is normally localized in the nucleus, and loss of nuclear TDP-43 staining is accompanied by TDP-43 positive cytoplasmic inclusions [36]. Over 30 dominant mutations in the *TARDBP* gene have been reported to date, accounting for 4% of FALS and 1.5% of SALS cases [38–41]. A mutant mouse model was recently described that develops a progressive and fatal neurodegenerative disease with both ALS and FTD features [42].

Mutations in the fused in sarcoma/translated in liposarcoma (*FUS/TLS*) gene were identified in 4% of FALS [43, 44] and in very few SALS cases [41, 45]. Interestingly, like *TDP-43*, this gene has a putative role in RNA regulation and is similarly translocated from the nucleus to the cytoplasm in spinal cord motor neurons.

Mutations in alsin (*ALS2*), vesicle associated membrane protein type B (*VAPB*) [46], the p150 subunit of dynactin [47], angiogenin (*ANG*) [48], senataxin (*SETX*) [49], optineurin (*OPTN*) [50], valosin-containing protein (*VCP*) [51] and most recently ubiquillin 2 (*UBQLN2*) [52] have also been reported in rare cases.

A complete list of ALS-associated genes is available along with analysis tools at the ALS Online Database [53, 54].

Until very recently only 25–30% of familial cases could be attributed to known genetic mutations. The new GGGGCC hexanucleotide repeat expansion in a non-coding region of *C9ORF72* gene on chromosome 9p21 has been recently identified as the most frequent cause of both FALS and familial FTD [55, 56]. This repeat expansion appears to be responsible for 23–46% of familial ALS cases, depending on the population studied [55–59]. Patients with ALS, FTD, or ALS-FTD have 700–1,600 repeats, whereas healthy individuals have less than 30 repeats [56].

This is for the first time that a repeat expansion mutation has been found to cause directly ALS, although an intermediate polyglutamine expansion in the ataxin 2 (*ATXN2*) gene was also recently identified as a contributor to the risk of developing ALS [60–62].

The *C9ORF72* mutation is more than twice as common as mutations in the *SOD1* gene in FALS, and more than three times as common as mutations in *TARDBP* (or *TDP43*), *FUS*, *OPTN* and *VCP* genes combined. This expansion was also found to be very frequent in patients with apparently sporadic ALS (5–7%) and sporadic FTD.

In spite of these major achievements during the last few years, the genetic defect causing ALS remains still unknown in half of the affected families and a yet unknown proportion of apparently sporadic patients harbor yet undetectable mutations.

Known genes and linkages associated with familial ALS, including rarely encountered genetic defects not mentioned above, are summarized in Table 1.

Table 1 Genes and linkages involved in familial ALS

Disease name	Genetic locus	Protein name (function)	Gene symbol	Onset	Inheritance	Incidence in familial ALS
ALS1	21q22.1	Cu-Zn superoxide dismutase 1	*SOD1*	Adult	Dominant	10–20%
ALS2	2q33	Alsin	*ALS2*	Juvenile	Recessive	Rare
ALS3	18q21	Unknown	Unknown	Adult	Dominant	Rare
ALS4	9q34	Senataxin	*SETX*	Juvenil	Dominant	Rare
ALS5	15q15-21	Spatacsin	*SPG11*	Juvenile	Recessive	Rare
ALS6	16p11.2	Fused in sarcoma	*FUS*	Adult	Dominant/recessive	4–5%
ALS7	20p13	Unknown	Unknown	Adult	Dominant	Rare
ALS8	20q13.3	Vesicle-associated membrane protein-associated protein B	*VAPB*	Adult	Dominant	Rare
ALS9	14q11.2	Angiogenin	*ANG*	Adult	Dominant	Rare
ALS10	1p36.2	TAR–DNA binding protein 43 (TDP-43)	*TARDBP*	Adult	Dominant/recessive (rare)	1–4%
ALS11	6q21	Polyphosphoinositide phosphatase	*FIG4*	Adult	Dominant	Rare
ALS12	10p13	Optineurin	*OPTN*	Adult	Dominant Recessive	1–4%
ALS?	9p13-p12	Valosin-containing protein	*VCP*	Adult	Dominant	1–2%
ALS-FTD	9q21-22	Chromosome 9 open reading frame 72	*C9ORF72*	Adult	Dominant	23–46%
ALS-FTD	9p13.3	Non-opioid intracellular receptor 1	*SIGMAR1*	Adult Juvenile	Dominant/recessive	Unknown
ALS-FTD with parkinsonism	17q21	Microtubule-associated protein tau	*MAPT*	Adult	Dominant	Rare
ALS-X	Xp11	Ubiquillin 2	*UBQLN2*	Adult	Recessive	Rare

Abbreviations: *ALS* amyotrophic lateral sclerosis, *FTD* Frontotemporal dementia

5.2 *Sporadic ALS*

The causes of SALS are still unknown. It is considered to be a multifactorial disease, in which both environmental and genetic factors interplay to trigger disease onset and modulate progression. Most ALS cases are probably due to oligogenic inheritance, but monogenic inheritance is also possible, as recently proven in the case of the newly identified hexanucleotide expansion in the gene *C9ORF72*. Mutations in *C9ORF72* appear to be the cause of SALS in about 20% of patients in the Finnish population that is genetically homogenous [55], 5% in the Flanders-Belgian and in the Irish population [57, 58]. In a very large cross-sectional study the repeat expansion was identified in 7% of sporadic white patients from the USA, Europe, and Australia, 4% of black individuals from the USA, and 8% of Hispanic individuals from the USA [59]. The mutation was not carried by patients from Asia or the Pacific Islands with sporadic ALS who were tested [59]. In a population-based incident cohort in Ireland the mean age of onset was 5 years younger and survival time significantly shorter in sporadic patients with the C9ORF72 mutation [58]. Cognitive and behavioral impairment on neuropsychological testing and a distinct pattern of non-motor cortex changes on high-resolution MRI were significantly associated with carriers of the C9ORF72 mutation [58].

The completion of a haplotype map of the human genome [63] and development of advanced genotyping technology have enabled the performance of genotype-wide association studies (GWAS), based on the theory that even sporadic ALS has a significant, but complex genetic contribution from common variants, each exerting individually a small effect. The genome-wide SNP association analysis is a powerful, relatively new method to identify, in an unbiased manner, genetic loci and genes underlying complex diseases. SNP arrays allow large-scale linkage analysis, association, and copy number studies. A number of GWAS have been performed in SALS, however to date this method has not revealed reproducible potential links to ALS [64–67]. This is due probably to a large degree of genetic heterogeneity, as well as to the possibility that disease susceptibility is influenced by additional environmental factors.

One of the most recent and largest GWAS analyses of SALS did not reveal any correlation with risk of disease; however it demonstrated for the first time an association between a gene (kinesin-associated protein 3 – *KIFAP3*) and survival [68]. Patients who were homozygotes for the favorable allele of the gene survived on average 14 months longer than heterozygotes and homozygotes for the unfavorable allele. Although the association between *KIFAP3* gene and survival in ALS has not been confirmed, genetic variation in *KIFAP3* was found to be associated with an upper motor neuron-predominant phenotype in ALS [69].

A specific variant in the *chromogranin B* (*CHGB*) gene has been reported to confer a 2.2-fold greater relative risk to develop ALS and was also associated with an earlier age at onset by almost 10 years in both SALS and FALS patients, therefore the CHGB gene was considered as apparently both a susceptibility and a phenotype modifier gene [70]. Further findings did not support the association between the chromogranin B variant and ALS susceptibility or age of onset [71, 72].

As mentioned above, intermediate-length polyglutamine repeats in *ATXN2* gene, encoding the ATXN2 protein have recently been identified as a significant risk factor for sporadic ALS in North American patients [60], a finding that has been confirmed in other populations [61, 62, 73, 74]. ATXN2 is a potent modifier of TDP-43 toxicity in animal and cellular models. ATXN2 and TDP-43 associate in a complex that depends on RNA and this interaction may be a target for therapeutic interventions in ALS [60].

It is important to mention that all genes mutated in FALS have also been found mutated in patients without a family history. Together with the fact that besides a lower mean of age of onset, FALS and SALS cases are clinically indistinguishable, and that similar histopathological inclusions have been found in both SALS and FALS, this implies that many cases labeled as sporadic may actually have a family history, and it is worth considering again if the classification of sporadic versus familial disease is biologically justified [30].

6 Individual Risk Factors

It should be remembered that association does not mean cause, but rather the first step in identifying a possible risk factor. When applying an evidence-based approach to the analysis of the epidemiological data, smoking exposure, including patients who stopped smoking, is the only established exogenous risk factor for occurrence of ALS [75, 76].

Head trauma, physical activity, residence in rural areas, and alcohol consumption were not proven as risk factors for ALS, although implicated in certain studies [77]. Specific occupational exposures are also mentioned as risk factors for ALS, although analyses of occupation as potential determinants of ALS did not reach a definitive conclusion [78].

Two putative risk factors that have gained recent attention include military service during the Gulf War [79, 80], and being a soccer player [81, 82]. However, the level of evidence supporting their role in triggering ALS is low [83, 84].

The use of statins has been mentioned as possibly associated with an increased incidence of ALS [85], but an analysis by the US Food and Drug Administration (FDA), undertaken after the agency received a higher than expected number of reports regarding occurrence of ALS among individuals on statins, failed to reveal an increased incidence of ALS in subjects treated with statins compared with placebo [86].

Individual metabolic traits encountered frequently in the general population were implicated recently as possible modulatory factors, able to influence prognosis in affected individuals. It was shown that an abnormally elevated low-density lipoprotein (LDL)/high-density lipoprotein (HDL) ratio increased mean patient survival by more than 12 months [87]. Another study demonstrated that ALS patients had lower mean serum uric acid levels than healthy individuals. The decreased uric acid levels were correlated to the rate of disease progression [88]. Pre-existent diabetes mellitus was shown to delay the onset of ALS-related symptoms by 4 years in another study [89]. Interestingly, all these metabolic factors are considered detrimental in the daily clinical practice, but for ALS patients they carried a favorable prognosis. Probably they are also interrelated, but were not analyzed as a group yet.

7 Pathophysiology and Pathogenesis

7.1 Involvement of Non-neuronal Cells

Many insights into the pathogenic mechanisms of ALS in the past decade came from studies in transgenic mice carrying the human SOD1 mutation. SOD1 is a ubiquitously expressed cytoplasmic enzyme that converts superoxide anions to hydrogen peroxide. However, mutant SOD1 confers its toxic properties without altering enzymatic activity [90–93].

A major contribution to understanding disease mechanisms was the discovery that ALS is not just cell-autonomous, but is influenced by the surrounding environment, namely by the interaction between motor neurons and non–neuronal cells, mainly glial cells. Initial experiments that selectively expressed mutant SOD1 in motor neurons or glial cells did not induce motor neuron degeneration or disease [94–96]. An elegant experiment in chimeric ALS mice demonstrated that wild-type non-neuronal cells extended survival of SOD1 mutant motor neurons in ALS mice, implicating that neighboring non-neuronal cells are important in determining survival [97]. The same study reported pathological abnormalities and cell death in wild-type motor neurons adjacent to mutant SOD1-expressing non-neuronal cells. Thus, the vulnerability of motor neurons to SOD1-mediated toxicity appears to be influenced by their cellular environment, created by their adjacent non-neuronal cells. Astrocytes that express mutant SOD1 secrete probably factors that are toxic to motor neurons, although such factors have not yet been identified [98]. Mutant SOD1 microglia produces and releases more superoxide and nitrite + nitrate and induce more neuronal death, compared to wild-type microglia [99].

It has also been demonstrated that selective reduction in mutant SOD1 expression within microglia [100] and astrocytes [101] significantly slows disease progression, but does not affect the timing of disease onset. These findings indicate that disease onset might have different genetic and cellular mechanisms. Disease onset is probably non-cell autonomous, and mutant SOD1 damage within cell types other than motor neurons and oligodendrocytes is a central contributor to initiation of motor neuron degeneration [102].

Mutant SOD1 mice and rats have been used extensively as a model to study motor neuron death in ALS and to study the efficacy of various drugs as possible therapeutic agents. However, emerging evidence shows that SOD1 mediated toxicity is not a major cause of SALS, therefore also the animal model based on SOD1 transgenicity reflects poorly the human disease. This fact is indicated clearly in the results of trials of potential drugs, in which therapeutic agents that were effective in this model failed in subsequent human clinical trials.

7.2 Glutamate Excitotoxicity

Glutamate-mediated excitotoxicity was the first proposed mechanism of motor neuron degeneration in ALS, following the finding of increased levels of glutamate in the cerebrospinal fluid of ALS patients [103]. ALS has been associated with reductions in the level of the glial cell excitatory amino acid transporter EAAT2 [104]. This defect might lead to an abnormal accumulation of extracellular glutamate and subsequently repetitive firing of motor neurons.

7.3 Free Radical Damage

The concept that oxidative damage is involved in ALS was based on the discovery that mutations in the *SOD1* gene cause FALS. Increased levels of 3-nitrityrosine, a marker for peroxinitrite – mediated oxidative damage, have been reported in ALS patients [105]. Peroxinitrite causes cell damage by nitration of tyrosine residues in target proteins. We have just recently reported that ALS patients have a significantly lower level of uric acid, a peroxinitrite specific antioxidant, compared to control subjects and that the concentration of uric acid was reversely correlated with the rate of disease progression [88]. Nevertheless, although oxidative damage has been linked to neurodegeneration, many antioxidants tested in clinical trials have failed to show benefit.

7.4 Intracellular Aggregates

In both FALS and SALS, affected neurons and glial cells contain abnormal inclusions of proteins, often labeled by anti-ubiquitin antibodies. Misfolded SOD1 protein is present in both FALS and SALS [106, 107]. It was recently demonstrated that wild-type SOD1 and mutant SOD1 share a conformational epitope prone to oxidation and therefore exert a strong neurotoxic effect [106, 107]. These data strongly support the hypothesis that either mutated or wild-type SOD1 misfolding contributes to disease pathogenesis in both familial and sporadic ALS.

Ubiquitinated inclusions containing pathologic forms of TAR DNA-binding protein-43 (TDP-43) were identified in the cytoplasm of motor neurons of patients with SALS, non-SOD1 FALS and patients with FTD [36, 37, 108, 109]. Cases with *SOD1* mutations had ubiquitin-positive neuronal inclusions; however, they were not immunoreactive for TDP-43. These findings implicate a common pathogenesis in SALS and non-SOD1 FALS. In contrast, SOD1 mutations and TDP-43 pathology probably involve different mechanisms [37, 110].

FUS/TLS-containing cytoplasmic inclusions are present in the brain and spinal cords of patients with FUS/TLS mutations [43, 44].

7.5 *Alteration in RNA Processing*

Both TDP-43 and FUS/TLS are genes that encode RNA processing proteins [44], reviewed in [41]. Patients with the FUS/TLS mutations have cytoplasmic inclusions containing FUS/TLS but not TDP-43. Although TDP-43 inclusions are not observed in FUS mutant cases, FUS is an RNA-binding protein that is found in complexes together with TDP-43, and ALS-associated TARDBP mutations promote the interaction between TDP43 and FUS [111].

TDP-43 and FUS/TLS have striking structural and functional similarities, implicating alterations in processing of RNA and microRNA a as a key event in ALS pathogenesis, although the underlying functional mechanisms remain unknown.

Furthermore, the hexanucleotide repeat expansion of the gene C9ORF72 leads to the loss of one alternatively spliced C9ORF72 transcript and to the formation of nuclear RNA foci [56].

Additional evidence of dysfunctional RNA metabolism in ALS emerges from the presence of mutations in angiogenin, which in addition to its role to promote angiogenesis during hypoxia also acts as a transfer RNA-specific ribonuclease and regulates ribosomal RNA transcription, and the DNA-RNA helicase senataxin. Mutations in *ANG* and *SEXT* are thought to result in loss of function of the encoded proteins in regulating transcription.

7.6 *Mitochondrial Dysfunction and Impaired Energy Metabolism*

Morphological and functional defects in mitochondria were found in both human patients and transgenic mice. Mutant SOD1 was found to be associated with mitochondria and subsequently to impair mitochondrial function. Recent studies suggest that axonal transport of mitochondria along microtubules and mitochondrial dynamics may be disrupted in ALS (reviewed in [112]). Creatine, an agent that improves mitochondrial function, has been shown to be neuroprotective in animal models of ALS [113]; however, in humans creatine showed only a trend toward beneficial effect in survival and did not influence motor, respiratory or functional measures [114].

ALS is associated with several defects in energy metabolism, including weight loss, hypermetabolism, and hyperlipidemia. Most of these abnormalities correlate with duration of survival, and available clinical evidence supports a negative contribution of defective energy metabolism to the overall pathogenic process (reviewed in [115]).

7.7 Dysregulation of Endosomal Vesicle Trafficking

Recessive mutations in the *ALS2* gene cause juvenile-onset ALS. Alsin is a guanine nucleotide change factor for the small GTPase protein Rab5 and is involved in endosomal fusion and trafficking [116]. Mutations in *VAPB*, which is primarily located in the endoplasmic reticulum and is involved in intracellular membrane transport, *OPTN*, a component of the endosomal sorting complex, and VCP, which forms a complex with the endocytic protein clathrin, further implicate dysregulation of endosomal trafficking as a probable important mechanism in the disease. Mutations in the *FIG4* gene have been found in a few familial cases [117]. *FIG4* encodes a polyphoshoinositide phosphatase that regulates the level of phosphatidylinositol 3,5-bisphosphate, a signaling lipid that mediates retrograde trafficking of endosomal vesicles to the Golgi apparatus [118].

Impaired axonal transport, lack of neurotrophic factors, caspase-mediated cell death (apoptosis), and neuroinflammation are also probably involved in ALS (reviewed in [119, 120]).

In spite of major scientific achievements in the field during the last years, the exact mechanisms responsible for the onset and progression of motor neuron degeneration remain largely unknown.

8 Spread of Disease

Neurodegeneration in ALS typically begins focally and asymmetrically in upper or lower limbs or with bulbar signs and then spreads spatiotemporally to contiguous groups of motor neurons. This pattern of spread implies propagation from an initial site of injury to neighboring healthy cells.

Prion-like domains in FUS and TDP-43 were just recently identified and proved to be required for aggregation [121]. Prion-like spread of SOD1 and TDP-43 misfolded proteins in ALS was recently proposed [122].

9 Biomarkers

The absence of biomarkers poses a significant problem to the diagnostic process of the disease in its early stages, thus limits the possibilities of correct clinical management of patients and delays time to initiation of adequate clinical care. The identification of disease-specific markers will also enable monitoring of disease progression and testing of drug efficacy in clinical trials, as well as improve understanding of disease mechanisms. In the absence of validated biomarkers for ALS, a combination of outcome measures is usually being used in clinical trials, including respiratory capacity, muscle strength, and the ALS functional rating scale, which scores a person's disability during daily activities.

Much of the efforts in finding biomarkers have concentrated in screening for protein markers, using mass spectroscopy in cerebrospinal fluid (CSF) and blood samples. Promising candidates include neurofilament heavy subunit [123], inflammatory molecules such as complement component C3 [124]; and other proteins such as cystatin C [125]. It was recently published that the combination of the phopsphorylated neurofilament heavy chain and complement compound C3 best distinguishes ALS from diseases that mimic ALS [126].

Metabolomics is a complementary technique to genomics, transcriptomics and proteomics. Recently, a proton nuclear magnetic resonance based metabolomic study reported that acetone, pyruvate, and ascorbate were increased, while acetate was decreased in the CSF of ALS patients compared to controls [127].

Disease-related changes in the CSF metabolome in ALS were just recently published. ALS patients with the SOD1 gene mutation appear to have a distinct metabolic profile in the CSF compared with ALS patients without mutations [128].

10 Treatment

The modern management of ALS is based on the recognition of the importance of multidisciplinary care. Although there is no cure for the disease, many of the symptoms that develop during the course of the disease are treatable and it is crucial to consider adjunctive measures to maintain function and independence and improve quality of life in patients with ALS. The improvement in symptomatic care of the patients during the last decade led to an improved rate of survival [2].

10.1 Non-drug Therapy

10.1.1 Ventilatory Management

Respiratory insufficiency is a common feature in patients with ALS, usually in advanced stages, although it may be the presenting clinical symptom in certain cases, and is the most frequent cause of death.

The respiratory status in ALS patients should be monitored serially by measurement of forced vital capacity (FVC) or sniff nasal pressure and questioning the patient about symptoms of respiratory insufficiency such as insomnia, daytime fatigue and morning headaches. Overnight pulse oximetry or polysomnography should be obtained as clinically indicated.

Respiratory rehabilitation therapy such as insufflation-exsufflation and chest physiotherapy can help maintain lung function and avoid secondary atelectasis and infections [129].

Non-invasive ventilatory support (NIV) should be considered in patients with symptoms of respiratory insufficiency when FVC (sitting or supine) is at or below 50% of predicted normal; or if there is evidence of nocturnal hypoventilation (O_2 saturation <90%), even if sitting and supine FVC are not significantly reduced.

NIV is usually provided by bi-level positive pressure devices (BiPAP). The rationale of using NIV is that non-invasive ventilation can improve survival and quality of life, and postpone the need for tracheostomy [130, 131]. In patients with severe bulbar impairment, NIV improves sleep-related symptoms, but it does not confer a large survival advantage [130].

During the last years it has been recognized that initiation of NIV might be indicated earlier than when FVC drops below 50% predicted. Patients treated early with NIV had an increased survival at 1 year and their median decline in FVC was slower than in patients who did not use NIV [132]. A Cochrane review analyzed recently the efficacy of NIV for ALS and concluded that NIV is able to prolong survival and to improve and maintain quality of life [133]. Similarly, a revision of the Practice Parameters of the American Academy of Neurology published in 2009 concluded that NIV is probably effective in prolonging survival and in slowing the rate of FVC decline, and possibly effective in improving quality of life. Therefore NIV should be considered to treat respiratory insufficiency in ALS [134].

Diaphragm pacing is a new technique using a device that has recently been approved by the U.S. Food and Drug Administration to treat respiratory insufficiency in ALS. The system is designed to supplement breathing efforts and preserve diaphragm muscle function; it may delay or abolish the need for invasive ventilation via tracheostomy and can be used in conjunction with non-invasive ventilation. This pacing system is indicated for patients with ALS who still have some preservation of diaphragm muscle fibers (a diaphragm still reactive to stimulation). The FDA humanitarian device exemption approval is based on the results of a multicenter clinical trial of patients with ALS and chronic hypoventilation [135, 136]. Results of this study show that ALS patients who used the diaphragm pacing system have been able to delay the need for mechanical ventilation for up to 24 months [135]. The laparoscopic surgery under general anesthesia, that is needed in order to implant the electrodes in the diaphragm, can be safely performed in patients with ALS so far the respiratory function is not very compromised [136].

10.1.2 Nutritional Management

In spite of decreasing movement abilities, patients with ALS have increased needs for energy in form of caloric intake [137, 138]. On the other hand the caloric intake is frequently low as a consequence of impaired swallowing. Dysphagia is common in ALS and leads to increased risk of aspiration, malnutrition, weight

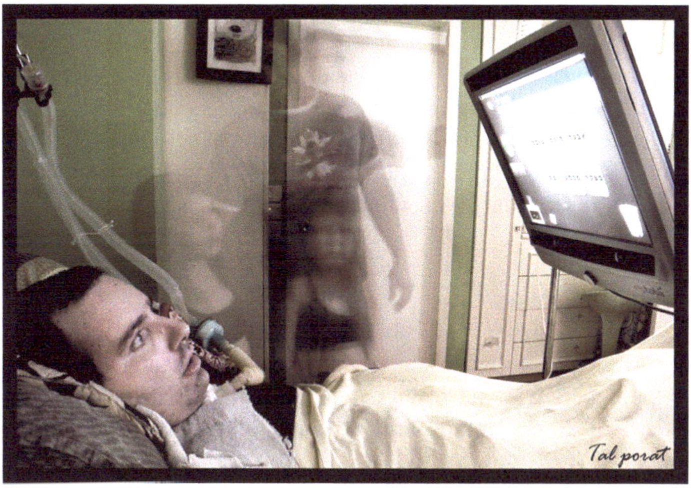

Fig. 2 This 40-year-old patient is completely paralyzed and mechanically ventilated. He chooses letters on a computer screen with light beams refracted by his eye balls. He wrote in this way (in Hebrew, on top of the screen): "I am completely paralyzed, there is no single part of my body that I can move"

loss and dehydration. Enteral tube feeding is commonly used in ALS to maintain adequate nutrition and hydration. This is most commonly performed by percutaneous endoscopic gastrostomy (PEG) and in fewer cases by radiologically inserted gastrostomy (RIG). Enteral nutrition should be initiated early after symptom onset in order to minimize complications and weight loss. Enteral nutrition via PEG is recommended by the revised Practice Parameters of the American Academy of Neurology, as it stabilizes body weight and improves survival in both bulbar and limb onset patients [134].

10.1.3 Communication

Verbal communication is for most humans one of the most important functions, as it determines the ability to initiate and maintain social interactions, and dominates the evaluation of percepted quality of life. Patients with bulbar ALS lose their ability to speak relatively early in the course of the disease. If the ability to write or use a keyboard is preserved, patients switch to written communication. But with progressing disease the hands become weak too, and more sophisticated communication aids are needed. Computer interfaces are being developed that can be activated by the patient through fixation with his eyes at a keyboard on a screen (Fig. 2), or even by patient's thought and registration of the pattern of his brain EEG activity [139].

10.1.4 Exercise and Physical Therapy

Performance of physical activity was considered in the past as possibly harmful for patients with ALS, but during the last decade there is rather robust evidence that regular moderate exercise is able to decrease the rate of deterioration in disability in ALS patients [140–142]. Physical therapy is indicated to preserve range of motion of joints and avoid secondary pain, contracture, or muscle spasms.

10.1.5 Multidisciplinary Care

Specialized ALS centers were established at many sites throughout US and Europe; they have the experience and resources to manage ALS patients and offer high level standard of care. Multidisciplinary care may have a survival benefit in ALS [143].

10.2 Drug Therapy

10.2.1 Contemporary Therapies

Inferences from animal models, including transgenic models of familial disease, to sporadic human disease are not straightforward. However, recognition of the role of glutamate excitotoxicity in sporadic disease and in animal models led to the development of riluzole, the only treatment to date that has been shown to ameliorate the course of sporadic ALS [144, 145]. The mechanism of action of riluzole includes interference with N-methyl-D-aspartate (NMDA) receptor mediated responses, stabilization of voltage-gated sodium channels, inhibition of glutamate release from pre-synaptic terminals, and increasing extracellular glutamate uptake [146].

Two randomized, controlled clinical trials have shown that riluzole prolongs survival in patients with ALS [144, 145]. A meta-analysis of the results indicates that riluzole at a dose of 100 mg daily is safe and probably prolongs median survival by approximately 2–3 months in patients with ALS [147].

Additional drug therapy exists to manage symptoms of ALS. Adequate symptomatic treatment of ALS can alleviate patient discomfort. Sialorrhea and emotional lability are bothersome and often socially embarrassing to ALS patients. Sialorrhea can be treated with amitriptyline, glycopyrrolate, or transdermal scopolamine patches; if sialorrhea is refractory, radiation therapy of the salivary glands or botulinum toxin injections of the salivary glands can be offered [148]. Thick mucus is amenable with propranolol or metoprolol. Pseudobulbar affect or emotional lability can be treated with the recently FDA approved combination of dextromethorphan/quinidine [149], amitriptyline or fluvoxamine.

Anxiety, depression, pain, and muscle discomfort impair quality of life of patients and often of their caregivers. Depression can be treated with tricyclic antidepressants such as amitriptyline or newer agents such as sertraline. Anxiety can be treated with anxiolytics such as benzodiazepines. Pain should be treated with anti-inflammatory and non-narcotic analgesic drugs or combinations; opioid analgesics and cannabis should be considered if pain is refractory or in later disease stages. Muscle stiffness and spasticity can be treated with muscle relaxants such as baclofen or tizanidine. A small study showed a positive effect of the anticonvulsant drug levetiracetam on muscle cramps and spasticity [150].

Fatigue may be a disabling symptom unrelated to muscle weakness or quality of sleep; in these patients modafinil may be indicated [151]. Insomnia causes nighttime distress and interferes with daytime functioning and well-being. Insomnia should be treated with non-invasive ventilation, pain management, antidepressants, or hypnotics, depending on the suspected cause.

10.2.2 Future Therapies and Directions

Despite the advances in understanding disease mechanisms and the performance of many clinical trials, no new therapy for ALS has been discovered during the last 15 years, although few recent clinical trials allow cautious optimism.

Stem Cells

Stem cell transplantation represents a new therapeutic approach with the opportunity of either cellular replacement, or trophic support for the existing cell population, or both. Potentially pluripotent and multipotent stem cells could be stimulated to develop into specialized cells that represent a renewable reservoir of cells for transplantation. A different approach is trophic and growth factor support, by stem-cell mediated delivery of growth factors at relevant sites.

There are a number of different sources for stem cells, ranging from embryonic, fetal, placental and cord blood to adult tissues. The adult tissue stem cells include bone marrow-derived cell populations such as the bone marrow mesenchymal stem cells and hematopoietic stem cells, blood hematopoietic stem cells and neuronal stem cells. Adult stem cells may be the best choice for cell therapy assuming that they can be harvested from the patient and grown in sufficient amounts. They offer the advantages of not raising any ethical problems, not inducing immune rejection, are less prone to promote tumors and are readily available.

All cell types can be administered by allogeneic or autologous delivery. Cells can be transplanted intracerebrally or directly into the spinal cord, or infused by an intravenous, intrathecal or intramuscular route. A further approach is mobilization of endogenous progenitor stem cells from the bone marrow into the peripheral blood by cytokines such as granulocyte colony-stimulating factor. Initial clinical trials to test stem cell treatment for ALS are ongoing.

The autologous transplantation of hematopoietic CD133(+) stem cells into the frontal motor cortex was safe and well-tolerated in ALS patients and delayed disease progression [152]. The autologous transplantation of mesenchymal stem cells into the spinal cord of ALS patients seems safe and might be useful for future ALS clinical trials [153].

The recent development of induced pluripotent stem cells (iPSC) by re-programming human skin fibroblasts is promising. Dimos et al. [154] demonstrated the feasibility of re-programming adult skin fibroblasts into embryonic stem cells that can subsequently yield patient-specific neural cells (neurons and astroglial cells). They not only demonstrated that human disease-related neural cells can be generated from adult fibroblasts via the generation of iPSC, but also that the method can be successfully applied to fibroblasts derived from elderly patients, a crucial issue for age-dependent diseases. The generation of pluripotent stem cells from an individual patient's somatic tissues would enable large scale production of the cell types affected by the patient's disease. These cells could in turn be used for better understanding of disease mechanisms, drug development and autologous cell replacement therapies.

High Fat Diet in ALS

There are some indications that a fat and calorie enriched nutrition may be beneficial to patients with ALS. An on-going clinical trial aims to study the safety, tolerability and preliminary efficacy of a high-fat/high-calorie diet versus both a high calorie diet and a normal diet. The estimated study completion date is by December 2012.

11 Conclusions and Prospects for the Future

The last decade of basic and clinical research was very fruitful for the ALS community, with plenty of exciting news that improve our understanding on the different aspects of this dreadful disease. We are at the edge of a new era, as we just begin to understand the basic processes that lead to individual differences in clinical presentation, course of disease and ultimately survival between ALS patients, either due to differences in genetic background, in the familial as well as in the much more common sporadic form, or due to other individual traits, as concomitant diseases and medications, life style and exogenous exposures.

Understanding these individual differences between patients will lead us to a more personalized approach in the treatment. The first hints for this approach are treatments being developed for patients carrying SOD1 mutations. As more genes are being discovered, treatments directed toward silencing these other genes will follow. Recommendations for changes in life style and environmental exposures for specific groups of patients and asymptomatic gene carriers are close to come. As opposed to the present clumping together of patients with various clinical forms and genetic and environmental backgrounds and offering them a common form of treatment, we can foresee for the next decade a splitting of the syndrome into subtypes, for which different diagnostic and treatment approaches will be used.

References

1. van der Graaff MM, de Jong JM, Baas F, de Visser M (2009) Upper motor neuron and extra-motor neuron involvement in amyotrophic lateral sclerosis: a clinical and brain imaging review. Neuromuscul Disord 19:53–58
2. Qureshi M, Schoenfeld DA, Paliwal Y, Shui A, Cudkowicz ME (2009) The natural history of ALS is changing: improved survival. Amyotroph Lateral Scler 10:324–331
3. Logroscino G, Traynor BJ, Hardiman O, Chiò A, Mitchell D, Swingler RJ, Millul A, Benn E, Beghi E; EURALS (2010) Incidence of amyotrophic lateral sclerosis in Europe. J Neurol Neurosurg Psychiatry 81:385–390
4. Steele JC, McGeer PL (2008) The ALS/PDC syndrome of Guam and the cycad hypothesis. Neurology 70:1984–1990
5. Kuzuhara S, Kokubo Y (2005) Atypical parkinsonism of Japan: amyotrophic lateral sclerosis – parkinsonism – dementia complex of the Kii peninsula of Japan (Muro disease): an update. Mov Disord 20:S108–S113
6. Worms PM (2001) The epidemiology of motor neuron diseases: a review of recent studies. J Neurol Sci 191:3–9
7. Zoccolella S, Beghi E, Palagano G, Fraddosio A, Guerra V, Samarelli V, Lepore V, Simone IL, Lamberti P, Serlenga L, Logroscino G; SLAP Registry (2008) Analysis of survival and prognostic factors in amyotrophic lateral sclerosis: a population based study. J Neurol Neurosurg Psychiatry 79:33–37
8. Sorenson EJ, Stalker AP, Kurland LT, Windebank AJ (2002) Amyotrophic lateral sclerosis in Olmsted County, Minnesota, 1925–1998. Neurology 59:280–282
9. Raaphorst J, de Visser M, Linssen WJP, de Haan R, Schmand B (2010) The cognitive profile of amyotrophic lateral sclerosis: a meta-analysis. Amyotroph Lateral Scler 11:38–45
10. Kurian KM, Forbes RB, Colville S, Swingler RJ (2009) Cause of death and clinical grading criteria in a cohort of amyotrophic lateral sclerosis cases undergoing autopsy from the Scottish Motor Neurone Disease Register. J Neurol Neurosurg Psychiatry 80:84–87
11. Zoccolella S, Beghi E, Palagano G, Fraddosio A, Guerra V, Samarelli V, Lepore V, Simone IL, Lamberti P, Serlenga L, Logroscino G (2008) Predictors of long survival in amyotrophic lateral sclerosis: a population-based study. J Neurol Sci 268:28–32
12. Brooks BR, Miller RG, Swash M, Munsat TL (2000) El Escorial revisited: revised criteria for the diagnosis of amyotrophic lateral sclerosis. Amyotroph Lateral Scler Other Motor Neuron Disord 1:293–299
13. Traynor BJ, Codd MB, Corr B, Forde C, Frost E, Hardiman OM (2000) Clinical features of amyotrophic lateral sclerosis according to the El Escorial and Airlie House criteria: a population-based study. Arch Neurol 57:1171–1176
14. de Carvalho M, Dengler R, Eisen A, England JD, Kaji R, Kimura J, Mills K, Mitsumoto H, Nodera H, Shefner J, Swash M (2008) Electrodiagnostic criteria for diagnosis of ALS. Clin Neurophysiol 119:497–503
15. Zhang L, Ulug AM, Zimmerman RD, Lin MT, Rubin M, Beal MF (2003) The diagnostic utility of FLAIR imaging in clinically verified amyotrophic lateral sclerosis. J Magn Reson Imaging 17:521–527
16. Lule D, Ludolph AC, Kassubek J (2009) MRI-based functional neuroimaging in ALS: an update. Amyotroph Lateral Scler 10:258–268
17. Senda J, Ito M, Watanabe H, Atsuta N, Kawai Y, Katsuno M, Tanaka F, Naganawa S, Fukatsu H, Sobue G (2009) Correlation between pyramidal tract degeneration and widespread white matter involvement in amyotrophic lateral sclerosis: a study with tractography and diffusion-tensor imaging. Amyotroph Lateral Scler 10:288–294
18. Filippini N, Douaud G, Mackay CE, Knight S, Talbot K, Turner MR (2010) Corpus callosum involvement is a consistent feature of amyotrophic lateral sclerosis. Neurology 75:1645–1652
19. Ben Bashat D, Artzi M, Tarrasch R, Nefussy B, Drory VE, Aizenstein O (2011) A potential tool for the diagnosis of ALS based on diffusion tensor imaging. Amyotroph Lateral Scler 12:398–405

20. Dabby R, Lange DJ, Trojaborg W, Hays AP, Lovelace RE, Brannagan TH, Rowland LP (2001) Inclusion body myositis mimicking motor neuron disease. Arch Neurol 58:1253–1256
21. Nobile-Orazio E, Carpo M, Meucci N (2001) Are there immunologically treatable motor neuron diseases? Amyotroph Lateral Scler Other Motor Neuron Disord 2:S23–S30
22. Parboosingh JS, Figlewicz DA, Krizus A, Meininger V, Azad NA, Newman DS, Rouleau GA (1997) Spinobulbar muscular atrophy can mimic ALS: the importance of genetic testing in male patients with atypical ALS. Neurology 49:568–572
23. Drory VE, Birnbaum M, Peleg L, Goldman B, Korczyn AD (2003) Hexosaminidase A deficiency is an uncommon cause of a syndrome mimicking amyotrophic lateral sclerosis. Muscle Nerve 28:109–112
24. Tashiro K, Kikuchi S, Itoyama Y, Tokumaru Y, Sobue G, Mukai E, Akiguchi I, Nakashima K, Kira J, Hirayama K (2006) Nationwide survey of juvenile muscular atrophy of distal upper extremity (Hirayama disease) in Japan. Amyotroph Lateral Scler 7:38–45
25. Byrne S, Walsh C, Lynch C, Bede P, Elamin M, Kenna K, McLaughlin R, Hardiman O (2011) Rate of familial amyotrophic lateral sclerosis: a systematic review and meta-analysis. J Neurol Neurosurg Psychiatry 82:623–627
26. Schymick JC, Talbot K, Traynor BJ (2007) Genetics of sporadic amyotrophic lateral sclerosis. Hum Mol Genet 16:R233–R242
27. Graham AJ, Macdonald AM, Hawkes CH (1997) British motor neuron disease twin study. J Neurol Neurosurg Psychiatry 62:562–569
28. Fang F, Kamel F, Lichtenstein P, Bellocco R, Sparén P, Sandler DP, Ye W (2009) Familial aggregation of amyotrophic lateral sclerosis. Ann Neurol 66:94–99
29. Hanby MF, Scott KM, Scotton W, Wijesekera L, Mole T, Ellis CE, Leigh PN, Shaw CE, Al-Chalabi A (2011) The risk to relatives of patients with sporadic amyotrophic lateral sclerosis. Brain 134:3454–3457
30. Andersen PM, Al-Chalabi A (2011) Clinical genetics of amyotrophic lateral sclerosis: what do we really know? Nat Rev Neurol 7:603–615
31. Beleza-Meireles A, Al-Chalabi A (2009) Genetic studies of amyotrophic lateral sclerosis: controversies and perspectives. Amyotroph Lateral Scler 10:1–14
32. Dion PA, Daoud H, Rouleau GA (2009) Genetics of motor neuron disorders: new insights into pathogenetic mechanisms. Nat Rev Genet 10:769–782
33. Rosen DR, Siddique T, Patterson D, Figlewicz DA, Sapp P, Hentati A, Donaldson D, Goto J, O'Regan JP, Deng HX, Rahmani Z, Krizus A, McKenna-Yasek D, Cayabyab A, Gaston SM, Berger R, Tanzi RE, Halperin JJ, Herzfeldt B, Van den Bergh R, Hung WY, Bird T, Deng G, Mulder DW, Smyth C, Laing NG, Soriano E, Pericak-Vance MA, Haines J, Rouleau GA, Gusella JS, Horvitz HR, Brown RH Jr (1993) Mutations in Cu/Zn superoxide dismutase gene are associated with familial amyotrophic lateral sclerosis. Nature 362:59–62
34. Shaw PJ (2005) Molecular and cellular pathways of neurodegeneration in motor neuron disease. J Neurol Neurosurg Psychiatry 76:1046–1057
35. Abe K, Aoki M, Ikeda M, Watanabe M, Hirai S, Itoyama Y (1996) Clinical characteristics of familial amyotrophic lateral sclerosis with Cu/Zn superoxide dismutase gene mutations. J Neurol Sci 136:108–116
36. Neumann M, Sampathu DM, Kwong LK, Truax AC, Micsenyi MC, Chou TT, Bruce J, Schuck T, Grossman M, Clark CM, McCluskey LF, Miller BL, Masliah E, Mackenzie IR, Feldman H, Feiden W, Kretzschmar HA, Trojanowski JQ, Lee VM (2006) Ubiquitinated TDP-43 in frontotemporal lobar degeneration and amyotrophic lateral sclerosis. Science 314:130–133
37. Mackenzie IR, Bigio EH, Ince PG, Geser F, Neumann M, Cairns NJ, Kwong LK, Forman MS, Ravits J, Stewart H, Eisen A, McClusky L, Kretzschmar HA, Monoranu CM, Highley JR, Kirby J, Siddique T, Shaw PJ, Lee VM, Trojanowski JQ (2007) Pathological TDP-43 distinguishes sporadic amyotrophic lateral sclerosis from amyotrophic lateral sclerosis with SOD1 mutations. Ann Neurol 61:427–434
38. Sreedharan J, Blair IP, Tripathi VB, Hu X, Vance C, Rogelj B, Ackerley S, Durnall JC, Williams KL, Buratti E, Baralle F, de Belleroche J, Mitchell JD, Leigh PN, Al-Chalabi A, Miller CC, Nicholson G, Shaw CE (2008) TDP-43 mutations in familial and sporadic amyotrophic lateral sclerosis. Science 319:1668–1672

39. Yokoseki A, Shiga A, Tan CF, Tagawa A, Kaneko H, Koyama A, Eguchi H, Tsujino A, Ikeuchi T, Kakita A, Okamoto K, Nishizawa M, Takahashi H, Onodera O (2008) TDP-43 mutation in familial amyotrophic lateral sclerosis. Ann Neurol 63:538–542
40. Kabashi E, Valdmanis PN, Dion P, Spiegelman D, McConkey BJ, Vande Velde C, Bouchard JP, Lacomblez L, Pochigaeva K, Salachas F, Pradat PF, Camu W, Meininger V, Dupre N, Rouleau GA (2008) TARDBP mutations in individuals with sporadic and familial amyotrophic lateral sclerosis. Nat Genet 40:572–574
41. Mackenzie IR, Rademakers R, Neumann M (2010) TDP-43 and FUS in amyotrophic lateral sclerosis and frontotemporal dementia. Lancet Neurol 9:995–1007
42. Wegorzewska I, Bell S, Cairns NJ, Miller TM, Baloh RH (2009) TDP-43 mutant transgenic mice develop features of ALS and frontotemporal degeneration. Proc Natl Acad Sci 106:18809–18814
43. Kwiatkowski TJ Jr, Bosco DA, Leclerc AL, Tamrazian E, Vanderburg CR, Russ C, Davis A, Gilchrist J, Kasarskis EJ, Munsat T, Valdmanis P, Rouleau GA, Hosler BA, Cortelli P, de Jong PJ, Yoshinaga Y, Haines JL, Pericak-Vance MA, Yan J, Ticozzi N, Siddique T, McKenna-Yasek D, Sapp PC, Horvitz HR, Landers JE, Brown RH Jr (2009) Mutations in the FUS/TLS gene on chromosome 16 cause familial amyotrophic lateral sclerosis. Science 323:1205–1208
44. Vance C, Rogelj B, Hortobagyi T, De Vos KJ, Nishimura AL, Sreedharan J, Hu X, Smith B, Ruddy D, Wright P, Ganesalingam J, Williams KL, Tripathi V, Al-Saraj S, Al-Chalabi A, Leigh PN, Blair IP, Nicholson G, de Belleroche J, Gallo JM, Miller CC, Shaw CE (2009) Mutations in FUS, an RNA processing protein, cause familial amyotrophic lateral sclerosis type 6. Science 323:1208–1211
45. Belzil VV, Valdmanis PN, Dion PA, Daoud H, Kabashi E, Noreau A, Gauthier J; S2D team, Hince P, Desjarlais A, Bouchard JP, Lacomblez L, Salachas F, Pradat PF, Camu W, Meininger V, Dupré N, Rouleau GA (2009) Mutations in *FUS* cause FALS and SALS in French and French Canadian populations. Neurology 73:1176–1179
46. Nishimura AL, Mitne-Neto M, Silva HC, Richieri-Costa A, Middleton S, Cascio D, Kok F, Oliveira JR, Gillingwater T, Webb J, Skehel P, Zatz M (2004) A mutation in the vesicle-trafficking protein VAPB causes late-onset spinal muscular atrophy and amyotrophic lateral sclerosis. Am J Hum Genet 75:822–831
47. Munch C, Sedlmeier R, Meyer T, Homberg V, Sperfeld AD, Kurt A, Prudlo J, Peraus G, Hanemann CO, Stumm G, Ludolph AC (2004) Point mutations in the p150 subunit of dynactin (DCTN1) gene in ALS. Neurology 63:724–726
48. Greenway MJ, Andersen PM, Russ C, Ennis S, Cashman S, Donaghy C, Patterson V, Swingler R, Kieran D, Prehn J, Morrison KE, Green A, Acharya KR, Brown RH Jr, Hardiman O (2006) ANG mutations segregate with familial and 'sporadic' amyotrophic lateral sclerosis. Nat Genet 38:411–413
49. Chen YZ, Bennett CL, Huynh HM, Blair IP, Puls I, Irobi J, Dierick I, Abel A, Kennerson ML, Rabin BA, Nicholson GA, Auer-Grumbach M, Wagner K, De Jonghe P, Griffin JW, Fischbeck KH, Timmerman V, Cornblath DR, Chance PF (2004) DNA/RNA helicase gene mutations in a form of juvenile amyotrophic lateral sclerosis (ALS4). Am J Hum Genet 74:1128–1135
50. Maruyama H, Morino H, Ito H, Izumi Y, Kato H, Watanabe Y, Kinoshita Y, Kamada M, Nodera H, Suzuki H, Komure O, Matsuura S, Kobatake K, Morimoto N, Abe K, Suzuki N, Aoki M, Kawata A, Hirai T, Kato T, Ogasawara K, Hirano A, Takumi T, Kusaka H, Hagiwara K, Kaji R, Kawakami H (2010) Mutations of optineurin in amyotrophic lateral sclerosis. Nature 456:223–226
51. Johnson JO, Mandrioli J, Benatar M, Abramzon Y, Van Deerlin VM, Trojanowski JQ, Gibbs JR, Brunetti M, Gronka S, Wuu J, Ding J, McCluskey L, Martinez-Lage M, Falcone D, Hernandez DG, Arepalli S, Chong S, Schymick JC, Rothstein J, Landi F, Wang YD, Calvo A, Mora G, Sabatelli M, Monsurrò MR, Battistini S, Salvi F, Spataro R, Sola P, Borghero G; ITALSGEN Consortium, Galassi G, Scholz SW, Taylor JP, Restagno G, Chiò A, Traynor BJ (2010) Exome sequencing reveals VCP mutations as a cause of familial ALS. Neuron 68:857–864

52. Deng HX, Chen W, Hong ST, Boycott KM, Gorrie GH, Siddique N, Yang Y, Fecto F, Shi Y, Zhai H, Jiang H, Hirano M, Rampersaud E, Jansen GH, Donkervoort S, Bigio EH, Brooks BR, Ajroud K, Sufit RL, Haines JL, Mugnaini E, Pericak-Vance MA, Siddique T (2011) Mutations in UBQLN2 cause dominant X-linked juvenile and adult-onset ALS and ALS/dementia. Nature 477:211–213
53. Wroe R, Wai-Ling Butler A, Andersen PM, Powell JF, Al-Chalabi A (2008) ALSOD: the amyotrophic lateral sclerosis online database. Amyotroph Lateral Scler 9:249–250
54. ALS Online Genetics database (online). http://alsod.iop.kcl.ac.uk
55. Renton AE, Majounie E, Waite A et al (2011) A hexanucleotide repeat expansion in C9ORF72 is the cause of chromosome 9p21-linked ALS-FTD. Neuron 72:257–268
56. DeJesus-Hernandez M, Mackenzie IR, Boeve BF, Boxer AL, Baker M, Rutherford NJ, Nicholson AM, Finch NA, Flynn H, Adamson J, Kouri N, Wojtas A, Sengdy P, Hsiung GY, Karydas A, Seeley WW, Josephs KA, Coppola G, Geschwind DH, Wszolek ZK, Feldman H, Knopman DS, Petersen RC, Miller BL, Dickson DW, Boylan KB, Graff-Radford NR, Rademakers R (2011) Expanded GGGGCC hexanucleotide repeat in a noncoding region of C9ORF72 causes chromosome 9p-linked FTD and ALS. Neuron 72:245–256
57. Gijselinck I, Van Langenhove T, van der Zee J et al (2012) A *C9orf72* promoter repeat expansion in a Flanders-Belgian cohort with disorders of the frontotemporal lobar degeneration-amyotrophic lateral sclerosis spectrum: a gene identification study. Lancet Neurol 11:54–65
58. Byrne S, Elamin M, Bede P, Shatunov A, Walsh C, Corr B, Heverin M, Jordan N, Kenna K, Lynch C, McLaughlin RL, Iyer PM, O'Brien C, Phukan J, Wynne B, Bokde AL, Bradley DG, Pender N, Al-Chalabi A, Hardiman O (2012) Cognitive and clinical characteristics of patients with amyotrophic lateral sclerosis carrying a C9orf72 repeat expansion: a population-based cohort study. Lancet Neurol 11:232–240
59. Majounie E, Renton AE, Mok K et al (2012) Frequency of the C9orf72 hexanucleotide repeat expansion in patients with amyotrophic lateral sclerosis and frontotemporal dementia: a cross-sectional study. Lancet Neurol 11:323–330
60. Elden A, Kim HJ, Hart MP, Chen-Plotkin AS, Johnson BS, Fang X, Armakola M, Geser F, Greene R, Lu MM, Padmanabhan A, Clay-Falcone D, McCluskey L, Elman L, Juhr D, Gruber PJ, Rüb U, Auburger G, Trojanowski JQ, Lee VM, Van Deerlin VM, Bonini NM, Gitler AD (2010) Ataxin 2 intermediate-length polyglutamine expansions are associated with increased risk for ALS. Nature 466:1069–1075
61. Lee T, Li YR, Ingre C, Weber M, Grehl T, Gredal O, de Carvalho M, Meyer T, Tysnes OB, Auburger G, Gispert S, Bonini NM, Andersen PM, Gitler AD (2011) Ataxin-2 intermediate-length polyglutamine expansions in European ALS patients. Hum Mol Genet 20:1697–1700
62. Daoud H, Belzil V, Martins S, Sabbagh M, Provencher P, Lacomblez L, Meininger V, Camu W, Dupré N, Dion PA, Rouleau GA (2011) Association of long *ATXN2* CAG repeats sizes with increased risk of amyotrophic lateral sclerosis. Arch Neurol 68:739–742
63. The International HapMap Consortium (2005) A haplotype map of the human genome. Nature 437:1299–1320
64. Dunckley T, Huentelman MJ, Craig DW et al (2007) Whole-genome analysis of sporadic amyotrophic lateral sclerosis. N Engl J Med 357:775–788
65. van Es MA, Veldink JH, Saris CG et al (2009) Genome-wide association study identifies 19p13.3 (UNC13A) and 9p21.2 as susceptibility loci for sporadic amyotrophic lateral sclerosis. Nat Genet 41:1083–1087
66. Blauw HM, Veldink JH, van Es MA, van Vught PW, Saris CG, van der Zwaag B, Franke L, Burbach JP, Wokke JH, Ophoff RA, van den Berg LH (2008) Copy-number variation in sporadic amyotrophic lateral sclerosis: a genome-wide screen. Lancet Neurol 7:319–326
67. Chio A, Schymick JC, Restagno G et al (2009) A two-stage genome-wide association study of sporadic amyotrophic lateral sclerosis. Hum Mol Genet 18:1524–1532
68. Landers JE, Melki J, Meininger V et al (2009) Reduced expression of the *Kinesin-associated protein 3* (*KIFAP3*) gene increases survival in sporadic amyotrophic lateral sclerosis. Proc Natl Acad Sci 106:9004–9009

69. Orsetti V, Pegoraro E, Cima V, D'Ascenzo C, Palmieri A, Querin G, Volpe M, Ermani M, Angelini C, Sorarù G (2011) Genetic variation in *KIFAP3* is associated with an upper motor neuron-predominant phenotype in amyotrophic lateral sclerosis. Neurodegener Dis 8: 491–495
70. Gros-Louis F, Andersen PM, Dupre N, Urushitani M, Dion P, Souchon F, D'Amour M, Camu W, Meininger V, Bouchard JP, Rouleau GA, Julien JP (2009) Chromogranin B P413L variant as risk factor and modifier of disease onset for amyotrophic lateral sclerosis. Proc Natl Acad Sci 106:21777–21782
71. van Vught PW, Veldink JH, van den Berg LH (2010) P413L CHGB is not associated with ALS susceptibility or age at onset in a Dutch population. Proc Natl Acad Sci U S A 107:E77
72. Blasco H, Corcia P, Veyrat-Durebex C, Coutadeur C, Fournier C, Camu W, Gordon P, Praline J, Andres CR, Vourc'h P; French ALS Study Group (2011) The P413L chromogranin B variation in French patients with sporadic amyotrophic lateral sclerosis. Amyotroph Lateral Scler 12:210–214
73. Van Damme P, Veldink JH, van Blitterswijk M, Corveleyn A, van Vught PW, Thijs V, Dubois B, Matthijs G, van den Berg LH, Robberecht W (2011) Expanded ATXN2 CAG repeat size in ALS identifies genetic overlap between ALS and SCA2. Neurology 76:2066–2072
74. Ross OA, Rutherford NJ, Baker M et al (2011) Ataxin-2 repeat-length variation and neurodegeneration. Hum Mol Genet 20:3207–3212
75. Gallo V, Bueno-De-Mesquita HB, Vermeulen R, Andersen PM, Kyrozis A, Linseisen J (2009) Smoking and risk for amyotrophic lateral sclerosis: analysis of the EPIC cohort. Ann Neurol 65:378–385
76. Armon C (2009) Smoking may be considered an established risk factor for sporadic ALS. Neurology 73:1693–1698
77. Armon C (2003) An evidence-based medicine approach to the evaluation of the role of exogenous risk factors in sporadic amyotrophic lateral sclerosis. Neuroepidemiology 22: 217–228
78. Sutedja NA, Fischer K, Veldink JH, van der Heijden GJ, Kromhout H, Heederik D, Huisman MH, Wokke JJ, van den Berg LH (2009) What we truly know about occupation as a risk factor for ALS: a critical and systematic review. Amyotroph Lateral Scler 10:295–301
79. Horner RD, Kamins KG, Feussner JR, Grambow SC, Hoff-Lindquist J, Harati Y, Mitsumoto H, Pascuzzi R, Spencer PS, Tim R, Howard D, Smith TC, Ryan MA, Coffman CJ, Kasarskis EJ (2003) Occurrence of amyotrophic lateral sclerosis among Gulf War veterans. Neurology 61:742–749
80. Haley RW (2003) Excess incidence of ALS in young Gulf War veterans. Neurology 61: 750–756
81. Chiò A, Benzi G, Dossena M, Mutani R, Mora G (2005) Severely increased risk of amyotrophic lateral sclerosis among Italian professional football players. Brain 128:472–476
82. Chio A, Calvo A, Dossena M, Ghiglione P, Mutani R, Mora G (2009) ALS in Italian professional soccer players: the risk is still present and could be soccer-specific. Amyotroph Lateral Scler 10:205–209
83. Hyams KC, Brown M, White DS (2005) Resolving disputes about toxicological risks during military conflict: the US Gulf War experience. Toxicol Rev 24:167–180
84. Armon C (2007) Sports and trauma in amyotrophic lateral sclerosis revisited. J Neurol Sci 262:45–53
85. Edwards IR, Star K, Kiuru A (2007) Statins, neuromuscular degenerative disease and an amyotrophic lateral sclerosis–like syndrome: an analysis of individual case safety reports from vigibase. Drug Saf 30:515–525
86. Colman E, Szarfman A, Wyeth J, Mosholder A, Jillapalli D, Levine J, Avigan M (2008) An evaluation of a data mining signal for amyotrophic lateral sclerosis and statins detected in FDA's spontaneous adverse event reporting system. Pharmacoepidemiol Drug Saf 17: 1068–1076
87. Dupuis L, Corcia P, Fergani A, Gonzalez De Aguilar JL, Bonnefont-Rousselot D, Bittar R, Seilhean D, Hauw JJ, Lacomblez L, Loeffler JP, Meininger V (2008) Dyslipidemia is a protective factor in amyotrophic lateral sclerosis. Neurology 70:1004–1009

88. Keizman D, Ish-Shalom M, Berliner S, Maimon N, Vered Y, Artamonov I, Tsehori J, Nefussy B, Drory VE (2009) Low uric acid levels in serum of patients with ALS: further evidence for oxidative stress? J Neurol Sci 285:95–99
89. Jawaid A, Salamone AR, Strutt AM, Murthy SB, Wheaton M, McDowell EJ, Simpson E, Appel SH, York MK, Schulz PE (2010) ALS disease onset may occur later in patients with pre-morbid diabetes mellitus. Eur J Neurol 17:733–739
90. Bruijn LI, Becher MW, Lee MK, Anderson KL, Jenkins NA, Copeland NG, Sisodia SS, Rothstein JD, Borchelt DR, Price DL, Cleveland DW (1997) ALS-linked SOD1 mutant G85R mediates damage to astrocytes and promotes rapidly progressive disease with SOD1-containing inclusions. Neuron 18:327–338
91. Gurney ME, Pu H, Chiu AY, Dal Canto MC, Polchow CY, Alexander DD, Caliendo J, Hentati A, Kwon YW, Deng HX, Chen W, Zhai P, Sufit RL, Siddique T (1994) Motor neuron degeneration in mice that express human Cu, Zn superoxide dismutase mutation. Science 264:1772–1775
92. Wong PC, Pardo CA, Borchelt DR, Lee MK, Copeland NG, Jenkins NA, Sisodia SS, Cleveland DW, Price DL (1995) An adverse property of a familial ALS-linked *SOD1* mutation causes motor neuron disease characterized by vacuolar degeneration of mitochondria. Neuron 14:1105–1116
93. Bruijn LI, Houseweart MK, Kato S, Anderson KL, Anderson SD, Ohama E, Reaume AG, Scott RW, Cleveland DW (1998) Aggregation and motor neuron toxicity of an ALS-linked *SOD1* mutant independent from wild-type *SOD1*. Science 281:1851–1854
94. Gong YH, Parsadanian AS, Andreeva A, Snider WD, Elliott JL (2000) Restricted expression of G86R Cu/Zn superoxide dismutase in astrocytes results in astrocytosis but does not cause motoneuron degeneration. J Neurosci 20:660–665
95. Pramatarova A, Laganiere J, Roussel J, Brisebois K, Rouleau GA (2001) Neuron-specific expression of mutant superoxide dismutase 1 in transgenic mice does not lead to motor neuron impairment. J Neurosci 21:3369–3374
96. Lino MM, Schneider C, Caroni P (2002) Accumulation of *SOD1* mutant in postnatal motor neurons does not cause motor neuron pathology of motoneuron disease. J Neurosci 22: 4825–4832
97. Clement AM, Nguyen MD, Roberts EA, Garcia ML, Boillée S, Rule M, McMahon AP, Doucette W, Siwek D, Ferrante RJ, Brown RH Jr, Julien JP, Goldstein LS, Cleveland DW (2003) Wild-type nonneuronal cells extended survival of SOD1 mutant motor neurons in ALS mice. Science 302:113–117
98. Nagai M, Re DB, Nagata T, Chalazonitis A, Jessell TM, Wichterle H, Przedborski S (2007) Astrocytes expressing ALS-linked mutated SOD1 release factors selectively toxic to motor neurons. Nat Neurosci 10:615–622
99. Beers DR, Henkel JS, Xiao Q, Zhao W, Wang J, Yen AA, Siklos L, McKercher SR, Appel SH (2006) Wild-type microglia extend survival in PU.1 knockout mice with familial amyotrophic lateral sclerosis. Proc Natl Acad Sci 103:16021–16026
100. Boillee S, Yamanaka K, Lobsiger CS, Copeland NG, Jenkins NA, Kassiotis G, Kollias G, Cleveland DW (2006) Onset and progression in inherited ALS determined by motor neurons and microglia. Science 312:1389–1392
101. Yamanaka K, Chun SJ, Boillee S, Fujimori-Tonou N, Yamashita H, Gutmann DH, Takahashi R, Misawa H, Cleveland DW (2008) Astrocytes as determinants of disease progression in inherited amyotrophic lateral sclerosis. Nat Neurosci 11:251–253
102. Yamanaka K, Boillee S, Roberts EA, Garcia ML, McAlonis-Downes M, Mikse OR, Cleveland DW, Goldstein LS (2008) Mutant *SOD1* in cell types other than motor neurons and oligodendrocytes accelerates onset of disease in ALS mice. Proc Natl Acad Sci 105: 7594–7599
103. Spreux-Varoquaux O, Bensimon G, Lacomblez L, Salachas F, Pradat PF, Le Forestier N, Marouan A, Dib M, Meininger V (2002) Glutamate levels in cerebrospinal fluid in amyotrophic lateral sclerosis: a reappraisal using a new HPLC method with colometric detection in a large cohort of patients. J Neurol Sci 193:73–78

104. Rothstein JD, Van Kammen M, Levey AL, Martin LJ, Kuncl RW (1995) Selective loss of glial glutamate transporter GLT-1 in amyotrophic lateral sclerosis. Ann Neurol 38:73–84
105. Beal MF, Ferrante RJ, Browne SE, Matthews RT, Kowall NW, Brown RH Jr (1997) Increased 3-nitrotyrosine in both sporadic and familial amyotrophic lateral sclerosis. Ann Neurol 42:644–654
106. Bosco DA, Morfini G, Karabacak NM, Song Y, Gros-Louis F, Pasinelli P, Goolsby H, Fontaine BA, Lemay N, McKenna-Yasek D, Frosch MP, Agar JN, Julien JP, Brady ST, Brown RH Jr (2010) Wild-type and mutated SOD1 share an aberrant conformation and a common pathogenic pathway in ALS. Nat Neurosci 13:1396–1403
107. Forsberg K, Jonsson PA, Andersen PM, Bergemalm D, Graffmo KS, Hultdin M, Jacobsson J, Rosquist R, Marklund SL, Brännström T (2010) Novel antibodies reveal inclusions containing non-native SOD1 in sporadic ALS patients. PLoS One 5:e11552
108. Arai T, Hasegawa M, Akiyama H, Ikeda K, Nonaka T, Mori H, Mann D, Tsuchiya K, Yoshida M, Hashizume Y, Oda T (2006) TDP-43 is a component of ubiquitin-positive tau-negative inclusions in frontotemporal lobar degeneration and amyotrophic lateral sclerosis. Biochem Biophys Res Commun 35:602–611
109. Tan CF, Eguchi H, Tagawa A, Onodera O, Iwasaki T, Tsujino A, Nishizawa M, Kakita A, Takahashi H (2007) TDP-43 immunoreactivity in neuronal inclusions in familial amyotrophic lateral sclerosis with or without SOD1 gene mutation. Acta Neuropathol 113:535–542
110. Kabashi E, Bercier V, Lissouba A, Liao M, Brustein E, Rouleau GA, Drapeau P (2011) FUS and TARDBP but not SOD1 interact in genetic models of amyotrophic lateral sclerosis. PLoS Genet 7:e1002214
111. Ling SC, Albuquerque CP, Han JS, Lagier-Tourenne C, Tokunaga S, Zhou H, Cleveland DW (2010) ALS-associated mutations in TDP-43 increase its stability and promote complexes with FUS/TLS. Proc Natl Acad Sci U S A 107:13318–13323
112. Shi P, Gal J, Kwinter DM, Liu X, Zhu H (2010) Mitochondrial dysfunction in amyotrophic lateral sclerosis. Biochim Biophys Acta 1802:45–51
113. Klivenyi P, Ferrante RJ, Matthews RT, Bogdanov MB, Klein AM, Andreassen OA, Mueller G, Wermer M, Kaddurah-Daouk R, Beal MF (1999) Neuroprotective effects of creatine in a transgenic animal model of amyotrophic lateral sclerosis. Nat Med 5:347–350
114. Rosenfeld J, King RM, Jackson CE, Bedlack RS, Barohn RJ, Dick A, Phillips LH, Chapin J, Gelinas DF, Lou JS (2008) Creatine monohydrate in ALS: effects on strength, fatigue, respiratory status and ALSFRS. Amyotroph Lateral Scler 9:266–272
115. Dupuis L, Pradat PF, Ludolph AC, Loeffler JP (2011) Energy metabolism in amyotrophic lateral sclerosis. Lancet Neurol 10:75–82
116. Kunita R, Otomo A, Mizumura H, Suzuki-Utsunomiya K, Hadano S, Ikeda JE (2007) The Rab5 activator ALS2/alsin acts as a novel Rac1 effector through Rac1-activated endocytocis. J Biol Chem 282:16599–16611
117. Chow CY, Landers JE, Bergren SK, Sapp PC, Grant AE, Jones JM, Everett L, Lenk GM, McKenna-Yasek DM, Weisman LS, Figlewicz D, Brown RH, Meisler MH (2009) Deleterious variants of FIG4, a phosphoinositide phosphatase, in patients with ALS. Am J Hum Genet 84:85–88
118. Michell RH, Dove SK (2009) A protein complex that regulates PtdIns(3,5)P_2 levels. EMBO J 28:86–87
119. Bruijn LI, Miller TM, Cleveland DW (2004) Unraveling the mechanisms involved in motor neuron degeneration in ALS. Annu Rev Neurosci 27:723–749
120. Rothstein JD (2009) Current hypotheses for the underlying biology of amyotrophic lateral sclerosis. Ann Neurol 65:S3–S9
121. Gitler AG, Shorter J (2011) RNA-binding proteins with prion-like domains in ALS and FTLD-U. Prion 5:179–189
122. Polymenidou M, Cleveland DW (2011) The seeds of neurodegeneration: prion-like spreading in ALS. Cell 147:498–506
123. Mendonça DM, Martins SC, Higashi R, Muscara MN, Neto VM, Chimelli L, Martinez AM (2011) Neurofilament heavy subunit in cerebrospinal fluid: a biomarker of amyotrophic lateral sclerosis? Amyotroph Lateral Scler 12:144–147

124. Goldknopf IL, Sheta EA, Bryson J, Folsom B, Wilson C, Duty J, Yen AA, Appel SH (2006) Complement C3c and related protein biomarkers in amyotrophic lateral sclerosis and Parkinson's disease. Biochem Biophys Res Commun 342:1034–1039
125. Wilson ME, Boumaza I, Lacomis D, Bowser R (2010) Cystatin C: a candidate biomarker for amyotrophic lateral sclerosis. PLoS One 5:e15133
126. Ganesalingam J, An J, Shaw CE, Shaw G, Lacomis D, Bowser R (2011) Combination of neurofilament heavy chain and complement C3 as CSF biomarkers for ALS. J Neurochem 117:528–537
127. Blasco H, Corcia P, Moreau C, Veau S, Fournier C, Vourc'h P, Emond P, Gordon P, Pradat PF, Praline J, Devos D, Nadal-Desbarats L, Andres CR (2010) H-1-NMR- based metabolomic profiling of CSF in early amyotrophic lateral sclerosis. PLoS One 5:e13223
128. Wuolikainen A, Moritz T, Marklund SL, Antti H, Andersen PM (2011) Disease- related changes in the cerebrospinal fluid metabolome in amyotrophic lateral sclerosis detected by GC/TOFMS. PLoS One 6:e17947
129. Cheah BC, Boland RA, Brodaty NE (2009) INSPIRATIonAL – INSPIRAtory muscle training in amyotrophic lateral sclerosis. Amyotroph Lateral Scler 10:384–392
130. Bourke SC, Tomlinson M, Williams TL, Bullock RE, Shaw PJ, Gibson GJ (2006) Effects of non-invasive ventilation on survival and quality of life in patients with amyotrophic lateral sclerosis. Lancet Neurol 5:140–157
131. Heffernan C, Jenkinson C, Holmes T, Macleod H, Kinnear W, Oliver D, Leigh N, Ampong MA (2006) Management of respiration in MND/ALS patients: an evidence-based review. Amyotroph Lateral Scler 7:5–15
132. Carratu P, Spicuzza L, Cassano A, Maniscalco M, Gadaleta F, Lacedonia D, Scoditti C, Boniello E, Di Maria G, Resta O (2009) Early treatment with noninvasive positive pressure ventilation prolongs survival in amyotrophic lateral sclerosis patients with nocturnal respiratory insufficiency. Orphanet J Rare Dis 10:4–10
133. Radunovic A, Annane D, Jewitt K, Mustfa N (2009) Mechanical ventilation for amyotrophic lateral sclerosis/motor neuron disease. Cochrane Database Syst Rev 4:CD004427
134. Miller RG, Jackson CE, Kasarskis EJ, England JD, Forshew D, Johnston W, Kalra S, Katz JS, Mitsumoto H, Rosenfeld J, Shoesmith C, Strong MJ, Woolley SC; Quality Standards Subcommittee of the American Academy of Neurology (2009) Practice Parameter update: the care of the patient with amyotrophic lateral sclerosis: drug, nutritional, and respiratory therapies (an evidence-based review). Neurology 73:1218–1226
135. Onders RP, Elmo M, Khansarinia S, Bowman B, Yee J, Road J, Bass B, Dunkin B, Ingvarsson PE, Oddsdóttir M (2009) Complete worldwide operative experience in laparoscopic diaphragm pacing: results and differences in spinal cord injured patients and amyotrophic lateral sclerosis patients. Surg Endosc 23:1433–1440
136. Onders RP, Carlin AM, Elmo M, Sivashankaran S, Katirji B, Schilz R (2009) Amyotrophic lateral sclerosis: the Midwestern surgical experience with the diaphragm pacing stimulation system shows that general anesthesia can be safely performed. Am J Surg 197:386–390
137. Vaisman N, Lusaus M, Nefussy B, Niv E, Comaneshter D, Hallack R, Drory VE (2009) Do patients with amyotrophic lateral sclerosis (ALS) have increased energy needs? J Neurol Sci 279:26–29
138. Funalot B, Desport JC, Sturtz F, Camu W, Couratier P (2009) High metabolic level in patients with familial amyotrophic lateral sclerosis. Amyotroph Lateral Scler 10:113–117
139. Daly JJ, Wolpaw JR (2008) Brain-computer interfaces in neurological rehabilitation. Lancet Neurol 7:1032–1043
140. Drory VE, Goltsman E, Goldman Reznik J, Mosek A, Korczyn AD (2001) The value of muscle exercise in patients with amyotrophic lateral sclerosis. J Neurol Sci 191:133–137
141. Dal Bello-Hass V, Florence JM, Kloos AD, Scheirbecker J, Lopate G, Hayes SM, Pioro EP, Mitsumoto H (2007) A randomized controlled trial of resistance exercise in individuals with ALS. Neurology 68:2003–2007
142. Dal Bello-Haas V, Florence JM, Krivickas LS (2008) Therapeutic exercise for people with amyotrophic lateral sclerosis or motor neuron disease. Cochrane Database Syst Rev 2:CD005229

143. Traynor BJ, Alexander M, Corr B, Frost E, Hardiman O (2003) Effect of a multidisciplinary amyotrophic lateral sclerosis (ALS) clinic on ALS survival: a population based study, 1996–2000. J Neurol Neurosurg Psychiatry 74:1258–1261
144. Bensimon G, Lacomblez L, Meininger VA (1994) A controlled trial of riluzole in amyotrophic lateral sclerosis. ALS/Riluzole Study Group. N Engl J Med 330:585–591
145. Lacomblez L, Bensimon G, Leigh PN, Guillet P, Meininger V (1996) Dose-ranging study of riluzole in amyotrophic lateral sclerosis. Amyotrophic Lateral Sclerosis/Riluzole Study Group II. Lancet 347:1425–1431
146. Bellingham MC (2011) A review of the neural mechanisms of action and clinical efficiency of riluzole in treating amyotrophic lateral sclerosis: what have we learned in the last decade? CNS Neurosci Ther 17:4–31
147. Miller RG, Mitchell JD, Lyon M, Moore DH (2007) Riluzole for amyotrophic lateral sclerosis (ALS)/motor neuron disease (MND). Cochrane Database Syst Rev 1:CD001447
148. Stone CA, O'Leary N (2009) Systematic review of the effectiveness of botulinum toxin or radiotherapy for sialorrhea in patients with amyotrophic lateral sclerosis. J Pain Symptom Manage 37:246–258
149. Pioro EP, Brooks BR, Cummings J, Schiffer R, Thisted RA, Wynn D, Hepner A, Kaye R; Safety, Tolerability, and Efficacy Results Trial of AVP-923 in PBA Investigators (2010) Dextromethorphan plus ultra low-dose quinidine reduces pseudobulbar affect. Ann Neurol 68:693–702
150. Bedlack RS, Pastula DM, Hawes J, Heydt D (2009) Open-label trial of levetiracetam for cramps and spasticity in patients with motor neuron disease. Amyotroph Lateral Scler 10:205–209
151. Rabkin JG, Gordon PH, McElhiney M, Rabkin R, Chew S, Mitsumoto H (2009) Modafinil treatment of fatigue in patients with ALS: a placebo-controlled study. Muscle Nerve 39: 297–303
152. Martinez HR, Gonzales-Garza MT, Moreno-Cuevas JE, Caro E, Gutierrez-Jimenez E, Segura JJ (2009) Stem-cell transplantation into the frontal motor cortex in amyotrophic lateral sclerosis patients. Cytotherapy 11:26–34
153. Mazzini L, Ferrero I, Luparello V, Rustichelli D, Gunetti M, Mareschi K, Testa L, Stecco A, Tarletti R, Miglioretti M, Fava E, Nasuelli N, Cisari C, Massara M, Vercelli R, Oggioni GD, Carriero A, Cantello R, Monaco F, Fagioli F (2010) Mesenchymal stem cell transplantation in amyotrophic lateral sclerosis: a phase I clinical trial. Exp Neurol 223:229–237
154. Dimos JT, Rodolfa KT, Niakan KK, Weisenthal LM, Mitsumoto H, Chung W, Croft GF, Saphier G, Leibel R, Goland R, Wichterle H, Henderson CE, Eggan K (2008) Induced pluripotent stem cells generated from patients with ALS can be differentiated into motor neurons. Science 321:1218–1221

Recent Advances in ALS Research: Perspectives for Personalized Clinical Application

Chen Benkler, Daniel Offen, Eldad Melamed, Tamar Amit, Silvia Mandel, Moussa B.H. Youdim, and Orly Weinreb

Abstract Treatment of amyotrophic lateral sclerosis (ALS) has been fueled, in part, by frustration over the shortcomings of the symptomatic drugs available, since these do not impede the progression of this disease. Currently, over 150 different potential therapeutic agents or strategies have been tested in preclinical models of ALS. Unfortunately, therapeutic modifiers of murine ALS have failed to be successfully translated into strategies for patients, probably because of differences in pharmacokinetics of the therapeutic agents, route of delivery, inefficiency of the agents to affect the distinct pathologies of the disease or inherent limitations of the available animal models. Given the multiplicity of the pathological mechanisms implicated in ALS, new therapies should consider the simultaneous manipulation of multiple targets. Additionally, a better management of ALS therapy should include understanding the interactions between potential risk factors, biomarkers and heterogeneous clinical features of the patients, aiming to manage their adverse events or personalize the safety profile of these agents. This review will discuss novel pharmacological approaches concerning adjusted therapy for ALS patients: iron-binding brain permeable multimodal compounds, genetic manipulation and cell-based treatment.

Keywords Amyotrophic lateral sclerosis • Iron chelating-multifunctional drugs • Stem cells • Genetic manipulation • Neurotrophic factors • Personalized medicine

---

C. Benkler • D. Offen • E. Melamed
Felsenstein Medical Research Center, Tel Aviv University, Tel-Aviv, Israel

T. Amit • M.B.H. Youdim • O. Weinreb (✉)
Eve Topf Centers of Excellence for Neurodegenerative Diseases Research and Department of Pharmacology, Rappaport Family Research Institute, Technion-Faculty of Medicine, P.O. Box 9649, Haifa 31096, Israel
e-mail: worly@tx.technion.ac.il

S. Mandel
Eve Topf Center for Neurodegenerative Diseases Research and Department of Molecular Pharmacology, Faculty of Medicine, Technion, Institute of Technology, Haifa, Israel

S. Mandel (ed.), *Neurodegenerative Diseases: Integrative PPPM Approach as the Medicine of the Future*, Advances in Predictive, Preventive and Personalised Medicine 2, DOI 10.1007/978-94-007-5866-7_11,

1 Introduction

Amyotrophic lateral sclerosis (ALS), commonly referred as 'Lou Gehrig's disease' is a relentlessly progressive, neurological disorder with an estimated uniform prevalence of 4–6 cases per 100,000 in Western countries, where it is 50–100 times higher than elsewhere in the world [1]. The onset of ALS is most common in midlife between 45 and 65 years of age, with a typical disease course of 1–5 years [1]. The progressive stage of this disease is characterized by specific degeneration of both upper and lower motor neurons in the brain, brainstem and spinal cord, resulting in paralysis due to muscle weakness and atrophy [2]. In a small percentage of cases, dementia is observed [3]. The majority of patients die within 3–5 years of symptom onset because of respiratory failure. Only 10% of all ALS cases are characterized as genetic familial ALS (fALS), whereas the vast majority is of unknown etiology and classed as sporadic ALS (sALS) [4, 5]. Both forms of ALS are clinically and pathologically similar, suggesting a possible common pathogenesis and final pathway of neurodegeneration [6] though sALS generally develops later than fALS [7].

However, the underlying cause of ALS remains elusive, as well as the development of the chain of events leading to this disease, contributing to the lack of effective therapy. Treatment of ALS has been fueled in part by frustration over the shortcomings of the symptomatic drugs available, since they are incapable of slowing down disease progression and neuronal degeneration. Currently, over 150 different potential therapeutic agents or strategies have been tested in transgenic ALS mice according to published reports [8]. This list involves pharmacotherapies, gene or antisense therapies, cell transplantations, immunizations and dietary or lifestyle regimens. The pharmacotherapy spectrum encompasses antioxidants, anti-excitotoxins, anti-aggregation compounds, anti-apoptotics, anti-inflammatories and neurotrophic agents. However, most of the candidates have been reported to have limited therapeutic benefits and only modest effects on survival of ALS patients. In addition, therapeutic modifiers of murine ALS have failed to be successfully translated into strategies for patients, probably because differences in pharmacokinetics of the therapeutic agents, route of delivery, inefficiency of the agents to affect the distinct pathologies of the disease or inherent limitations of the available animal models [9]. For example, antioxidant treatments such as vitamin E, N-acetylcysteine, and L-methionine demonstrated to delay the disease onset and progression in an ALS animal model [10], show a lack of clinical efficacy in randomized controlled trials in 830 ALS patients [11]. Similarly, systemic administration of the antiapoptotic compound, minocycline, a second-generation tetracycline antibiotic, that can delay the symptom onset in ALS animal model [12], has been reported to fail in a multicenter, phase III, randomized, controlled trial in more than 400 ALS patients [13]. Regrettably, riluzole, an anti-glutamatergic agent, demonstrated to inhibit the presynaptic release of glutamate and the only drug for the treatment of ALS approved by the US Food and Drug Administration (FDA), is reported to have limited therapeutic benefits and only modest effects on survival of ALS patients [14].

Given the devastating nature of ALS, the disease management and therapy should include understanding of the interactions between potential risk factors, biomarkers and heterogeneous clinical features of the patients, with the aim of improving management of their adverse events or personalizing the safety profile of these agents. This knowledge may offer a superior distinction of the pathological profiles among individuals or groups of ALS patients and in turn lead to new-focused therapeutic clinical trials.

The clinical aspects and advanced therapies in personalized treatment of ALS patients are beyond the scope of this chapter and thoroughly discussed by Nefussy and Drory in the current book. The actual chapter will discuss novel therapeutic approaches employed in preclinical models of ALS, including iron-binding brain permeable compounds possessing multimodal neuroprotective activities and genetic and cell-based therapies. These therapeutic approaches take into account the various challenging features of this disease and the interplay of genetic and environmental risk factors, to provide superior efficacy in translation from preclinical studies to the clinic.

2 Multifunctional-Iron Chelating Therapeutic Strategies in ALS

Current therapeutic strategies suggest that drugs acting at a single target may be insufficient for the treatment of multifactorial neurodegenerative diseases, such as Parkinson's disease (PD), Alzheimer's disease (AD) and ALS, all characterized by the coexistence of multiple pathologies. Indeed, a small number of phase 3 studies in ALS patients based on one therapeutic tested target, are currently in progress, for example: ceftriaxone, a β-lactam antibiotic, which reported to increase glutamate transport (ClinicalTrials.gov, number NCT00349622), and dexpramipexole, the optical enantiomer of the dopamine agonist pramipexole (ClinicalTrials.gov, number NCT01281189). The molecular pathogenesis of ALS is poorly understood, contributing to the lack of effective system-based therapies to treat this disease. Investigations have inferred that ALS is a multifactorial and multisystemic disease that arises from combination of several mechanisms, which act through concurring damage inside motor neurons and their neighboring non-motor cells. These include iron accumulation, oxidative stress damage, protein misfolding and aggregation, mitochondrial malfunction, neuroinflammation, impaired axonal transport and vesicle release, genetic factors, increased glutamate excitotoxicity and deficits in neurotrophic factors (Fig. 1) [1, 4, 15].

The effect of OS within neurons has been studied mainly in the G93A-superoxide dismutase (SOD)1 transgenic mouse model-related fALS (which develops symptoms and pathology similar to those of ALS patients); however, the roles of OS in sALS remain obscure. Recently, a role of iron in OS-induced motor neuron damage has been suggested. It is well established that ionic iron, participates in the Fenton chemistry and reacts with hydrogen peroxide to produce reactive oxygen species

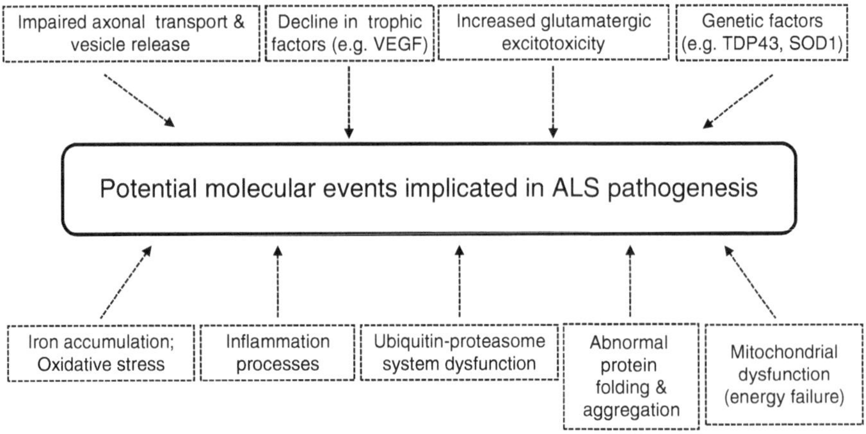

Fig. 1 A schematic illustration of the potential diverse molecular events implicated in the CNS pathogenesis of both sporadic and familial ALS forms that may involve in motor neuronal degeneration. ALS is a multifactorial and multisystemic disease that arises from combination of several mechanisms, which act through concurring damage inside motor neurons and their neighboring non-motor cells. These include iron accumulation, oxidative stress damage, protein misfolding and aggregation, mitochondrial malfunction, neuroinflammation, impaired axonal transport and vesicle release, genetic factors, increased glutamate excitotoxicity and deficits in neurotrophic factors [1]

(ROS), displays elevated iron levels in the central nervous system (CNS) of both sporadic and familial forms have been reported [16–18]. Indeed, a recent clinical report described a patient, who was diagnosed with neurodegeneration and brain iron accumulation, but had clinical criteria for frontotemporal dementia and picture of ALS [3]. Recently, a defect in the Hfe gene, which has been previously associated with iron overload diseases, hemochromatosis and Alzheimer's disease (AD), is currently associated with ALS [19]. The protein normally made by the Hfe gene is thought to limit the uptake of iron by cells, to protect against OS and possibly to dampen inflammatory reactions. An increased incidence of the Hfe mutation has been reported in ALS patients [20]. The presence of this mutation was shown to disrupt expression of tubulin and actin at the protein levels potentially consistent with the disruption of axonal transport seen in ALS and associated with a decrease in SOD1 expression [20]. In support for the involvement of iron in ALS pathology, the iron-responsive protein transferrin, was localized in Bunina bodies of spinal cord neurons from ALS patients [21, 22], suggesting its involvement in the formation of these inclusions.

In accordance with the human studies, in the G93A-SOD1 transgenic mouse model, the expression of ferritin was induced at late stages of the disease, indicating high iron concentrations [4]. Interestingly, in transgenic mice expressing the wild type SOD1 or SOD1-active mutant enzyme, G93A-SOD1, the expression of transferrin receptor (TfR) and iron regulatory proteins 1(IRP1), a positive transcriptional regulator of TfR, were positively modulated in response to increased SOD1 mutation [21]. Jeong and collaborators [23] have recently described dysregulation of

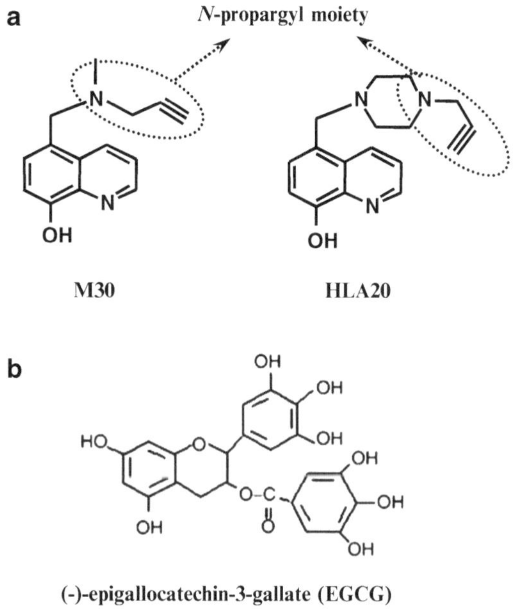

Fig. 2 Chemical structure of the novel brain permeable, multifunctional iron chelating compounds and natural iron chelator, EGCG. (**a**) M30 (5-[N-methyl-N-propargylaminomethyl]-8-hydroxyquinoline) and HLA20 (5-[4-propargylpiperazin-1-ylmethyl]-8-hydroxyquinoline), restrained the propargyl moiety. (**b**) The major polyphenolic compound of green tea, (−)-epigallocatechin-3-gallate, EGCG

iron homeostasis mechanism in the CNS in the G37R-SOD1 transgenic mice model of ALS, suggesting that iron chelating therapy might be useful for the treatment of ALS. In support, iron-binding porphyrin (a catalytic antioxidant) treatment has extended survival and increased neuronal survival in the spinal cord in G93A-SOD1 mice [24].

Noticeably, drug therapy of neurodegenerative diseases with multifunctional compounds embracing diverse biological properties with single bioavailability and pharmacokinetic metabolism, will have a pronounced advantage over individual-target drug or cocktail of drugs. Based on this reasoning, Youdim and collaborators [25] have designed and synthesized several chimeric compounds, by amalgamating the neuroprotective propargyl moiety of the anti-Parkinson's drug/irreversible selective monoamine oxidase (MAO)-B inhibitor, rasagiline [N-propargyl-1R(+)-aminoindan], Azilect (Teva Pharmaceutical Netanya Industries, Ltd. Israel) into the antioxidant-iron chelating skeleton of the 8-hydroxyquinoline derivative of VK-28, to produce several multiomodal compounds for combating neurodegenerative diseases, such as ALS.

In searching for the superlative neuroprotective agents in a series of multifunctional iron chelators, the compound M30 (5-[N-methyl-N-propargylaminomethyl]-8-hydroxyquinoline) (Fig. 2) was found to be a most potent iron chelator, non-toxic, brain permeable, displaying highly effective inhibition of MAO-A and -B activities in vitro and in vivo and iron-induced membrane lipid peroxidation [25–29]. Additionally, M30 was found to invoke a wide range of pharmacological activities, including a neurorescue response, a protective potency against OS insults and regulatory action on neuronal differentiation and neurite outgrowth in various neuronal cell lines [30–32]. In in vivo studies, M30 possesses a significant neuroprotective as well as neurorescue activity against N-methyl-4-phenyl-1,2,3,6-tetrahydropyridine (MPTP) neurotoxicity in mice [29, 33]. In addition, both M30 and VK28 were found to prevent the loss of mice tyrosine hydroxylase (TH) positive neurons, induced by post-intra-nigral injection of lactacystin, significantly improve behavioral performances and attenuate inhibition of ubiquitin-proteasome activity, iron increase and microglial activation in the ipsilateral substantia nigra (SN) [34].

Given the diversity of pathological mechanisms implicated in ALS, which suggested that multiple pathways may contribute to this disease [35], new ALS therapies may consider a simultaneous manipulation of multiple targets. Indeed, combination treatments or polypharmacy, targeting different disease mechanisms have consistently shown superior efficacy in transgenic ALS mice, such as the combination of rasagiline with riluzole [36].

2.1 Multimodal Iron Chelating Compounds, M30 and HLA20

2.1.1 Neuroprotective Effects on Survival and Signaling Cascades in Cell Cultures

In a recent experimental study of ALS, M30 and an additional newly designed multifunctional compound, HLA20 (5-[4-propargylpiperazin-1-ylmethyl]-8-hydroxyquinoline) [37] (Fig. 2) were demonstrated to possess multiple pharmacological activities in NSC-34 cells, a widely used mouse motor neuron-like cell line [32]. Initially, M30 and HLA20 were demonstrated to acquire a protective potency after exposure of cultured NSC-34 cells to the OS insults, hydrogen peroxide (H_2O_2) and peroxynitrite generator, 3-morpholinosydnonimine (SIN-1) [32], previously shown to be associated with motoneuron degeneration in ALS [38, 39]. This observation is consistent with previous studies showing that M30 decreased apoptosis of SH-SY5Y neuroblastoma cells in a neurorescue, serum deprivation model, via various protective mechanisms, including reduction of the pro-apoptotic proteins, Bad and Bax and inhibition of the apoptosis-associated phosphorylated H2A.X protein (Ser 139) and caspase-3 activation [30]. Since N-propargylamine and rasagiline confer neuroprotective/neurorescue effect via activation of protein kinase C (PKC) and mitogen-activated protein kinase (MAPK) signaling pathways, coupled to pro-survival Bcl-2 family members and mitochondrial membrane stabilization [40–42], it can be assumed that the neuroprotective effect of the novel multifunctional drugs, demonstrated in NSC-34 cells, may be partially associated with their propargyl moiety.

2.1.2 Effect on Iron Homeostasis-Related Proteins and Hypoxia-Inducible Factor (HIF) Pathway in Cell Cultures

The iron complexing moiety, embedded in the novel multifunctional iron chelating drugs may favorably influence cell survival by reducing the levels of ROS and

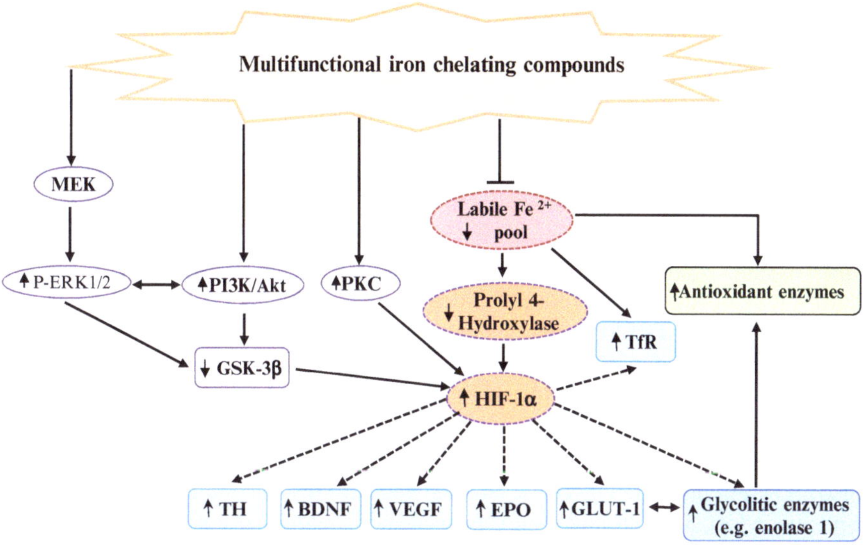

Fig. 3 Neuroprotective targets involved in mechanism of action of the multifunctional iron chelating compounds (e.g. M30, HLA20). Full explanation is discussed in the text. *Abbreviations*; *BDNF* brain derived neurotrophic factor, *EPO* erythropoietin, *ERK* extracellular signal-regulated kinase, *HIF* hypoxia-inducible factor, *GLUT-1* glucose transporter 1, *GSK-3* glycogen synthase kinase 3 β, *PKC* PI3K, phosphatidylinositol 3-kinase protein kinase C, *TfR* transferrin receptor, *TH* tyrosine hydroxylase, *VEGF* vascular endothelial growth factor

reactive nitrogen species (RNS), due to neutralization of excessive free-reactive Fe^{2+}. An alternative pathway of protection by iron chelators may include the inhibition of the iron-dependent hypoxia-inducible factor (HIF)- prolyl 4-hydroxylases (PHDs), enzymes that regulates HIF stability (Fig. 3). Indeed, PHDs have been suggested as an additional target for neuroprotection in various neurodegenerative diseases. Inhibition of HIF-PHDs prevents the hydroxylation and subsequent degradation of HIF-1α [43]. Stabilization and nuclear localization of HIF-1α results in heterodimerization with its partner HIF-1β, binding to the hypoxia-response element in the gene regulatory regions, and subsequent transcriptional up-regulation of established protective genes, such as erythropoietin, vascular endothelial growth factor (VEGF), p21waf1/cip1, glycolytic enzymes (e.g. aldolase and enolase 1) as well as iron homeostasis-related genes (e.g., TfR) [44]. Siddiq et al. [43] used an in vitro model of OS to correlate the protective effects of iron chelators and small molecules and peptides that don't bind iron, but do inhibit the PHDs and activate HIF-1α. This model has been further supported by the observation that the neuroprotective effects of the prototype iron chelating drug, desferrioxamine (DFO), in deprived oxygen and glucose hippocampal neuronal cultures were prevented by blockade of HIF-1α with antisense oligonucleotide transfection [45].

In accordance, we have demonstrated that M30 and HLA20 induced mRNA expression of HIF-1α and enhanced the protein levels of HIF-1α and its nuclear translocation in NSC-34 cells [32]. Furthermore, M30 and HLA20 significantly increased the levels of the endogenous HIF-1-dependent genes, enolase 1, VEGF, brain derived neurotrophic factor (BDNF) and TfR in NSC-34 cells (Fig. 3) [32]. Although VEGF was once considered to be only a specific angiogenic factor, emerging evidence indicates that it also has direct effects on neuronal cells and protects motoneurons from cell death induced by various insults, such as OS, hypoxia/hypoglycemia, glutamate-excitoxicity and serum deprivation [46]. Deletion of the hypoxia-responsive element in the promotor region of the VEGF gene can cause motor degeneration in mice and low-VEGF producing alleles of the VEGF gene are associated with motoneuron degeneration in human ALS, suggesting that VEGF is a modifier of motoneuron degeneration in human ALS [47–49]. Although there is evidence for and against the role of VEGF in ALS etiopathogenesis, the literature has widespread interest in developing VEGF-based therapies for motoneuron degenerative disorders, raising new hope for the treatment of ALS and other neurodegenerative diseases. Besides regulating VEGF, HIF-1 has been also shown to promote glycolytic enzyme gene expression and consequent aerobic glycolysis [50, 51]. Recent studies have demonstrated that a shift in energy generation from glucose oxidation in mitochondria to aerobic glycolysis is associated with resistance to OS [52]. Thus, by inducing glycolytic enzyme expression, iron chelators may reduce the ambient free radical burden of neurons by enabling the cell to generate more energy glycolytically and minimize deleterious consequences of mitochondrial glucose oxidation. Taken together, it can be suggested that regulation of HIF-1α expression and its related genes may constitute an additional pathway underlying the neuroprotective effect of M30 and HLA20.

In addition, it has been reported that phosphatidylinositol-3-kinase (PI3K)/ protein kinase B (AKT) signaling plays an important role in regulating HIF-1α expression [53, 54]. Indeed, both M30 and HLA20 markedly increased the amount of phospho-AKT and phosphor- glycogen synthase kinase 3 (GSK-3)β (at ser-9) in NSC-34 cells, suggesting that both drugs induced activation of AKT followed by phosphorylation (inactivation) of GSK-3β (Fig. 3) [32].

2.1.3 Induction of Neurite Outgrowth and Differentiation in Cell Cultures

Another important characteristic aspect of M30 and HLA20 is their ability to induce differentiation, cell elongation and stimulate neurite outgrowth of NSC-34 motoneuron cells [32]. These morphological modifications were accompanied by an increase in the immunoreactivity of the neuronal marker growth-associated protein, GAP-43, and decrease in cyclin D1 expression [32], in accordance with results of previous studies, demonstrating that M30 induced a neuritogenic effect and triggered cell cycle arrest in G0/G1 in rat pheochromocytoma PC12 and human SH-SY5Y neuroblastoma cell lines [30]. Indeed, many cell cycle regulating factors require iron for their function [55, 56]. Thus, an inhibitory effect of the novel iron-chelator compounds on cell cycle reentry [30], may trigger the process of

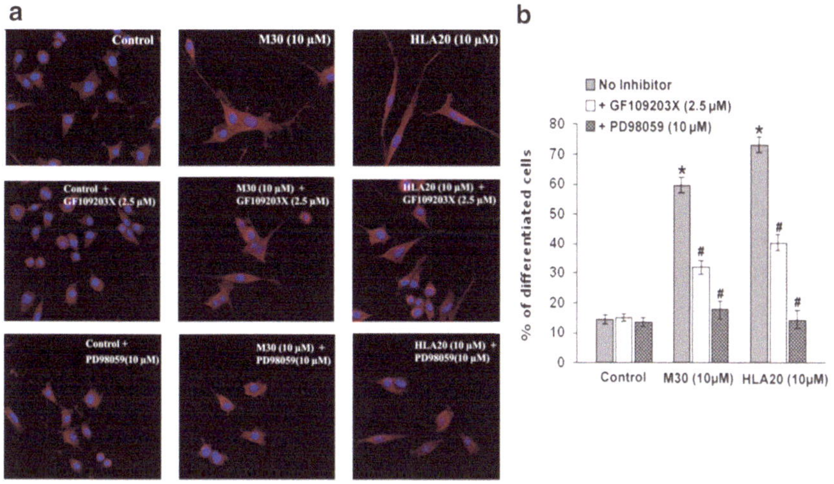

Fig. 4 Attenuation of the neuritogenic effect of M30 and HLA20 by specific inhibitors of PKC and MAPK/ERK kinase (MEK). NSC-34 cells were incubated with PKC inhibitor GF109203X (2.5 μM) or MEK inhibitor PD98059 (10 μM) for 1 h before administration of M30 (10 μM) or HLA20 (10 μM) for a further 24 h. The cells were fixed and permeabilized for GAP-43 detection. (**a**) The images are representative fields from three independent experiments. (**b**) The histogram represents averages of the differentiated cell percentages (±SEM). One-way ANOVA followed by Student's *t*-test was used for statistic analysis. * $p < 0.001$ vs. respective controls; # $p < 0.001$ vs. M30 and HLA20 only (without inhibitors) treated cells (reproduced from [32])

differentiation through various iron-associated biological events [57]. In addition, the effect of the multifunctional drugs on motoneuron differentiation may be associated with their propargyl moiety, since N-propargylamine and rasagiline were shown to up-regulate BDNF and glial cell line-derived neurotrophic factor (GDNF) gene expression in PC12 cells [41, 42]. In in vitro studies in NSC-34 cell, M30 and HLA20 were demonstrated to induce mRNA levels of BDNF, which is a well recognized neurotrophic factor for motoneurons [32, 58–60]. Motor neuron differentiation, induced by M30 and HLA20, was shown to be modulated by inhibitors of ERK/MAPK and PKC signaling pathways (Fig. 4) [32]. In results, complementary to inhibition studies, we found that the drugs significantly increased the immunoreactivity of phosphorylated MAPK and PKC in NSC-34 cells, further indicating the involvement of both ERK and PKC activation [32].

2.1.4 Neuroprotective Effects in Mouse Model of ALS

Treatment with M30 provided clear benefits in the fast-progressing strain of G93A-SOD1 mutant ALS transgenic mice [61] even when the treatment was initiated at relatively advanced stage of the disease (day 70th post-natal) [32]. Oral administration of M30 (1 mg/kg, 4Xwk) resulted in delaying the onset of motor dysfunction (the average age of onset was 107 ± 3 days in the control group, while 112 ± 4 days in the M30-treated group; n = 14–16/group; $p < 0.001$; log -rank Mantel-Cox test) and in significantly increasing their survival (an increase in survival by M30 treatment from 124 ± 6 to 134 ± 12 days; n = 14–16/group; $p < 0.025$; log -rank Mantel-Cox test) (Fig. 5) [32]. In a supplementary study performed in G93A-SOD1 mice, initiated at presymptomatic stage of the neurological dysfunctions, M30 was demonstrated to reduce spinal motor neuronal loss [62]. M30 also inhibited the inflammatory-related factor, tumor necrosis factor-α (TNFα) by suppressing the glial activation, as well as reduced TAR-DNA binding protein (TDP-43) [62], which is a major ubiquitinated inclusion of ALS inclusions [63, 64].

2.2 *Natural Multifunctional Iron Chelating Compound, EGCG*

Naturally multifunctional occurring polyphenols, in particular, the major component of green tea, (−)-epigallocatechin-3-gallate (EGCG) (Fig. 2) has been indicated as novel and promising therapeutic approach for treating neurodegenerative disease, including AD, PD and ALS. Indeed, several studies of acute and chronic EGCG administration in different cellular and animal models of neurodegeneration, demonstrated various novel molecular targets, including, the MAPK/ PKC and the PI3K/AKT signaling pathways; survival genes and proteins associated with mitochondrial function [65]. EGCG was demonstrated to exert neuroprotective effects against cytotoxicity in ventral spinal cord 4.1 cells (VSC 4.1), motoneuron–neuroblastoma hybrid cell line, wild-type and G93A mutant cells and affect upstream signaling through the PI3K/Akt, GSK-3 pathway, as well as downstream cascades involved in cytochrome c release and caspase-3 activation [66]. Oral administration of EGCG to G93A-SOD1 mutant transgenic mice, beginning from a pre-symptomatic stage, significantly prolonged the symptom onset, delayed the onset of disease and extended life span [66–68]. Additionally, EGCG increased the number of motor neurons, diminished microglial activation, reduced immunohistochemical reaction of NF-kappaB and cleaved caspase-3, as well as reduced protein level of iNOS and NF-kappaB in the spinal cords of G93A-SOD1 mutant transgenic mice [68]. Regarding glutamate-mediated neurotoxicity in ALS, it was demonstrates that EGCG prevented glutamate excitotoxicity induced by threohydroxyaspartate, an inhibitor of glutamate transporter, in organotypic culture of rat spinal cord [69].

3 Genetic and Cellular Therapeutic Approaches in ALS Patients and Animal Models

The vast majority of ALS cases have not been associated with genetic factors. However, approximately 2% of all known ALS cases are attributed to mutations in the zinc/copper superoxide dismutase (SOD1) gene [5]. To date, over 100 mutations

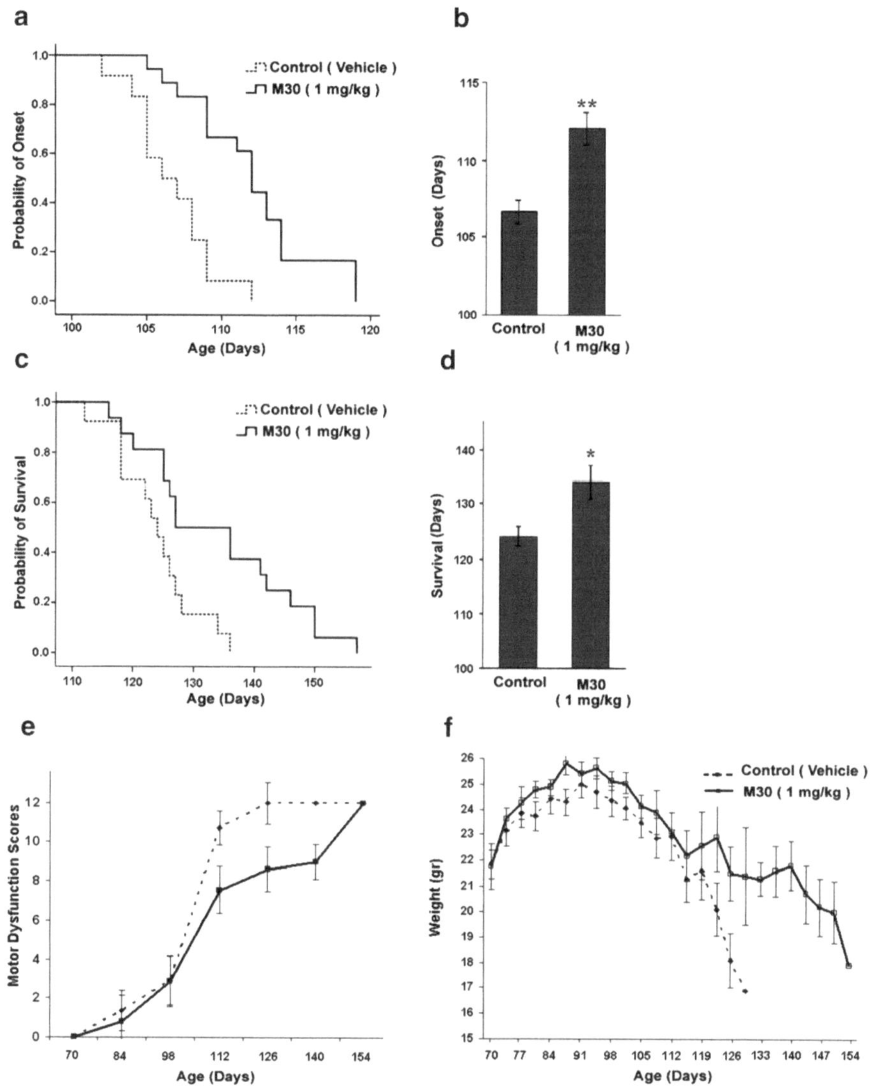

Fig. 5 Effect of M30 treatment on motor dysfunction onset, survival time, motor deficits and weight in G93A-SOD1 mutant ALS transgenic mice. Mutant G93A-SOD1 mice were treated by the oral gavage method with vehicle (control) or M30 (1 mg/kg) four times a week starting from the 70th day after birth and continuing until death. Plots present cumulative probability of (**a**) the symptoms onset ($n = 13–16$ per group; $p < 0.001$; log -rank Mantel-Cox test) and (**c**) overall survival ($n = 13–16$ per group; $p < 0.025$; log -rank Mantel-Cox test) against the age of the mutant mice. Histograms present (**b**) mean onset (days) and (**d**) mean survival (days) of vehicle- or M30-treated G93A-SOD1 mice. Values are means ± SEM ($n = 13–16$ per group; *$p < 0.05$, **$p < 0.001$ vs. control group; one-way ANOVA). (**e**) Overall neurological deficit scores vs. the age of animals. The total neurological deficits were determined from four independent tests (rotarod performance, postural reflex, screen grasping and tail suspension behavior). Total score of 12 represent a complete loss of motor function. Values are means ± SEM ($n = 13–16$ per group). (**f**) Weight vs. the age of animals. Values are means ± SEM (reproduced from [32])

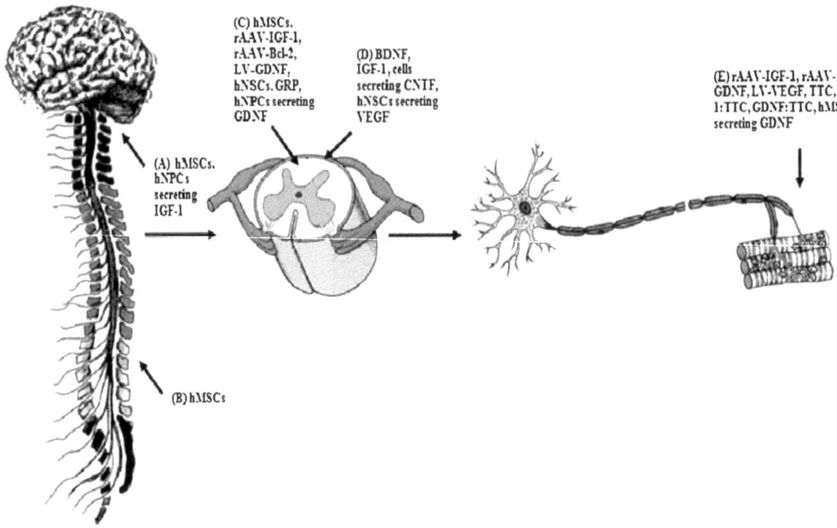

Fig. 6 A schematic representation of methods allowing local delivery of neurotrophic factors and cellular transplantations performed in ALS animal models and patients. (**a**) Represents the cisterna magna. (**b**) Refers to the CSF. (**c**) Concerns intraspinal treatment. (**d**) Indicates the spinal intrathecal space. (**e**) Pertains to the intramuscular space. *Abbreviations*; *hMSCs* human mesenchymal stem cells, *hNPCs* human neuronal progenitor cells, *rAAV* recombinant Adeno-associated virus, *LV* lentivirus, *hNSCs* human neuronal stem cells, *GRP* glial restricted precursors, *TTC* tetanus toxin heavy chain, *IGF-1:TTC* fusion protein of IGF-1 and TTC, *GDNF:TTC* fusion protein of GDNF and TTC

associated with fALS have been mapped to the SOD1 gene. One of the most prevalent mutations is the G93A variant, in which the amino acid glycine is replaced by alanine at position 93. This information was used in developing animal models of ALS; the first and most commonly used model is of transgenic rodents harboring the G93A mutation. These animals develop a pathophysiological condition resembling ALS, presenting with hindlimb weakness at 3–4 months of age and progressive paralysis culminating in death at the age of 4–5 months. The pathological features of the disease resemble human neurologic diseases, including loss of large spinal cord motoneurons and robust spinal astrogliosis and microgliosis [4, 61]. Other genes linking to fALS include alsin (ALS2), senataxin (ALS4), vesicle associated membrane protein (VAPB, ALS8), angiogenin and a mutation in the p150 subunit of dynactin [70]. Recently, mutations in TARDBP gene, encoded the TAR-DNA binding protein TDP-43 and located on chromosome 1p36.22, has been reported to be associated with fALS and sALS [70].

In the next section, we will focus on genetic therapies, cell replacement therapies and their combination, and we will review experiments, performed predominantly in the SOD1 rodent ALS models and later adapted for clinical trials (Fig. 6; Table 1).

Table 1 An overview of several genetic and cellular therapeutic approaches relating to ALS patients and animal models

Method of treatment	Treated group	Method of delivery	Effect	Reference
Recombinant methionyl BDNF	Clinical trails ALS patients	Intrathecal administration	No clinical or survival benefits	[71–73]
Recombinant IGF-1	Clinical trails ALS patients	Subcutaneous administration	American trial: slower progression of functional impairment European trial: no clinical benefits	[74, 75]
Recombinant IGF-1	Clinical trails ALS patients	Intrathecal infusion	Modest beneficial effects	[76]
Recombinant human CNTF	Clinical trails ALS patients	Subcutaneous injections	No beneficial clinical results, high dose CNTF administration was accompanied by severe adverse reactions and an increased number of deaths	[77–79]
Human MSCs	Clinical trails ALS patients	Intraparenchymal autologous transplantation at the thoracic level of the spinal cord	Contradicting results of ether slight clinical improvement or no clinical benefits	[80, 81]
Human HSCs	Clinical trails sporadic ALS patients	Post-irradiation transplantation with a sibling's HSCs	No clinical benefits	[82]
Baby hamster kidney cells overexpressing CNTF	Clinical trails ALS patients	Intrathecal transplantation of membrane encapsulated cells	Clinical benefits have not been reported	[83–85]
rAAV-IGF-1	SOD1 G93A mice	rAAV-IGF-1 at respiratory and motor limb muscles or deep cerebellar nuclei	Delayed disease onset, prolonged life and moderated disease progression	[86, 87]
rAAV-IGF-1	SOD1 G93A mice	rAAV-IGF-1 Spinal cord infection	Delayed disease onset, weight loss, decline in hindlimb grip strength and increased survival was observed in male but not female mice	[88, 89]
rAAV-IGF-1 or rAAV-VEGF	SOD1 G93A mice	rAAV-IGF-1 or rAAV-VEGF intracerebroventricular (ICV) administration	Delayed motor decline and moderately extended the survival. Combined IGF-1 and VEGF administration yielded no add neuroprotection	[90]

(continued)

Table 1 (continued)

Method of treatment	Treated group	Method of delivery	Effect	Reference
rAAV-GDNF	SOD1 G93A mice	rAAV-GDNF at respiratory and motor limb muscles	Delayed disease onset, prolonged life and moderated disease progression	[91]
rAAV-GDNF	SOD1 G93A mice	rAAV-GDNF intramuscularly injecting	Delayed disease onset, amelioration of disease progression and prolongation of life expectancy	[91]
rAAV-G-CSF (or direct G-CSF administration)	SOD1 G93A mice	rAAV-G-CSF intramuscular injection or subcutaneous G-CSF administration	Protected motor neurons, ameliorated functional outcome and prolonged life	[92, 93]
rAAV-G-CSF	SOD1 G93A mice	rAAV-G-CSF intraspinal administration	Improved motor function, delayed disease progression and increased survival by 10% more then when G-CSF was subcutaneously administrated or rAAV-G-CSF was intramuscularly injected	[92]
rAAV-Bcl-2	SOD1 G93A mice	Lumber spinal cord rAAV-Bcl-2 infection	Delayed disease onset and preservation of spinal motoneurons	[94]
LV-GDNF	SOD1 G93A mice and facial motoneurons axotomy model	LV-GDNF lumber spinal cord and facial nucleus injections	Rescue of motoneurons in the facial nucleus. No effect on spinal motoneurons	[95, 96]
LV-VEGF	SOD1 G93A mice	LV-VEGF infection at muscle terminals	Delayed disease onset, slower disease progression and prolonged survival	[97]
TTC	SOD1 G93A mice	Intramuscular injections of naked TTC DNA	Delayed onset of symptoms and functional deficits, improved spinal motor neuron survival, and prolonged lifespan	[98]
IGF-1:TTC	SOD1 G93A mice	Intramuscular and intrathecal injections of soluble IGF-1:TTC protein	No significant clinical effect	[99]

GDNF:TTC	SOD1 G93A mice	Intramuscular injections of soluble GDNF:TTC protein	Mild survival prolongation	[100]
GDNF:TTC	SOD1 G93A mice	Intramuscular injection of naked DNA encoding GDNF or GDNF:TTC fusion	Delayed symptom onset and functional deficits. Prolonged lifespan	[101]
(GDNF alone)				
BDNF:TTC	SOD1 G93A mice	Intramuscular injection of BDNF:TTC of TTC plasmid DNA	Fusing BDNF to the TTC molecule did not result in enhanced synergistic effect when compared to TTC alone	[102]
(TTC alone)				
Human NSCs	SOD1 G93A rats	Lumbar spinal cord transplantation	Delayed disease onset and prolongation of life expectancy	[103, 104]
Glial restricted precursors	SOD1 G93A rats	GRP transplantation at cervical spinal cord respiratory motor neuron pools	Extended survival and disease duration, attenuated motor neuron loss and moderated decline in forelimb motor and respiratory functions	[105, 106]
Human MSCs	SOD1 G93A mice and rats	Lumber spinal cord, cisterna magna and CSF transplantations	Extended survival, reduced neuroinflammation; astrogliosis and microgliosis and ameliorated disease symptoms	[107–109]
Human MSCs	SOD1 G93A mice	Post irradiation intravenous administration	Delayed disease onset, delayed disease progression and increased lifespan	[110]
Lin-c-kit positive bone marrow cells	SOD1 G93A mice	Systemic intravascular injection	Delayed motor degeneration, reduced microgliosis delayed onset of the early symptomatic disease stage, slowed disease progression and extended the survival of SOD1-G93A mice by 16–17 days	[111]

(continued)

Table 1 (continued)

Method of treatment	Treated group	Method of delivery	Effect	Reference
Bone marrow cells	SOD1 G93A mice	Post-irradiation intra-bone marrow grafts	Slowed disease progression and extended survival	[113]
Human umbilical cord blood cells	SOD1 G93A mice	Retro-ocular transplantation	Delayed disease onset and extended survival	[114]
Human umbilical cord blood cells	SOD1 G93A mice	Intraspinal administration	Transplantation before symptom onset (day 40) slightly enhanced survival and preserved motor function in a gender bias manor. transplantation after symptom onset (day 90) had no such effect	[115]
Human cord blood mononuclear cells	SOD1 G93A mice	Intracerebroventricular (ICV) administration	Significantly ameliorated symptoms progression and motor decline as well as significantly prolonged survival	[116]
Human NPCs secreting GDNF	SOD1 G93A mice and rats	Spinal cord grafts	Delayed disease onset but did not extended survival	[117–119]
Human MSCs secreting GDNF	SOD1 G93A rats	Muscle transplantations	Delayed disease progression and extended overall life span	[120]
Human NSCs overexpressing VEGF	SOD1 G93A mice	Intrathecal transplantation into the lumber spinal cord	Delayed disease onset, slower decline of motor function and prolonged survival	[121]
Human NPCs secreting IGF-1	SOD1 G93A mice	Intrathecal transplantation into the cisterna magna	Enhanced motor neuron survival, but with no functional improvement or extension of life span	[118]
Bone marrow cells	A crossbred of PU.1−/− mice and SOD1 G93A mice	Bone marrow transplantation	Reduced motoneuron degeneration, ameliorated disease progression and prolonged survival	[112]

Recombinant adenoviruses with anti SOD1 RNAi properties	SOD1 G93A mice	Direct nerve injection	Extended the life span of SOD1 G93A mice	[122]
LV with anti SOD1 RNAi properties	SOD1 G93A mice	Intraspinal injections	Delayed disease onset and progression rate	[123]
LV carrying RNAi molecules specifically targeting the SOD1	SOD1 G93A mice	Intramuscular injections	Preserved motor function, delayed disease onset and prolonged survival	[124]
Antisense peptide nucleic acid targeting p75	SOD1 G93A mice	Systemic intraperitoneal delivery	Delayed motor decline and prolonged survival	[125]
Anti-Fas RNAi	SOD1 G93A mice	Intrathecal administration	Delayed motor injury and extension of life span	[126]

Abbreviations: MSCs Mesenchymal stem cells, *NPCs* neuronal progenitor cells, *NSCs* neuronal stem cells, *GRP* glial restricted precursors, *rAAV* recombinant Adeno-associated virus, *LV* lentivirus, *TTC* tetanus toxin heavy chain, *IGF-1:TTC* fusion protein of IGF-1 and TTC, *GDNF:TTC* fusion protein of GDNF and TTC

3.1 Adeno-Associated Viruses

Adeno-associated viruses (AAVs) are small single strand DNA genome encapsidated viruses, which require helper viruses for successful infection and are therefore not themselves pathogenic [127, 128]. Recombinant AAVs (rAAVs) are produced by replacing their pathogenic viral DNA with the gene of interest.

Injecting rAAVs containing either IGF-1 or GDNF into respiratory and motor limb muscles of SOD1 G93A mice delayed disease onset, prolonged life and moderated disease progression. IGF-1 was found to perform more potently than GDNF. It was further established that the rAAVs selectively transfected motoneurons without infecting neighboring glia cells. To assess the importance of rAAVs retrograde delivery into the central nervous system (CNS), SOD1 mice were transfected with recombinant lentiviruses (rLVs) containing IGF-1 which were not retrogradely transported and were found to increase survival to a lesser extent than rAAV-IGF-1 [86]. Similar results were obtained when rAAVs containing IGF-1 were delivered to the deep cerebellar nuclei [87]. Spinal cord infection was accompanied by a sexual bias predominantly effecting male mice [88, 89]. More recently, Dodge et al. [90] expanded their experiments with IGF-1 in SOD1 G93A mice and evaluated the effect of intracerebroventricular (ICV) injections of AAVs encoding not only for IGF-1 gene but also for vascular endothelial growth factor (VEGF) another potentially promising neurotrophic factor. Interestingly, they found that injection of IGF-1 as well as VEGF delayed motor decline and moderately extended the survival of SOD1 G93A mice. Both factors proved to be equally effective, however the combination of IGF-1 and VEGF expressing AAVs resulted in no additional protection. This suggests that these two neurotrophic factors are likely to influence common therapeutic signaling pathways [90]. Additional research on rAAVs carrying GDNF, injected intramuscularly, further emphasized its therapeutic potential, as long term GDNF expression in the CNS, a delay in disease onset, an amelioration in disease progression and a prolongation of life expectancy, were obtained [91, 129].

Another factor with neuroprotective properties rose from an-unexpected source. Granulocyte-colony stimulating factor (G-CSF) was originally described as a potent hematopoietic growth factor that enhances the survival and stimulates proliferation and differentiation of myeloid lineage cells [130, 131]. It was later discovered to possess potent neuroprotective properties through anti-apoptotic activity as well as through activation of survival pathways in neural progenitor cells, adult neural stem cells and in mature neurons [130]. Direct G-CSF subcutaneous administration and intraspinal injection of rAAV-G-CSF improved motor function, delayed disease progression and increased the survival of SOD1 G93A mice [92, 93]. However, intramuscular injection of rAAV-G-CSF did not retrogradely transducer spinal cord cells but yielded high G-CSF serum levels. As G-CSF crosses the blood brain barrier it yielded similar results to those obtained by subcutaneous administration [92].

Neurotrophic factors are not the only genes that can be delivered using viral vectors. Bcl-2 an anti-apoptotic protein has previously been delivered to the lumber spinal cord of SOD1 G93A mice using a rAAV construct. Intraspinal injection of these rAAVs resulted in sustained motoneuron Bcl-2 expression, delayed disease onset and significantly increased the number of surviving motoneurons at disease end-stage, but was ultimately unable to prolong survival [94]. This represents an interesting alternative for gene selection in AAV based gene therapy in ALS. However, when considering this approach we must be extremely vigilant regarding the transfected cells, as the effect of Bcl-2 might be a double edged sword, protective in non-dividing, degeneration prone motoneurons, yet potentially oncogenic in other cells.

rAAVs are emerging as a prominent vector for neurotrophic factor gene delivery in ALS, probably due to their many advantages, which include rare vector sequence integration into host genome, efficient transfection of both dividing and non-dividing cells, long-term gene expression, low risk of toxicity, and minimal stimulation of the host immune response. The existence of several serotypes allows differential targeting of specific CNS cell populations [89]. However, they are limited by the relatively small amount of DNA they can encapsulate, thus not suitable for larger genetic constructs [127].

3.2 Lentiviruses

Lentiviruses (LVs) are the most complex viruses in the retroviral family and are particularly suitable for CNS related genetic therapy. LV vectors have relatively large RNA packing capacity. They can efficiently transfect not only dividing but also non-dividing cells, such as neurons. LVs can reverse transcribe their RNA into the host cell's DNA, thus promoting stable and long term expression of the gene of interest [128, 132–134].

Recombinant LVs for gene therapy are produced similarly to rAAVs. First, removal of as much of the viral genes as possible occurs, then the desired expression cassette is inserted. rLVs incapable of producing a functional envelope can be pseudotyped with different viruses to selectively direct the infection to the desired cellular type and prevent secondary infection [135, 136].

Lumber spinal cord and facial nucleus injections of rLVs incorporating the GDNF gene were found to significantly rescue motoneurons in the facial nucleus, but did not prevent spinal motoneuron loss, despite robust intraspinal GDNF expression in SOD1 G93A mice and facial motoneurons axotomy mice models [95, 96].

Injection of a recombinant rabies-G pseudotyped lentiviral vector based on the equine infectious anemia virus encoding the human vascular endothelial growth factor (hVEGF) gene at several muscle terminals in SOD1 G93A mice, promoted hVEGF expression in spinal cord motoneurons, delayed disease onset, slowed disease progression and prolonged survival [97].

3.3 Tetanus Toxin Heavy Chain

Gene therapy can also be based on non-viral delivery of naked DNA. It has been suggested that the non-toxic carboxy-terminal fragment of the tetanus toxin heavy chain (TTC) can act as a neuroprotective agent through the downregulation of caspases 1 and 3. Intramuscular injections of naked DNA encoding the TTC gene in SOD1 G93A mice delays the onset of symptoms and functional deficits, improved spinal motor neuron survival, and prolonged lifespan [98]. TTC has also been used as a platform for delivering IGF-1 and GDNF to motoneurons. Intrathecal and intramuscular injections of a soluble IGF-1: TTC fusion protein had no significant effect on disease progression or survival in SOD1 G93A mice [99]. However, a slight increase in survival was noted when similar intramuscular injections of a GDNF:TTC fusion protein in SOD1 G93A mice were administered [100]. Later research by the same group also found that intramuscular administration of naked GDNF:TTC DNA rather than protein yielded similar results. It is noteworthy that the fusion of GDNF to TTC did not augment the therapeutic effect when compared to GDNF alone [101]. On the other hand, intramuscular administration of plasmid DNA encoding the fused BDNF:TTC molecule did not enhance the therapeutic effect conferred by TTC alone [102]. In this experiment, unlike the situation in the GDNF experiment, the neuroprotective outcome was conferred by the TTC protein and not by the neurotrophic factor.

Genetic therapy approaches offer promising strategies for delivering factors that might enhance the survival chances of degeneration prone neurons in ALS. However, it is not the only option: transplantation of whole cells presents a viable and potent non-viral alternative.

3.4 Motor Neuron Replacement Therapies

Motor neurons can be effectively derived from embryonic stem cells (ESCs), neuronal stem cells (NSCs) and induced pluripotent stem (iPS) cells [137–139]. The strategy for treatment is to use these cells to replace damaged neurons in ALS disease.

In several studies of neuronal transplantations, performed in mice and rat models of ALS, a delay in disease onset and a mild prolongation of life expectancy have been observed [139–141].

It has been further shown that human NSCs grafted into the lumbar spinal cord of SOD1 G93A rats, survive after transplantation and they present with predominantly neuronal differentiation at the transplantation site. The grafts secrete BDNF and GDNF, while forming functional synapses with host motor and non-motor neurons. However, the differentiated neurons do not develop into large motor neurons, but are predisposed to a GABAergic phenotype. A mild delay in disease onset and prolongation of life expectancy has been observed [103, 104].

Another potential source for motor neuron replacement is human embryonic stem cell-derived motor neuron progenitors (hMNPs). Wyatt et al. [142] transplanted hMNPs and found that they secrete physiologically functional levels of the neurotrophic factors nerve growth factor (NGF) and NT-3. This was accompanied by preservation of motor neurons in three models of motor neuron loss, including ALS. However, the cells differentiate mostly into young motor neurons and non of the transplanted cells were found to differentiate into mature motor neurons were found impaling that the therapeutic effect observed was quite possibly attributed to factors other than motor neuron replacement.

The limited in vivo axonal growth rate of grafted motoneurons derived from ESC [138] hinders their viability as candidates for replacement therapy in neurodegenerative diseases such as ALS. The observed therapeutic effects of neuronal stem cells in these experiments could be attributed to their neuron to neuron trophic functions, such as buffering glutamate levels and neurotrophic factor delivery [104, 143, 144]. In this case, whether motor neurons are best suited to provide such effects, or whether other cell types, such as astrocytes or microglia might provide a more potent tool, must be discussed.

3.5 Astroglial Replacement Therapies

Astrocytic abnormalities participate in the degeneration occurring in ALS [145–148]. Furthermore, astrocytes possess potent neuroprotective potential [149]. Taken together, these attributes render astrocytes interesting targets in ALS related therapeutic strategies.

Transplantation of healthy rat and mice glial restricted precursors (GRPs) around cervical spinal cord respiratory motor neuron pools of SOD1 G93A rats, extended survival and disease duration, attenuated motor neuron loss and moderated the decline in forelimb motor and respiratory functions. These effects correlated with reduced microgliosis at the transplantation site and were severely hindered when there was an absence of GLT1, the glutamate transporter, in the cells grafted [105, 106].

These results suggest some potential for glial replacement therapies that should be further investigated.

Another potential cellular source for motor neuron protection is bone marrow derived mesenchymal stem cells (MSCs). MSCs are self-renewing precursors that can be differentiated in vitro to give rise to cells from multiple lineages, including but not limited to neuron, oligodendrocyte and astrocyte like cells [150–152]. The use of MSCs as the cellular source for stem cell based treatments not only circumvents ethical concerns but also are candidates for autologous transplantation.

Several studies have shown that human MSCs (hMSCs), survive after transplantation into the lumber spinal cord. The cisterna magna and the CSF for long periods of time, show limited migration into and within the spinal cord, extend survival, reduce neuroinflammation; astrogliosis and microgliosis, perform moderate differentiation into neurons and astrocytes and ameliorate disease symptoms in mice and rat SOD1 G93A models [107–109].

A different approach utilizes irradiation in order to improve hMSC engraftment into the brain and muscles [153]. Intravenous administration of hMSC in irradiated pre-symptomatic SOD1-G93A mice has yielded similar results to those previously mentioned, with additional aspects of neuroglial differentiation and migration into the brain and spinal cord parenchyma [110]. It was later found that even in non-irradiated mice, systemic intravascular injection of the Lin-c-kit positive stem/progenitor cell population derived from bone marrow extended the survival of SOD1-G93A mice by 16–17 days and delayed motor degeneration. Transplanted cells were widely distributed throughout the CNS, they expressed high levels of GLT-1 and reduced microgliosis in their microenvironment. Extended survival resulted from delayed onset of the early symptomatic disease stage and a slower progression of later disease [111].

Mazzini et al. performed two clinical trials [80, 81] in which MSC were autologously transplanted intraparenchymally at the thoracic level of the spinal cord of ALS patients. A total of 19 patients have been grafted in both experiments. Patients were monitored for up to 4 years and no significant acute or late side effects were evident. In most cases, no modifications of spinal cord volume or other signs of spinal or brain abnormal cell proliferation, were observed. However, in four of the cases the spinal cord was tethered anteriorly and posteriorly by postoperative scarring, producing marked distortion of the cord due to traction. In the first trial, the linear decline of the forced vital capacity was significantly moderated in four of the nine transplanted patients. However, in second trial, slower deterioration was observed in only two of the ten grafted patients. The results suggest that hMSC might be suggested as candidates for stem cell based treatment of ALS. Moreover, intraspinal injection of MSCs showed no long term adverse outcome. Additional clinical trials have recently been launched.

Several mechanisms have been proposed to explain the neuroprotective effect observed in transplantations of native and differentiated MSC including cell replacement, immunomodulation and trophic factor delivery. hMSC can be further utilized as a platform for delivering neurotrophic factors to target areas within the brain.

When considering the use of any stem cell grafts in the future treatment of ALS, it must be noted that there is a risk of tumor formation, non-specific differentiation and long term toxic effect in these cells. One of the most prominent advantages in selecting MSC in this respect is their relative safety [154]. However, MSC derived from ALS patients might sustain disease related damages [155, 156]. Thus, autologous transplantations could be contaminated with cells impaired in their ability to counteract disease pathophysiology and result in loss of treatment effectivity. The limited migration performed by hMSC means that transplantations must take place in the already afflicted spinal cord of ALS patients, posing a difficulty. These problems must be addressed to be overcome in future treatments.

3.6 Microglia Replacement Therapies

Neurons and astrocytes are not the only cells suspected as impaired in ALS. Microglia specifically, and the immune system in general have been implicated in disease pathophysiology [157–161]. In fact, exposure of primary rat spinal cord cultures to immunoglobulin Gs (IgGs) isolated from ALS patient lead to motor neuron apoptosis [158]. When intraperitoneally injected into healthy mice, ALS-IgGs caused spinal cord microglia recruitment and motor neuron degeneration [162, 163]. PU.1$^{-/-}$ mice are impaired in their ability to develop myeloid and lymphoid cells and therefore lack CNS microglia. These mice were crossbred with SOD1 G93A mice and subsequently transplanted with donor wild type bone marrow cells. Donor cells differentiated into microglia and integrated into the CNS. Grafted mice benefited from reduced motoneuron degeneration, ameliorated disease progression and prolonged survival compared to mice transplanted with SOD1 G93A bone marrow cells and SOD1 G93A non-crossbred and non-transplanted mice [112]. Similar results were obtained when SOD1 G93A mice were irradiated and then intra-bone marrow grafted with wild type bone marrow cells [113] and when SOD1 mice retro-ocularly received human umbilical cord blood cells [114]. Direct intraspinal administration of human umbilical cord blood cells in SOD1-G93A mice before symptom onset (day 40) slightly enhanced their survival in a gender bias manor (an average of 6 days prolongation in survival in mixed gender evaluation and 12 days when evaluating the female fraction alone). Similarly a slight and gender bias preservation of motor function was observed. However, when the cells were transplanted after symptom onset (day 90) no such effect was observed [115]. In attempt to improve the potential therapeutic effect of transplanted human umbilical cord blood cells in models of motor neuron disease, Bigini et al. [116] examined the effect and distribution of human cord blood mononuclear cells after a single bilateral ICV injection in two models of motor neuron degeneration including the transgenic SOD1-G93A mouse model of ALS. This type of administration significantly ameliorated symptoms progression and motor decline as well as significantly prolonged their survival. Even 4 months following ICV injection the cells were localized to the lateral ventricles, but were not found in the spinal cord ventral horns. Unlike the results described here of previous reports the results of this study indicate a protective effect that is the result of the production and release of circulating protective factors rather than direct cell replacement. These findings prompted a clinical trial where 6 sporadic ALS patients underwent full body irradiation and were then infused with a siblings' hematopoietic stem cells (HSCs). Unfortunately, clinical benefits were not evident. Nevertheless, the experiment was not completely unsuccessful: grafted cells were found in post mortem autopsy brain and spinal cord samples. Biases in HSCs engraftment were observed, indicating that the cells gravitate towards areas of motoneuron injury and neuroinflammation [82].

Yet, it is possible that with alternative grafting techniques or cellular genetic engineering microglial cells therapeutic potential might be enhanced.

3.7 Genetically Altered Cells

Using gene therapy technology to enhance the neuroprotective performance of stem cells prior to their transplantation might offer a potential therapeutic tool.

Neuronal progenitor cells (NPCs), NSCs and hMSCs are only a small portion of the cellular sources that can be utilized for this purpose [117–121, 164, 165]. A vast variety of genes could also be considered as genetic targets for cellular enhancement, the most prominent of which are neurotrophic factors and glutamate transporters.

Neurotrophic factors are a group of proteins essential for neuronal development and survival. Delivery of appropriate neurotrophic factors to the immediate environment of afflicted neurons in ALS might improve their survival and thus ameliorate disease progression and symptoms.

3.8 Glial Cell Line Derived Neurotrophic Factor (GDNF)

Treatment with GDNF has shown interesting results in several different neurodegenerative diseases models [165–168]. However, spinal cord grafting of GDNF secreting cells in ALS animal models appears to influence predominantly disease onset rather than progression [117–119]. Therefore, other neurotrophic factors might be better suited for ALS. On the other hand, preliminary results suggest that direct muscle delivery of GDNF, by genetically engineered hMSC, delayed disease progression and improved the overall life span [120]. In light of these results, the delivery site of GDNF rather than the neurotrophic factor itself, must be investigated as a factor in clinical outcome [169]. GDNF based treatments might be suitable as complementary to other neurotrophic factors.

3.9 Vascular Endothelial Growth Factor (VEGF)

Deletion of the hypoxia response element sequence in the vascular endothelial growth factor (VEGF) promoter in otherwise healthy mice results in adult-onset muscle weakness due to the degeneration of lower motor neurons innervating skeletal muscle fibers, reminiscent of the clinical symptoms and neuropathology observed in ALS [170]. Furthermore, crossbreeding of SOD1 G93A mice with mice overexpressing VEGF in their motor neurons attenuates motor impairment, delays motoneuron loss and prolongs the survival of SOD1 G93A mice [171]. This experiment was followed by an evaluation of the effect of intramuscular administration of a plasmid encoding a zinc finger protein transcription factor engineered to induce VEGF expression on the SOD1 G93A ALS rat model. Weekly intramuscular injections managed to preserve motor function and delay motor decline in SOD1 G93A rats [172]. These findings closely link ALS with VEGF and highlight its therapeutic potential. Human NSCs (hNSCs) overexpressing VEGF were intrathecally transplanted into the lumber spinal cord of SOD1 G93A mice. Grafted cells survive for at least 4 weeks in the diseased spinal cord and migrate into the gray matter of the spinal cord. A small portion of grafted cells differentiated into motoneurons. Transplanted animals benefited from delayed disease onset, slower decline of motor function and prolonged survival [121]. The manner by which these cells exert their therapeutic effect remains to be determined. However, several mechanisms might be proposed. VEGF's neurotrophic function is one possibility, which includes anti-apoptotic effect on motoneurons, encouragement of neo-vascularization [173–175] and VEGF's protection against glutamate excitotoxicity [176, 177]. While only few cells differentiated into motoneurons, the possibility of neuronal replacement should not be discounted altogether.

3.10 Insulin Like Growth Factor Type 1 (IGF-1)

Despite promising results in subcutaneous administration of insulin like growth factor type 1 (IGF-1) in animal models of ALS [178, 179], similar clinical trials have been found wanting [74, 75, 180]. Yet, modest beneficial effects were obtained in high dose intrathecal administration of IGF-1 in ALS patients [76].

Localized and continues delivery of IGF-1, via genetically enhanced cells could lower the necessary IGF-1 dosage thus reducing side effects and improving the therapeutic outcome. Recently, human NPCs (hNPS) have been adenovirally infected with the IGF-1 gene, to induce IGF-1 secretion. Transplantation of these cells into the cisterna magna of SOD1 G93A mice, showed enhanced motoneuron survival, but no functional improvement or extension of life span [118].

3.11 Ciliary Neurotrophic Factor (CNTF)

Another factor that has been examined for therapy is the ciliary neurotrophic factor (CNTF). Unfortunately, clinical trials systemically delivering CNTF caused severe adverse reactions in ALS patients [77–79]. In an attempt to reduce unwanted CNTF related side effects, baby hamster kidney cells were genetically engineered to overexpress CNTF, encapsulated in a semispherical membrane and intrathecally implanted in the lumber spinal cord of ALS patients. The membrane allows for passage of nutrients and cell-released therapeutic agents, but restricts diffusion of larger molecules and cells, thus facilitating graft survival and enabling the retrieval of the transplanted cells in the event of severe side effects or oncogenic developments. A robust CNTF presence could be detected for several weeks following transplantation in the CSF of most patients. Grafts survive for up to 20 weeks in the absence of immunosuppression, but an antigenic reaction to bovine fetuin, a component of pre-implantation medium was observed. However, neither

CNTF nor the presence of antibodies against bovine fetuin, elicited any adverse side effects in the implanted patients. However, clinical benefits have not been reported [83–85].

3.12 Glutamate Transporters

A different approach to genetically engineered cells considers increasing glutamate removal from the neuronal environment rather than infusion of neurotrophic factors. Glutamate is the main excitatory amino acid neurotransmitter in the human central nervous system (CNS). It plays a major role in synaptic plasticity, learning, development, cognitive functions and human behavior [181–185]. However, when glutamate is not properly controlled, it may lead to detrimental results. Prolonged exposure to glutamate can promote a process culminating in neuronal cell death [79, 164, 181, 182, 184, 186–188]. Accumulating evidence implicates glutamate excitotoxicity in the pathophysiology of ALS. Furthermore, RNA editing impairments and a significant reduction in the expression and activity of the glutamate transporter GLT1 in animal models of ALS and its human counterpart EAAT2 in patients has been noted [189–194]. We can thus propose that reducing extracellular glutamate levels around the susceptible neurons affected by glutamate toxicity in ALS may mitigate the neurodegenerative progression of the disease. So far, several experiments investigated this possibility using a pharmacological approach. In a screening study of 1040 pharmacological substances ceftriaxone was found to increase astrocyte mediated glutamate transport by stimulating the expression of GLT-1 [195]. Furthermore, the use of ceftriaxone in the SOD1G93A mouse model of ALS prolonged overall survival and upregulated GLT-1 mRNA levels [195]. More recently, a similar screening study, identified harmine, a naturally occurring beta-carboline alkaloid, as a very potent activator of the GLT-1 promoter. In vivo evaluation of harmine yielded similar results to those obtained with ceftriaxone [196]. In an effort to find compounds that can induce translation of EAAT2 transcripts, high-throughput screening was used to evaluate approximately 140,000 compounds. This study identified 61 compounds with the ability to produce a dose-dependent increase in EAAT2 protein levels [197]. Crossbreeding of SOD1 G93A mice with astrocyte overexpressing human EAAT2 mice, improved motor performance, protected motoneurons from glutamate induced excitotoxicity but did not delay the onset of paralysis nor did it prolong life span [198]. In addition to the pharmacological approach, transplantation of cells over-expressing GLT1 in animal models of ALS should be investigated to maximize the therapeutic potential of this approach.

3.13 Gene Silencing

Gene therapy technologies can be used not only for delivery of potential therapeutic genes but could further be used for silencing of potentially pathogenic genes

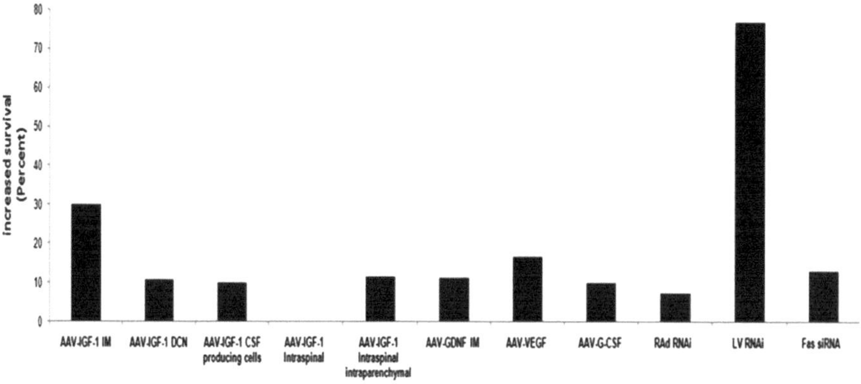

Fig. 7 Survival effect of gene therapy in SOD1 transgenic mice. Graph demonstrates the effect on lifespan of several gene therapy techniques. AAV-IGF-1 IM (intramuscular, [86]), AAV-IGF-1 DCN (deep cerebellar nuclei, [87]), AAV-IGF-1 CSF producing cells [90], AAV-IGF-1 Intraspinal [88], AAV-IGF-1 Intraspinal intraparenchymal (in males only, [89]), AAV-GDNF IM [104, 161], AAV-VEGF CSF producing cells [90], LV-VEGF (IM, [97]), AAV-G-CSF [92], recombinant adenovirus (RAd) RNAi [122], LV RNAi [124], Fas siRNA [126]

(Fig. 7). RNA interference (RNAi) is a process by which non coding RNA mediates sequence specific gene regulation and silencing. Several forms of fALS are caused by dominant, gain of function genetic mutations. In these cases silencing the mutant gene might reduces its toxic effect and thus delay the disease progression. A potent proof of concept emerged from the development of transgenic mice harboring an anti-SOD1 small interference RNA gene, crossbreeding of these mice with SOD1 G93A mice prevented the clinical development of the disease [199]. Several studies further demonstrated the ability to suppress mutant SOD1 products and in most cases delay clinical disease progression in rodent models of ALS. There are several available methods of RNAi delivery, injecting RNAi carrying recombinant adenoviruses directly into large nerve trunks such as the sciatic nerve exposes the viruses to a large number of axons thus reaching a large population of motor neuron. Wu et al. [122] found that such injections extended the life span of SOD1 G93A mice. Intraspinal injections of a lentiviral vector that results in specific RNAi mediated silencing of SOD1 substantially delayed disease onset and progression rate in SOD1 G93A mice [123]. Similarly, intramuscular injections of lentiviral vectors carrying RNAi molecules specifically targeting the SOD1 gene preserved motor function, considerably delayed disease onset and significantly prolonged the survival of SOD1 G93A mice [124]. Other molecular pathogenic targets, which could also apply to the sporadic patients, have also yielded positive results. Silencing the death signaling p75 neurotrophin receptor through systemic intraperitoneal administration as well as intrathecal administration of anti-Fas RNAi significantly delayed motor injury and prolonged the survival of SOD1 G93A mice [125, 126]. To realize the full therapeutic potential of siRNA for neurodegenerative diseases, researchers need to overcome many hurdles and challenges such as selecting suitable tissue-specific delivery vectors, minimizing the off-target effects and reducing side effects. Furthermore, with RNAi therapies safety and efficiency must be demonstrated for every desired mode of delivery individually.

4 Conclusions and Future Perspectives

To date there is no treatment that can substantially prolong life expectancy in individuals afflicted with ALS. Nevertheless where current drug treatments have failed, gene therapy, cell replacement therapies or a combination of both, could offer improved alternatives. Also, trophic factors, known for their role in the survival, development and maintenance of motor neurons [200–206] have been considered as potential therapeutic agents in ALS. Encouraging preliminary results using neurotrophic factors in animal models has prompted further research. Unfortunately, in clinical trials subcutaneous delivery showed no clinical or survival benefits. However, more recent clinical trial infusing IGF-1 and BDNF intrathecally found modest beneficial effects in ALS patients with no severe adverse reactions which indicate that further studies should be design.

Transplantation of stem cells genetically engineered or otherwise induced to secrete such neurotrophic factors is another viable therapeutic approach. Applying gene therapy on stem cells prior to their transplantation to selectively combine the therapeutic effect of neurotrophic factor secretion and glutamate uptake might prove to provide the best survival chances of degeneration prone neurons, to ameliorate symptoms and reduce degeneration in ALS patients. In the case of fALS, a limited subset of ALS patients, the genetic bases of the disease is known and a more personalized treatment regime, such as RNA based silencing of the mutant SOD1 or other ALS related genes (e.g. TARDBP gene), should be considered.

A different therapeutic approach that may have the potential to modify the progression of ALS is a design of iron chelating multifunctional molecules. Indeed, preclinical studies with the novel multimodal iron chelating compound, M30, provide clear beneficial neuroprotective effects at the pre-symptomatic phase and at relatively advanced stage of the neurological dysfunctions. Further development of efficient neuroprotective drugs for early treatment along with early diagnostic of ALS patients is significant, since up-to-date almost all ALS patients are diagnosed after symptom onset. In view of the neuritogenic effect of M30 and its regulation of neurotrophic factors in motoneuron cells, it is possible that in vivo pro-survival effect of M30 is mediated through the regeneration process of motor nerves, inducing neuro-differentiation and sprouting of axons, leading to reinnervation of muscle fibers. Getting insight the molecular mechanism of the drug will assist in the translation of its protective effect from animal preclinical studies into human ALS. Since the process of motoneuron degeneration in ALS is complex and multifactorial, future studies should consider the use of multifunctional iron chelating drugs, such as M30 and HLA20, alone or in combination with other compounds. These compounds should be directed against distinct pathological mechanisms (e.g. anti-glutamergic drug, riluzole) and non-pharmaceutical approaches (e.g. highly energetic diet, moderated exercises) in other models of motoneuron degeneration, to develop the optimal neuroprotective treatment for ALS. The implication of multifunctional drug or "cocktail" therapies for ALS, together with a pathology-specific and patient-specific manner, may provide efficient tools against the significant heterogeneity at the molecular level and the population of ALS patients, to make progress in overcoming this disease.

Preclinical neuroprotective studies are planned with the intent of designing personalized medicine in which clinicians can tailor therapies to maximize benefit and minimize risk. Taken together, the therapeutic approaches outlined in this review, including multifunctional molecules, gene therapy and stem cell-based therapies could possibly form a beacon of hope for those patients suffering from ALS.

Acknowledgments The authors gratefully acknowledged the support of the Israeli ALS Research Association (Haifa, Israel) and the Technion-Research and Development and Rappaport Family Research Institute, Technion-Israel Institute of Technology (Haifa, Israel). The work was also supported, in part, by the Norma and Alan Aufzein Chair for Parkinson's Disease Research and the Devora Eleonora Kirshman Fund for Research of Parkinson's Disease, Tel Aviv University.

References

1. Cronin S, Berger S, Ding J, Schymick JC, Washecka N, Hernandez DG, Greenway MJ, Bradley DG, Traynor BJ, Hardiman O (2008) A genome-wide association study of sporadic ALS in a homogenous Irish population. Hum Mol Genet 17:768–774
2. Gonzalez de Aguilar JL, Echaniz-Laguna A, Fergani A, Rene F, Meininger V, Loeffler JP, Dupuis L (2007) Amyotrophic lateral sclerosis: all roads lead to Rome. J Neurochem 101:1153–1160
3. Santillo AF, Skoglund L, Lindau M, Eeg-Olofsson KE, Tovi M, Engler H, Brundin RM, Ingvast S, Lannfelt L, Glaser A, Kilander L (2009) Frontotemporal dementia-amyotrophic lateral sclerosis complex is simulated by neurodegeneration with brain iron accumulation. Alzheimer Dis Assoc Disord 23:298–300
4. Cozzolino M, Ferri A, Carri MT (2008) Amyotrophic lateral sclerosis: from current developments in the laboratory to clinical implications. Antioxid Redox Signal 10:405–443
5. Valdmanis PN, Daoud H, Dion PA, Rouleau GA (2009) Recent advances in the genetics of amyotrophic lateral sclerosis. Curr Neurol Neurosci Rep 9:198–205
6. Strong MJ, Gordon PH (2005) Primary lateral sclerosis, hereditary spastic paraplegia and amyotrophic lateral sclerosis: discrete entities or spectrum? Amyotroph Lateral Scler Other Motor Neuron Disord 6:8–16
7. Valdmanis PN, Rouleau GA (2008) Genetics of familial amyotrophic lateral sclerosis. Neurology 70:144–152
8. Turner BJ, Talbot K (2008) Transgenics, toxicity and therapeutics in rodent models of mutant SOD1-mediated familial ALS. Prog Neurobiol 85:94–134
9. Brooks BR (2009) Managing amyotrophic lateral sclerosis: slowing disease progression and improving patient quality of life. Ann Neurol 65(Suppl 1):S17–S23
10. Pioro EP (2000) Antioxidant therapy in ALS. Amyotroph Later Scler Other Motor Neuron Disord 1(Suppl 4):5–12, discussion 13–15

11. Orrell RW, Lane RJ, Ross M (2007) Antioxidant treatment for amyotrophic lateral sclerosis / motor neuron disease. Cochrane Database Syst Rev 1:CD002829
12. Van Den Bosch L, Tilkin P, Lemmens G, Robberecht W (2002) Minocycline delays disease onset and mortality in a transgenic model of ALS. Neuroreport 13:1067–1070
13. Gordon PH, Moore DH, Miller RG, Florence JM, Verheijde JL, Doorish C, Hilton JF, Spitalny GM, MacArthur RB, Mitsumoto H, Neville HE, Boylan K, Mozaffar T, Belsh JM, Ravits J, Bedlack RS, Graves MC, McCluskey LF, Barohn RJ, Tandan R (2007) Efficacy of minocycline in patients with amyotrophic lateral sclerosis: a phase III randomised trial. Lancet Neurol 6:1045–1053
14. Miller RG, Mitchell JD, Lyon M, Moore DH (2007) Riluzole for amyotrophic lateral sclerosis (ALS)/motor neuron disease (MND). Cochrane Database Syst Rev 1:CD001447
15. Urushitani M, Ezzi SA, Julien JP (2007) Therapeutic effects of immunization with mutant superoxide dismutase in mice models of amyotrophic lateral sclerosis. Proc Natl Acad Sci U S A 104:2495–2500
16. Oba H, Araki T, Ohtomo K, Monzawa S, Uchiyama G, Koizumi K, Nogata Y, Kachi K, Shiozawa Z, Kobayashi M (1993) Amyotrophic lateral sclerosis: T2 shortening in motor cortex at MR imaging. Radiology 189:843–846
17. Imon Y, Yamaguchi S, Yamamura Y, Tsuji S, Kajima T, Ito K, Nakamura S (1995) Low intensity areas observed on T2-weighted magnetic resonance imaging of the cerebral cortex in various neurological diseases. J Neurol Sci 134(Suppl):27–32
18. Kasarskis EJ, Tandon L, Lovell MA, Ehmann WD (1995) Aluminum, calcium, and iron in the spinal cord of patients with sporadic amyotrophic lateral sclerosis using laser microprobe mass spectroscopy: a preliminary study. J Neurol Sci 130:203–208
19. Sutedja NA, Sinke RJ, Van Vught PW, Van der Linden MW, Wokke JH, Van Duijn CM, Njajou OT, Van der Schouw YT, Veldink JH, Van den Berg LH (2007) The association between H63D mutations in HFE and amyotrophic lateral sclerosis in a Dutch population. Arch Neurol 64:63–67
20. Wang XS, Lee S, Simmons Z, Boyer P, Scott K, Liu W, Connor J (2004) Increased incidence of the Hfe mutation in amyotrophic lateral sclerosis and related cellular consequences. J Neurol Sci 227:27–33
21. Danzeisen R, Achsel T, Bederke U, Cozzolino M, Crosio C, Ferri A, Frenzel M, Gralla EB, Huber L, Ludolph A, Nencini M, Rotilio G, Valentine JS, Carri MT (2006) Superoxide dismutase 1 modulates expression of transferrin receptor. J Biol Inorg Chem 11:489–498
22. Mizuno Y, Amari M, Takatama M, Aizawa H, Mihara B, Okamoto K (2006) Transferrin localizes in Bunina bodies in amyotrophic lateral sclerosis. Acta Neuropathol 112:597–603
23. Jeong SY, Rathore KI, Schulz K, Ponka P, Arosio P, David S (2009) Dysregulation of iron homeostasis in the CNS contributes to disease progression in a mouse model of amyotrophic lateral sclerosis. J Neurosci 29:610–619
24. Wu AS, Kiaei M, Aguirre N, Crow JP, Calingasan NY, Browne SE, Beal MF (2003) Iron porphyrin treatment extends survival in a transgenic animal model of amyotrophic lateral sclerosis. J Neurochem 85:142–150
25. Zheng H, Youdim MB, Weiner LM, Fridkin M (2005) Synthesis and evaluation of peptidic metal chelators for neuroprotection in neurodegenerative diseases. J Pept Res 66:190–203
26. Zheng H, Gal S, Weiner LM, Bar-Am O, Warshawsky A, Fridkin M, Youdim MB (2005) Novel multifunctional neuroprotective iron chelator-monoamine oxidase inhibitor drugs for neurodegenerative diseases: in vitro studies on antioxidant activity, prevention of lipid peroxide formation and monoamine oxidase inhibition. J Neurochem 95:68–78
27. Zheng H, Youdim MB, Weiner LM, Fridkin M (2005) Novel potential neuroprotective agents with both iron chelating and amino acid-based derivatives targeting central nervous system neurons. Biochem Pharmacol 70:1642–1652
28. Gal S, Fridkin M, Amit T, Zheng H, Youdim MB (2006) M30, a novel multifunctional neuroprotective drug with potent iron chelating and brain selective monoamine oxidase-ab inhibitory activity for Parkinson's disease. J Neural Transm Suppl 70:447–456

29. Gal S, Zheng H, Fridkin M, Youdim MB (2005) Novel multifunctional neuroprotective iron chelator-monoamine oxidase inhibitor drugs for neurodegenerative diseases. In vivo selective brain monoamine oxidase inhibition and prevention of MPTP-induced striatal dopamine depletion. J Neurochem 95:79–88
30. Avramovich-Tirosh Y, Amit T, Bar-Am O, Zheng H, Fridkin M, Youdim MB (2007) Therapeutic targets and potential of the novel brain- permeable multifunctional iron chelator-monoamine oxidase inhibitor drug, M-30, for the treatment of Alzheimer's disease. J Neurochem 100:490–502
31. Avramovich-Tirosh Y, Reznichenko L, Mit T, Zheng H, Fridkin M, Weinreb O, Mandel S, Youdim MB (2007) Neurorescue activity, APP regulation and amyloid-beta peptide reduction by novel multi-functional brain permeable iron- chelating- antioxidants, M-30 and green tea polyphenol, EGCG. Curr Alzheimer Res 4:403–411
32. Kupershmidt L, Weinreb O, Amit T, Mandel S, Carri MT, Youdim MB (2009) Neuroprotective and neuritogenic activities of novel multimodal iron-chelating drugs in motor-neuron-like NSC-34 cells and transgenic mouse model of amyotrophic lateral sclerosis. FASEB J 23:3766–3779
33. Gal S, Zheng H, Fridkin M, Youdim MB (2010) Restoration of nigrostriatal dopamine neurons in post-MPTP treatment by the novel multifunctional brain-permeable iron chelator-monoamine oxidase inhibitor drug, M30. Neurotox Res 17:15–27
34. Zhu W, Xie W, Pan T, Xu P, Fridkin M, Zheng H, Jankovic J, Youdim MB, Le W (2007) Prevention and restoration of lactacystin-induced nigrostriatal dopamine neuron degeneration by novel brain-permeable iron chelators. FASEB J 21:3835–3844
35. Kriz J, Nguyen MD, Julien JP (2002) Minocycline slows disease progression in a mouse model of amyotrophic lateral sclerosis. Neurobiol Dis 10:268–278
36. Waibel S, Reuter A, Malessa S, Blaugrund E, Ludolph AC (2004) Rasagiline alone and in combination with riluzole prolongs survival in an ALS mouse model. J Neurol 251:1080–1084
37. Youdim MB, Fridkin M, Zheng H (2005) Bifunctional drug derivatives of MAO-B inhibitor rasagiline and iron chelator VK-28 as a more effective approach to treatment of brain ageing and ageing neurodegenerative diseases. Mech Ageing Dev 126:317–326
38. Martin LJ (2007) Transgenic mice with human mutant genes causing Parkinson's disease and amyotrophic lateral sclerosis provide common insight into mechanisms of motor neuron selective vulnerability to degeneration. Rev Neurosci 18:115–136
39. Cassina P, Peluffo H, Pehar M, Martinez-Palma L, Ressia A, Beckman JS, Estevez AG, Barbeito L (2002) Peroxynitrite triggers a phenotypic transformation in spinal cord astrocytes that induces motor neuron apoptosis. J Neurosci Res 67:21–29
40. Bar-Am O, Yogev-Falach M, Amit T, Sagi Y, Youdim MBH (2004) Regulation of protein kinase C by the anti-Parkinson drug, MAO-B inhibitor, rasagiline and its derivatives, in vivo. J Neurochem 89:1119–1125
41. Bar-Am O, Weinreb O, Amit T, Youdim MB (2005) Regulation of Bcl-2 family proteins, neurotrophic factors, and APP processing in the neurorescue activity of propargylamine. FASEB J 19:1899–1901
42. Weinreb O, Bar-Am O, Amit T, Chillag-Talmor O, Youdim MBH (2004) Neuroprotection via pro-survival protein kinase C isoforms associated with Bcl-2 family members. FASEB J 18:1471–1473
43. Siddiq A, Ayoub IA, Chavez JC, Aminova L, Shah S, LaManna JC, Patton SM, Connor JR, Cherny RA, Volitakis I, Bush AI, Langsetmo I, Seeley T, Gunzler V, Ratan RR (2005) Hypoxia-inducible factor prolyl 4-hydroxylase inhibition. A target for neuroprotection in the central nervous system. J Biol Chem 280:41732–41743
44. Semenza GL (2000) Expression of hypoxia-inducible factor 1: mechanisms and consequences. Biochem Pharmacol 59:47–53
45. Almli LM, Hamrick SE, Koshy AA, Tauber MG, Ferriero DM (2001) Multiple pathways of neuroprotection against oxidative stress and excitotoxic injury in immature primary hippocampal neurons. Brain Res Dev Brain Res 132:121–129

46. Tolosa L, Mir M, Olmos G, Llado J (2009) Vascular endothelial growth factor protects motoneurons from serum deprivation-induced cell death through phosphatidylinositol 3-kinase-mediated p38 mitogen-activated protein kinase inhibition. Neuroscience 158: 1348–1355
47. Van Den Bosch L, Storkebaum E, Vleminckx V, Moons L, Vanopdenbosch L, Scheveneels W, Carmeliet P, Robberecht W (2004) Effects of vascular endothelial growth factor (VEGF) on motor neuron degeneration. Neurobiol Dis 17:21–28
48. Lambrechts D, Storkebaum E, Morimoto M, Del-Favero J, Desmet F, Marklund SL, Wyns S, Thijs V, Andersson J, van Marion I, Al-Chalabi A, Bornes S, Musson R, Hansen V, Beckman L, Adolfsson R, Pall HS, Prats H, Vermeire S, Rutgeerts P, Katayama S, Awata T, Leigh N, Lang-Lazdunski L, Dewerchin M, Shaw C, Moons L, Vlietinck R, Morrison KE, Robberecht W, Van Broeckhoven C, Collen D, Andersen PM, Carmeliet P (2003) VEGF is a modifier of amyotrophic lateral sclerosis in mice and humans and protects motoneurons against ischemic death. Nat Genet 34:383–394
49. Moreau C, Devos D, Brunaud-Danel V, Defebvre L, Perez T, Destee A, Tonnel AB, Lassalle P, Just N (2006) Paradoxical response of VEGF expression to hypoxia in CSF of patients with ALS. J Neurol Neurosurg Psychiatry 77:255–257
50. Semenza GL, Roth PH, Fang HM, Wang GL (1994) Transcriptional regulation of genes encoding glycolytic enzymes by hypoxia-inducible factor 1. J Biol Chem 269:23757–23763
51. Chen C, Pore N, Behrooz A, Ismail-Beigi F, Maity A (2001) Regulation of glut1 mRNA by hypoxia-inducible factor-1. Interaction between H-ras and hypoxia. J Biol Chem 276: 9519–9525
52. Brand K (1997) Aerobic glycolysis by proliferating cells: protection against oxidative stress at the expense of energy yield. J Bioenerg Biomembr 29:355–364
53. Blancher C, Moore JW, Robertson N, Harris AL (2001) Effects of ras and von Hippel-Lindau (VHL) gene mutations on hypoxia-inducible factor (HIF)-1alpha, HIF-2alpha, and vascular endothelial growth factor expression and their regulation by the phosphatidylinositol 3'-kinase/Akt signaling pathway. Cancer Res 61:7349–7355
54. Jiang BH, Jiang G, Zheng JZ, Lu Z, Hunter T, Vogt PK (2001) Phosphatidylinositol 3-kinase signaling controls levels of hypoxia-inducible factor 1. Cell Growth Differ 12:363–369
55. Nurtjahja-Tjendraputra E, Fu D, Phang JM, Richardson DR (2007) Iron chelation regulates cyclin D1 expression via the proteasome: a link to iron deficiency-mediated growth suppression. Blood 109:4045–4054
56. Kulp KS, Green SL, Vulliet PR (1996) Iron deprivation inhibits cyclin-dependent kinase activity and decreases cyclin D/CDK4 protein levels in asynchronous MDA-MB-453 human breast cancer cells. Exp Cell Res 229:60–68
57. Amit T, Avramovich-Tirosh Y, Youdim MB, Mandel S (2008) Targeting multiple Alzheimer's disease etiologies with multimodal neuroprotective and neurorestorative iron chelators. FASEB J 22:1296–1305
58. Buck CR, Seburn KL, Cope TC (2000) Neurotrophin expression by spinal motoneurons in adult and developing rats. J Comp Neurol 416:309–318
59. Namiki J, Kojima A, Tator CH (2000) Effect of brain-derived neurotrophic factor, nerve growth factor, and neurotrophin-3 on functional recovery and regeneration after spinal cord injury in adult rats. J Neurotrauma 17:1219–1231
60. Qin DX, Zou XL, Luo W, Zhang W, Zhang HT, Li XL, Zhang H, Wang XY, Wang TH (2006) Expression of some neurotrophins in the spinal motoneurons after cord hemisection in adult rats. Neurosci Lett 410:222–227
61. Bendotti C, Carri MT (2004) Lessons from models of SOD1-linked familial ALS. Trends Mol Med 10:393–400
62. Wang Q, Zhang X, Chen S, Zhang X, Zhang S, Youdium M, Le W (2011) Prevention of motor neuron degeneration by novel iron chelators in SOD1G93A transgenic mice of amyotrophic lateral sclerosis. Neurodegener Dis 8:310–321
63. Arai T, Hasegawa M, Akiyama H, Ikeda K, Nonaka T, Mori H, Mann D, Tsuchiya K, Yoshida M, Hashizume Y, Oda T (2006) TDP-43 is a component of ubiquitin-positive tau-negative

inclusions in frontotemporal lobar degeneration and amyotrophic lateral sclerosis. Biochem Biophys Res Commun 351:602–611
64. Sumi H, Kato S, Mochimaru Y, Fujimura H, Etoh M, Sakoda S (2009) Nuclear TAR DNA binding protein 43 expression in spinal cord neurons correlates with the clinical course in amyotrophic lateral sclerosis. J Neuropathol Exp Neurol 68:37–47
65. Weinreb O, Amit T, Mandel S, Youdim MB (2009) Neuroprotective molecular mechanisms of (−)-epigallocatechin-3-gallate: a reflective outcome of its antioxidant, iron chelating and neuritogenic properties. Genes Nutr 4:283–296
66. Koh SH, Kwon H, Kim KS, Kim J, Kim MH, Yu HJ, Kim M, Lee KW, Do BR, Jung HK, Yang KW, Appel SH, Kim SH (2004) Epigallocatechin gallate prevents oxidative-stress-induced death of mutant Cu/Zn-superoxide dismutase (G93A) motoneuron cells by alteration of cell survival and death signals. Toxicology 202:213–225
67. Koh SH, Lee SM, Kim HY, Lee KY, Lee YJ, Kim HT, Kim J, Kim MH, Hwang MS, Song C, Yang KW, Lee KW, Kim SH, Kim OH (2006) The effect of epigallocatechin gallate on suppressing disease progression of ALS model mice. Neurosci Lett 395:103–107
68. Xu Z, Chen S, Li X, Luo G, Li L, Le W (2006) Neuroprotective effects of (−)-epigallocatechin-3-gallate in a transgenic mouse model of amyotrophic lateral sclerosis. Neurochem Res 31:1263–1269
69. Yu J, Jia Y, Guo Y, Chang G, Duan W, Sun M, Li B, Li C (2010) Epigallocatechin-3-gallate protects motor neurons and regulates glutamate level. FEBS Lett 584:2921–2925
70. Wijesekera LC, Leigh PN (2009) Amyotrophic lateral sclerosis. Orphanet J Rare Dis 4:1–22
71. Beck M, Flachenecker P, Magnus T, Giess R, Reiners K, Toyka KV, Naumann M (2005) Autonomic dysfunction in ALS: a preliminary study on the effects of intrathecal BDNF. Amyotroph Later Scler Other Motor Neuron Disord 6:100–103
72. Ochs G, Penn RD, York M, Giess R, Beck M, Tonn J, Haigh J, Malta E, Traub M, Sendtner M, Toyka KV (2000) A phase I/II trial of recombinant methionyl human brain derived neurotrophic factor administered by intrathecal infusion to patients with amyotrophic lateral sclerosis. Amyotroph Later Scler Other Motor Neuron Disord 1:201–206
73. The BDNF Study Group (1999) A controlled trial of recombinant methionyl human BDNF in ALS: the BDNF Study Group (phase III). Neurology 52:1427–1433
74. Borasio GD, Robberecht W, Leigh PN, Emile J, Guiloff RJ, Jerusalem F, Silani V, Vos PE, Wokke JH, Dobbins T (1998) A placebo-controlled trial of insulin-like growth factor-I in amyotrophic lateral sclerosis. European ALS/IGF-I Study Group. Neurology 51:583–586
75. Lai EC, Felice KJ, Festoff BW, Gawel MJ, Gelinas DF, Kratz R, Murphy MF, Natter HM, Norris FH, Rudnicki SA (1997) Effect of recombinant human insulin-like growth factor-I on progression of ALS. A placebo-controlled study. The North America ALS/IGF-I Study Group. Neurology 49:1621–1630
76. Nagano I, Shiote M, Murakami T, Kamada H, Hamakawa Y, Matsubara E, Yokoyama M, Moritaz K, Shoji M, Abe K (2005) Beneficial effects of intrathecal IGF-1 administration in patients with amyotrophic lateral sclerosis. Neurol Res 27:768–772
77. Group ACTS (1996) A double-blind placebo-controlled clinical trial of subcutaneous recombinant human ciliary neurotrophic factor (rHCNTF) in amyotrophic lateral sclerosis. ALS CNTF Treatment Study Group. Neurology 46:1244–1249
78. Miller RG, Bryan WW, Dietz MA, Munsat TL, Petajan JH, Smith SA, Goodpasture JC (1996) Toxicity and tolerability of recombinant human ciliary neurotrophic factor in patients with amyotrophic lateral sclerosis. Neurology 47:1329–1331
79. Miller RG, Petajan JH, Bryan WW, Armon C, Barohn RJ, Goodpasture JC, Hoagland RJ, Parry GJ, Ross MA, Stromatt SC (1996) A placebo-controlled trial of recombinant human ciliary neurotrophic (rhCNTF) factor in amyotrophic lateral sclerosis. RhCNTF ALS Study Group. Ann Neurol 39:256–260
80. Mazzini L, Ferrero I, Luparello V, Rustichelli D, Gunetti M, Mareschi K, Testa L, Stecco A, Tarletti R, Miglioretti M, Fava E, Nasuelli N, Cisari C, Massara M, Vercelli R, Oggioni GD, Carriero A, Cantello R, Monaco F, Fagioli F (2009) Mesenchymal stem cell transplantation in amyotrophic lateral sclerosis: a phase I clinical trial. Exp Neurol 223:229–237

81. Mazzini L, Mareschi K, Ferrero I, Vassallo E, Oliveri G, Nasuelli N, Oggioni GD, Testa L, Fagioli F (2008) Stem cell treatment in amyotrophic lateral sclerosis. J Neurol Sci 265:78–83
82. Appel SH, Engelhardt JI, Henkel JS, Siklos L, Beers DR, Yen AA, Simpson EP, Luo Y, Carrum G, Heslop HE, Brenner MK, Popat U (2008) Hematopoietic stem cell transplantation in patients with sporadic amyotrophic lateral sclerosis. Neurology 71:1326–1334
83. Aebischer P, Pochon NA, Heyd B, Deglon N, Joseph JM, Zurn AD, Baetge EE, Hammang JP, Goddard M, Lysaght M, Kaplan F, Kato AC, Schluep M, Hirt L, Regli F, Porchet F, De Tribolet N (1996) Gene therapy for amyotrophic lateral sclerosis (ALS) using a polymer encapsulated xenogenic cell line engineered to secrete hCNTF. Hum Gene Ther 7:851–860
84. Aebischer P, Schluep M, Deglon N, Joseph JM, Hirt L, Heyd B, Goddard M, Hammang JP, Zurn AD, Kato AC, Regli F, Baetge EE (1996) Intrathecal delivery of CNTF using encapsulated genetically modified xenogeneic cells in amyotrophic lateral sclerosis patients. Nat Med 2:696–699
85. Zurn AD, Henry H, Schluep M, Aubert V, Winkel L, Eilers B, Bachmann C, Aebischer P (2000) Evaluation of an intrathecal immune response in amyotrophic lateral sclerosis patients implanted with encapsulated genetically engineered xenogeneic cells. Cell Transplant 9: 471–484
86. Kaspar BK, Llado J, Sherkat N, Rothstein JD, Gage FH (2003) Retrograde viral delivery of IGF-1 prolongs survival in a mouse ALS model. Science 301:839–842
87. Dodge JC, Haidet AM, Yang W, Passini MA, Hester M, Clarke J, Roskelley EM, Treleaven CM, Rizo L, Martin H, Kim SH, Kaspar R, Taksir TV, Griffiths DA, Cheng SH, Shihabuddin LS, Kaspar BK (2008) Delivery of AAV-IGF-1 to the CNS extends survival in ALS mice through modification of aberrant glial cell activity. Mol Ther 16:1056–1064
88. Franz CK, Federici T, Yang J, Backus C, Oh SS, Teng Q, Carlton E, Bishop KM, Gasmi M, Bartus RT, Feldman EL, Boulis NM (2009) Intraspinal cord delivery of IGF-I mediated by adeno-associated virus 2 is neuroprotective in a rat model of familial ALS. Neurobiol Dis 33:473–481
89. Lepore AC, Haenggeli C, Gasmi M, Bishop KM, Bartus RT, Maragakis NJ, Rothstein JD (2007) Intraparenchymal spinal cord delivery of adeno-associated virus IGF-1 is protective in the SOD1G93A model of ALS. Brain Res 1185:256–265
90. Dodge JC, Treleaven CM, Fidler JA, Hester M, Haidet A, Handy C, Rao M, Eagle A, Matthews JC, Taksir TV, Cheng SH, Shihabuddin LS, Kaspar BK (2010) AAV4-Mediated expression of IGF-1 and VEGF within cellular components of the ventricular system improves survival outcome in familial ALS mice. Mol Ther 18:2075–2084
91. Wang LJ, Lu YY, Muramatsu S, Ikeguchi K, Fujimoto K, Okada T, Mizukami H, Matsushita T, Hanazono Y, Kume A, Nagatsu T, Ozawa K, Nakano I (2002) Neuroprotective effects of glial cell line-derived neurotrophic factor mediated by an adeno-associated virus vector in a transgenic animal model of amyotrophic lateral sclerosis. J Neurosci 22:6920–6928
92. Henriques A, Pitzer C, Dittgen T, Klugmann M, Dupuis L, Schneider A (2011) CNS-targeted viral delivery of G-CSF in an animal model for ALS: improved efficacy and preservation of the neuromuscular unit. Mol Ther 19:284–292
93. Pitzer C, Kruger C, Plaas C, Kirsch F, Dittgen T, Muller R, Laage R, Kastner S, Suess S, Spoelgen R, Henriques A, Ehrenreich H, Schabitz WR, Bach A, Schneider A (2008) Granulocyte-colony stimulating factor improves outcome in a mouse model of amyotrophic lateral sclerosis. Brain 131:3335–3347
94. Azzouz M, Hottinger A, Paterna JC, Zurn AD, Aebischer P, Bueler H (2000) Increased motoneuron survival and improved neuromuscular function in transgenic ALS mice after intraspinal injection of an adeno-associated virus encoding Bcl-2. Hum Mol Genet 9: 803–811
95. Guillot S, Azzouz M, Deglon N, Zurn A, Aebischer P (2004) Local GDNF expression mediated by lentiviral vector protects facial nerve motoneurons but not spinal motoneurons in SOD1(G93A) transgenic mice. Neurobiol Dis 16:139–149
96. Hottinger AF, Azzouz M, Deglon N, Aebischer P, Zurn AD (2000) Complete and long-term rescue of lesioned adult motoneurons by lentiviral-mediated expression of glial cell line-derived neurotrophic factor in the facial nucleus. J Neurosci 20:5587–5593

97. Azzouz M, Ralph GS, Storkebaum E, Walmsley LE, Mitrophanous KA, Kingsman SM, Carmeliet P, Mazarakis ND (2004) VEGF delivery with retrogradely transported lentivector prolongs survival in a mouse ALS model. Nature 429:413–417
98. Moreno-Igoa M, Calvo AC, Penas C, Manzano R, Olivan S, Munoz MJ, Mancuso R, Zaragoza P, Aguilera J, Navarro X, Osta Pinzolas R (2010) Fragment C of tetanus toxin, more than a carrier. Novel perspectives in non-viral ALS gene therapy. J Mol Med (Berl) 88:297–308
99. Chian RJ, Li J, Ay I, Celia SA, Kashi BB, Tamrazian E, Matthews JC, Bronson RT, Rossomando A, Pepinsky RB, Fishman PS, Brown RH Jr, Francis JW (2009) IGF-1:tetanus toxin fragment C fusion protein improves delivery of IGF-1 to spinal cord but fails to prolong survival of ALS mice. Brain Res 1287:1–19
100. Ciriza J, Moreno-Igoa M, Calvo AC, Yague G, Palacio J, Miana-Mena FJ, Munoz MJ, Zaragoza P, Brulet P, Osta R (2008) A genetic fusion GDNF-C fragment of tetanus toxin prolongs survival in a symptomatic mouse ALS model. Restor Neurol Neurosci 26:459–465
101. Moreno-Igoa M, Calvo AC, Ciriza J, Munoz MJ, Zaragoza P, Osta R (2011) Non-viral gene delivery of the GDNF, either alone or fused to the C-fragment of tetanus toxin protein, prolongs survival in a mouse ALS model. Restor Neurol Neurosci 30:69–80
102. Calvo AC, Moreno-Igoa M, Mancuso R, Manzano R, Olivan S, Munoz MJ, Penas C, Zaragoza P, Navarro X, Osta R (2011) Lack of a synergistic effect of a non-viral ALS gene therapy based on BDNF and a TTC fusion molecule. Orphanet J Rare Dis 6:10–26
103. Xu L, Ryugo DK, Pongstaporn T, Johe K, Koliatsos VE (2009) Human neural stem cell grafts in the spinal cord of SOD1 transgenic rats: differentiation and structural integration into the segmental motor circuitry. J Comp Neurol 514:297–309
104. Xu L, Yan J, Chen D, Welsh AM, Hazel T, Johe K, Hatfield G, Koliatsos VE (2006) Human neural stem cell grafts ameliorate motor neuron disease in SOD-1 transgenic rats. Transplantation 82:865–875
105. Lepore AC, Rauck B, Dejea C, Pardo AC, Rao MS, Rothstein JD, Maragakis NJ (2008) Focal transplantation-based astrocyte replacement is neuroprotective in a model of motor neuron disease. Nat Neurosci 11:1294–1301
106. Maragakis NJ, Rao MS, Llado J, Wong V, Xue H, Pardo A, Herring J, Kerr D, Coccia C, Rothstein JD (2005) Glial restricted precursors protect against chronic glutamate neurotoxicity of motor neurons in vitro. Glia 50:145–159
107. Boucherie C, Schafer S, Lavand'homme P, Maloteaux JM, Hermans E (2009) Chimerization of astroglial population in the lumbar spinal cord after mesenchymal stem cell transplantation prolongs survival in a rat model of amyotrophic lateral sclerosis. J Neurosci Res 87: 2034–2046
108. Kim H, Kim HY, Choi MR, Hwang S, Nam KH, Kim HC, Han JS, Kim KS, Yoon HS, Kim SH (2010) Dose-dependent efficacy of ALS-human mesenchymal stem cells transplantation into Cisterna magna in SOD1-G93A ALS mice. Neurosci Lett 468:190–194
109. Vercelli A, Mereuta OM, Garbossa D, Muraca G, Mareschi K, Rustichelli D, Ferrero I, Mazzini L, Madon E, Fagioli F (2008) Human mesenchymal stem cell transplantation extends survival, improves motor performance and decreases neuroinflammation in mouse model of amyotrophic lateral sclerosis. Neurobiol Dis 31:395–405
110. Zhao CP, Zhang C, Zhou SN, Xie YM, Wang YH, Huang H, Shang YC, Li WY, Zhou C, Yu MJ, Feng SW (2007) Human mesenchymal stromal cells ameliorate the phenotype of SOD1-G93A ALS mice. Cytotherapy 9:414–426
111. Corti S, Nizzardo M, Nardini M, Donadoni C, Salani S, Simone C, Falcone M, Riboldi G, Govoni A, Bresolin N, Comi GP (2010) Systemic transplantation of c-kit + cells exerts a therapeutic effect in a model of amyotrophic lateral sclerosis. Hum Mol Genet 19:3782–3796
112. Beers DR, Henkel JS, Xiao Q, Zhao W, Wang J, Yen AA, Siklos L, McKercher SR, Appel SH (2006) Wild-type microglia extend survival in PU.1 Knockout mice with familial amyotrophic lateral sclerosis. Proc Natl Acad Sci U S A 103:16021–16026
113. Ohnishi S, Ito H, Suzuki Y, Adachi Y, Wate R, Zhang J, Nakano S, Kusaka H, Ikehara S (2009) Intra-bone marrow-bone marrow transplantation slows disease progression and prolongs survival in G93A mutant SOD1 transgenic mice, an animal model mouse for amyotrophic lateral sclerosis. Brain Res 1296:216–224

114. Ende N, Weinstein F, Chen R, Ende M (2000) Human umbilical cord blood effect on sod mice (amyotrophic lateral sclerosis). Life Sci 67:53–59
115. Knippenberg S, Thau N, Schwabe K, Dengler R, Schambach A, Hass R, Petri S (2012) Intraspinal injection of human umbilical cord blood-derived cells is neuroprotective in a transgenic mouse model of amyotrophic lateral sclerosis. Neurodegener Dis 9:107–120
116. Bigini P, Veglianese P, Andriolo G, Cova L, Grignaschi G, Caron I, Daleno C, Barbera S, Ottolina A, Calzarossa C, Lazzari L, Mennini T, Bendotti C, Silani V (2011) Intracerebroventricular administration of human umbilical cord blood cells delays disease progression in two murine models of motor neuron degeneration. Rejuvenation Res 14:623–639
117. Klein SM, Behrstock S, McHugh J, Hoffmann K, Wallace K, Suzuki M, Aebischer P, Svendsen CN (2005) GDNF delivery using human neural progenitor cells in a rat model of ALS. Hum Gene Ther 16:509–521
118. Park S, Kim HT, Yun S, Kim IS, Lee J, Lee IS, Park KI (2009) Growth factor-expressing human neural progenitor cell grafts protect motor neurons but do not ameliorate motor performance and survival in ALS mice. Exp Mol Med 41:487–500
119. Suzuki M, McHugh J, Tork C, Shelley B, Klein SM, Aebischer P, Svendsen CN (2007) GDNF secreting human neural progenitor cells protect dying motor neurons, but not their projection to muscle, in a rat model of familial ALS. PLoS One 2:e689
120. Suzuki M, McHugh J, Tork C, Shelley B, Hayes A, Bellantuono I, Aebischer P, Svendsen CN (2008) Direct muscle delivery of GDNF with human mesenchymal stem cells improves motor neuron survival and function in a rat model of familial ALS. Mol Ther 16:2002–2010
121. Hwang DH, Lee HJ, Park IH, Seok JI, Kim BG, Joo IS, Kim SU (2009) Intrathecal transplantation of human neural stem cells overexpressing VEGF provide behavioral improvement, disease onset delay and survival extension in transgenic ALS mice. Gene Ther 16:1234–1244
122. Wu R, Wang H, Xia X, Zhou H, Liu C, Castro M, Xu Z (2009) Nerve injection of viral vectors efficiently transfers transgenes into motor neurons and delivers RNAi therapy against ALS. Antioxid Redox Signal 11:1523–1534
123. Raoul C, Abbas-Terki T, Bensadoun JC, Guillot S, Haase G, Szulc J, Henderson CE, Aebischer P (2005) Lentiviral-mediated silencing of SOD1 through RNA interference retards disease onset and progression in a mouse model of ALS. Nat Med 11:423–428
124. Ralph GS, Radcliffe PA, Day DM, Carthy JM, Leroux MA, Lee DC, Wong LF, Bilsland LG, Greensmith L, Kingsman SM, Mitrophanous KA, Mazarakis ND, Azzouz M (2005) Silencing mutant SOD1 using RNAi protects against neurodegeneration and extends survival in an ALS model. Nat Med 11:429–433
125. Turner BJ, Cheah IK, Macfarlane KJ, Lopes EC, Petratos S, Langford SJ, Cheema SS (2003) Antisense peptide nucleic acid-mediated knockdown of the p75 neurotrophin receptor delays motor neuron disease in mutant SOD1 transgenic mice. J Neurochem 87:752–763
126. Locatelli F, Corti S, Papadimitriou D, Fortunato F, Del Bo R, Donadoni C, Nizzardo M, Nardini M, Salani S, Ghezzi S, Strazzer S, Bresolin N, Comi GP (2007) Fas small interfering RNA reduces motoneuron death in amyotrophic lateral sclerosis mice. Ann Neurol 62:81–92
127. Berns KI (1990) Parvovirus replication. Microbiol Rev 54:316–329
128. Alisky JM, Davidson BL (2000) Gene therapy for amyotrophic lateral sclerosis and other motor neuron diseases. Hum Gene Ther 11:2315–2329
129. Lu YY, Wang LJ, Muramatsu S, Ikeguchi K, Fujimoto K, Okada T, Mizukami H, Matsushita T, Hanazono Y, Kume A, Nagatsu T, Ozawa K, Nakano I (2003) Intramuscular injection of AAV-GDNF results in sustained expression of transgenic GDNF, and its delivery to spinal motoneurons by retrograde transport. Neurosci Res 45:33–40
130. Schneider A, Kruger C, Steigleder T, Weber D, Pitzer C, Laage R, Aronowski J, Maurer MH, Gassler N, Mier W, Hasselblatt M, Kollmar R, Schwab S, Sommer C, Bach A, Kuhn HG, Schabitz WR (2005) The hematopoietic factor G-CSF is a neuronal ligand that counteracts programmed cell death and drives neurogenesis. J Clin Invest 115:2083–2098
131. Welte K, Platzer E, Gabrilove JL, Lu L, Levi E, Polivka A, Mertelsmann R, Moore MA (1985) Purification to apparent homogeneity and biochemical characterization of human pluripotent hematopoietic colony-stimulating factor. Haematol Blood Transfus 29:398–401

132. Cockrell AS, Kafri T (2007) Gene delivery by lentivirus vectors. Mol Biotechnol 36:184–204
133. Naldini L, Blomer U, Gallay P, Ory D, Mulligan R, Gage FH, Verma IM, Trono D (1996) In vivo gene delivery and stable transduction of nondividing cells by a lentiviral vector. Science 272:263–267
134. Lewis P, Hensel M, Emerman M (1992) Human immunodeficiency virus infection of cells arrested in the cell cycle. EMBO J 11:3053–3058
135. Azzouz M (2006) Gene therapy for ALS: progress and prospects. Biochim Biophys Acta 1762:1122–1127
136. Cronin J, Zhang XY, Reiser J (2005) Altering the tropism of lentiviral vectors through pseudotyping. Curr Gene Ther 5:387–398
137. Dimos JT, Rodolfa KT, Niakan KK, Weisenthal LM, Mitsumoto H, Chung W, Croft GF, Saphier G, Leibel R, Goland R, Wichterle H, Henderson CE, Eggan K (2008) Induced pluripotent stem cells generated from patients with ALS can be differentiated into motor neurons. Science 321:1218–1221
138. Harper JM, Krishnan C, Darman JS, Deshpande DM, Peck S, Shats I, Backovic S, Rothstein JD, Kerr DA (2004) Axonal growth of embryonic stem cell-derived motoneurons in vitro and in motoneuron-injured adult rats. Proc Natl Acad Sci U S A 101:7123–7128
139. Lopez-Gonzalez R, Kunckles P, Velasco I (2009) Transient recovery in a rat model of familial amyotrophic lateral sclerosis after transplantation of motor neurons derived from mouse embryonic stem cells. Cell Transplant 18:1171–1181
140. Garbuzova-Davis S, Willing AE, Milliken M, Saporta S, Zigova T, Cahill DW, Sanberg PR (2002) Positive effect of transplantation of hNT neurons (NTera 2/D1 cell-line) in a model of familial amyotrophic lateral sclerosis. Exp Neurol 174:169–180
141. Willing AE, Garbuzova-Davis S, Saporta S, Milliken M, Cahill DW, Sanberg PR (2001) hNT neurons delay onset of motor deficits in a model of amyotrophic lateral sclerosis. Brain Res Bull 56:525–530
142. Wyatt TJ, Rossi SL, Siegenthaler MM, Frame J, Robles R, Nistor G, Keirstead HS (2011) Human motor neuron progenitor transplantation leads to endogenous neuronal sparing in 3 models of motor neuron loss. Stem Cells Int 2011:207230
143. Sharma HS (2007) Interaction between amino acid neurotransmitters and opioid receptors in hyperthermia-induced brain pathology. Prog Brain Res 162:295–317
144. Woo RS, Li XM, Tao Y, Carpenter-Hyland E, Huang YZ, Weber J, Neiswender H, Dong XP, Wu J, Gassmann M, Lai C, Xiong WC, Gao TM, Mei L (2007) Neuregulin-1 enhances depolarization-induced GABA release. Neuron 54:599–610
145. Luo Y, Xue H, Pardo AC, Mattson MP, Rao MS, Maragakis NJ (2007) Impaired SDF1/CXCR4 signaling in glial progenitors derived from SOD1(G93A) mice. J Neurosci Res 85:2422–2432
146. Nagai M, Re DB, Nagata T, Chalazonitis A, Jessell TM, Wichterle H, Przedborski S (2007) Astrocytes expressing ALS-linked mutated SOD1 release factors selectively toxic to motor neurons. Nat Neurosci 10:615–622
147. Pramatarova A, Laganiere J, Roussel J, Brisebois K, Rouleau GA (2001) Neuron-specific expression of mutant superoxide dismutase 1 in transgenic mice does not lead to motor impairment. J Neurosci 21:3369–3374
148. Yamanaka K, Chun SJ, Boillee S, Fujimori-Tonou N, Yamashita H, Gutmann DH, Takahashi R, Misawa H, Cleveland DW (2008) Astrocytes as determinants of disease progression in inherited amyotrophic lateral sclerosis. Nat Neurosci 11:251–253
149. Maragakis NJ, Rothstein JD (2006) Mechanisms of disease: astrocytes in neurodegenerative disease. Nat Clin Pract Neurol 2:679–689
150. Bahat-Stroomza M, Barhum Y, Levy YS, Karpov O, Bulvik S, Melamed E, Offen D (2009) Induction of adult human bone marrow mesenchymal stromal cells into functional astrocyte-like cells: potential for restorative treatment in Parkinson's disease. J Mol Neurosci 39: 199–210

151. Suzuki H, Taguchi T, Tanaka H, Kataoka H, Li Z, Muramatsu K, Gondo T, Kawai S (2004) Neurospheres induced from bone marrow stromal cells are multipotent for differentiation into neuron, astrocyte, and oligodendrocyte phenotypes. Biochem Biophys Res Commun 322:918–922
152. Sanchez-Ramos J, Song S, Cardozo-Pelaez F, Hazzi C, Stedeford T, Willing A, Freeman TB, Saporta S, Janssen W, Patel N, Cooper DR, Sanberg PR (2000) Adult bone marrow stromal cells differentiate into neural cells in vitro. Exp Neurol 164:247–256
153. Francois S, Bensidhoum M, Mouiseddine M, Mazurier C, Allenet B, Semont A, Frick J, Sache A, Bouchet S, Thierry D, Gourmelon P, Gorin NC, Chapel A (2006) Local irradiation not only induces homing of human mesenchymal stem cells at exposed sites but promotes their widespread engraftment to multiple organs: a study of their quantitative distribution after irradiation damage. Stem Cells 24:1020–1029
154. Battiwalla M, Hematti P (2009) Mesenchymal stem cells in hematopoietic stem cell transplantation. Cytotherapy 11:503–515
155. Bossolasco P, Cova L, Calzarossa C, Servida F, Mencacci NE, Onida F, Polli E, Lambertenghi Deliliers G, Silani V (2010) Metalloproteinase alterations in the bone marrow of ALS patients. J Mol Med (Berl) 88:553–564
156. Cho GW, Noh MY, Kim HY, Koh SH, Kim KS, Kim SH (2010) Bone marrow-derived stromal cells from amyotrophic lateral sclerosis patients have diminished stem cell capacity. Stem Cell Dev 19:1035–1042
157. Engelhardt JI, Soos J, Obal I, Vigh L, Siklos L (2005) Subcellular localization of IgG from the sera of ALS patients in the nervous system. Acta Neurol Scand 112:126–133
158. Demestre M, Pullen A, Orrell RW, Orth M (2005) ALS-IgG-induced selective motor neurone apoptosis in rat mixed primary spinal cord cultures. J Neurochem 94:268–275
159. Henkel JS, Beers DR, Zhao W, Appel SH (2009) Microglia in ALS: the good, the bad, and the resting. J Neuroimmune Pharmacol 4:389–398
160. Sargsyan SA, Monk PN, Shaw PJ (2005) Microglia as potential contributors to motor neuron injury in amyotrophic lateral sclerosis. Glia 51:241–253
161. Hall ED, Oostveen JA, Gurney ME (1998) Relationship of microglial and astrocytic activation to disease onset and progression in a transgenic model of familial ALS. Glia 23:249–256
162. Obal I, Jakab JS, Siklos L, Engelhardt JI (2001) Recruitment of activated microglia cells in the spinal cord of mice by ALS IgG. Neuroreport 12:2449–2452
163. Pullen AH, Demestre M, Howard RS, Orrell RW (2004) Passive transfer of purified IgG from patients with amyotrophic lateral sclerosis to mice results in degeneration of motor neurons accompanied by $Ca2+$ enhancement. Acta Neuropathol 107:35–46
164. de Hemptinne I, Vermeiren C, Maloteaux JM, Hermans E (2004) Induction of glial glutamate transporters in adult mesenchymal stem cells. J Neurochem 91:155–166
165. Pineda JR, Rubio N, Akerud P, Urban N, Badimon L, Arenas E, Alberch J, Blanco J, Canals JM (2007) Neuroprotection by GDNF-secreting stem cells in a Huntington's disease model: optical neuroimage tracking of brain-grafted cells. Gene Ther 14:118–128
166. Cunningham LA, Su C (2002) Astrocyte delivery of glial cell line-derived neurotrophic factor in a mouse model of Parkinson's disease. Exp Neurol 174:230–242
167. Patel NK, Bunnage M, Plaha P, Svendsen CN, Heywood P, Gill SS (2005) Intraputamenal infusion of glial cell line-derived neurotrophic factor in PD: a two-year outcome study. Ann Neurol 57:298–302
168. Slevin JT, Gash DM, Smith CD, Gerhardt GA, Kryscio R, Chebrolu H, Walton A, Wagner R, Young AB (2007) Unilateral intraputamenal glial cell line-derived neurotrophic factor in patients with Parkinson disease: response to 1 year of treatment and 1 year of withdrawal. J Neurosurg 106:614–620
169. Li W, Brakefield D, Pan Y, Hunter D, Myckatyn TM, Parsadanian A (2007) Muscle-derived but not centrally derived transgene GDNF is neuroprotective in G93A-SOD1 mouse model of ALS. Exp Neurol 203:457–471

170. Oosthuyse B, Moons L, Storkebaum E, Beck H, Nuyens D, Brusselmans K, Van Dorpe J, Hellings P, Gorselink M, Heymans S, Theilmeier G, Dewerchin M, Laudenbach V, Vermylen P, Raat H, Acker T, Vleminckx V, Van Den Bosch L, Cashman N, Fujisawa H, Drost MR, Sciot R, Bruyninckx F, Hicklin DJ, Ince C, Gressens P, Lupu F, Plate KH, Robberecht W, Herbert JM, Collen D, Carmeliet P (2001) Deletion of the hypoxia-response element in the vascular endothelial growth factor promoter causes motor neuron degeneration. Nat Genet 28:131–138
171. Wang Y, Mao XO, Xie L, Banwait S, Marti HH, Greenberg DA, Jin K (2007) Vascular endothelial growth factor overexpression delays neurodegeneration and prolongs survival in amyotrophic lateral sclerosis mice. J Neurosci 27:304–307
172. Kliem MA, Heeke BL, Franz CK, Radovitskiy I, Raore B, Barrow E, Snyder BR, Federici T, Kaye Spratt S, Boulis NM (2011) Intramuscular administration of a VEGF zinc finger transcription factor activator (VEGF-ZFP-TF) improves functional outcomes in SOD1 rats. Amyotroph LaterScler 12:331–339
173. Lambrechts D, Carmeliet P (2006) VEGF at the neurovascular interface: therapeutic implications for motor neuron disease. Biochim Biophys Acta 1762:1109–1121
174. Rizvanov AA, Kiyasov AP, Gaziziov IM, Yilmaz TS, Kaligin MS, Andreeva DI, Shafigullina AK, Guseva DS, Kiselev SL, Matin K, Palotas A, Islamov RR (2008) Human umbilical cord blood cells transfected with VEGF and L(1)CAM do not differentiate into neurons but transform into vascular endothelial cells and secrete neuro-trophic factors to support neurogenesis-a novel approach in stem cell therapy. Neurochem Int 53:389–394
175. Zheng C, Nennesmo I, Fadeel B, Henter JI (2004) Vascular endothelial growth factor prolongs survival in a transgenic mouse model of ALS. Ann Neurol 56:564–567
176. Bogaert E, Van Damme P, Poesen K, Dhondt J, Hersmus N, Kiraly D, Scheveneels W, Robberecht W, Van Den Bosch L (2009) VEGF protects motor neurons against excitotoxicity by upregulation of GluR2. Neurobiol Aging 31:2185–2191
177. Tolosa L, Mir M, Asensio VJ, Olmos G, Llado J (2008) Vascular endothelial growth factor protects spinal cord motoneurons against glutamate-induced excitotoxicity via phosphatidylinositol 3-kinase. J Neurochem 105:1080–1090
178. Vergani L, Finco C, Di Giulio AM, Muller EE, Gorio A (1997) Effects of low doses of glycosaminoglycans and insulin-like growth factor-I on motor neuron disease in wobbler mouse. Neurosci Lett 228:41–44
179. Vergani L, Losa M, Lesma E, Di Giulio AM, Torsello A, Muller EE, Gorio A (1999) Glycosaminoglycans boost insulin-like growth factor-I-promoted neuroprotection: blockade of motor neuron death in the wobbler mouse. Neuroscience 93:565–572
180. Sorenson EJ, Windbank AJ, Mandrekar JN, Bamlet WR, Appel SH, Armon C, Barkhaus PE, Bosch P, Boylan K, David WS, Feldman E, Glass J, Gutmann L, Katz J, King W, Luciano CA, McCluskey LF, Nash S, Newman DS, Pascuzzi RM, Pioro E, Sams LJ, Scelsa S, Simpson EP, Subramony SH, Tiryaki E, Thornton CA (2008) Subcutaneous IGF-1 is not beneficial in 2-year ALS trial. Neurology 71:1770–1775
181. Danbolt NC, Chaudhry FA, Dehnes Y, Lehre KP, Levy LM, Ullensvang K, Storm-Mathisen J (1998) Properties and localization of glutamate transporters. Prog Brain Res 116:23–43
182. Danbolt NC (2001) Glutamate uptake. Prog Neurobiol 65:1–105
183. Maragakis NJ, Rothstein JD (2001) Glutamate transporters in neurologic disease. Arch Neurol 58:365–370
184. Mattson MP (2008) Glutamate and neurotrophic factors in neuronal plasticity and disease. Ann N Y Acad Sci 1144:97–112
185. Molz S, Decker H, Dal-Cim T, Cremonez C, Cordova FM, Leal RB, Tasca CI (2008) Glutamate-induced toxicity in hippocampal slices involves apoptotic features and p38 MAPK signaling. Neurochem Res 33:27–36
186. Lee Y, Park HW, Park SG, Cho S, Myung PK, Park BC (2007) Proteomic analysis of glutamate-induced toxicity in HT22 cells. Proteomics 7:185–193
187. Sheldon AL, Robinson MB (2007) The role of glutamate transporters in neurodegenerative diseases and potential opportunities for intervention. Neurochem Int 51:333–355

188. Sonnewald U, Qu H, Aschner M (2002) Pharmacology and toxicology of astrocyte-neuron glutamate transport and cycling. J Pharmacol Exp Ther 301:1–6
189. Bendotti C, Tortarolo M, Suchak SK, Calvaresi N, Carvelli L, Bastone A, Rizzi M, Rattray M, Mennini T (2001) Transgenic SOD1 G93A mice develop reduced GLT-1 in spinal cord without alterations in cerebrospinal fluid glutamate levels. J Neurochem 79:737–746
190. Howland DS, Liu J, She Y, Goad B, Maragakis NJ, Kim B, Erickson J, Kulik J, DeVito L, Psaltis G, DeGennaro LJ, Cleveland DW, Rothstein JD (2002) Focal loss of the glutamate transporter EAAT2 in a transgenic rat model of SOD1 mutant-mediated amyotrophic lateral sclerosis (ALS). Proc Natl Acad Sci U S A 99:1604–1609
191. Lin CL, Bristol LA, Jin L, Dykes-Hoberg M, Crawford T, Clawson L, Rothstein JD (1998) Aberrant RNA processing in a neurodegenerative disease: the cause for absent EAAT2, a glutamate transporter, in amyotrophic lateral sclerosis. Neuron 20:589–602
192. Meyer T, Fromm A, Munch C, Schwalenstocker B, Fray AE, Ince PG, Stamm S, Gron G, Ludolph AC, Shaw PJ (1999) The RNA of the glutamate transporter EAAT2 is variably spliced in amyotrophic lateral sclerosis and normal individuals. J Neurol Sci 170:45–50
193. Rothstein JD, Van Kammen M, Levey AI, Martin LJ, Kuncl RW (1995) Selective loss of glial glutamate transporter GLT-1 in amyotrophic lateral sclerosis. Ann Neurol 38:73–84
194. Flomen R, Makoff A (2011) Increased RNA editing in EAAT2 pre-mRNA from amyotrophic lateral sclerosis patients: involvement of a cryptic polyadenylation site. Neurosci Lett 497:139–143
195. Rothstein JD, Patel S, Regan MR, Haenggeli C, Huang YH, Bergles DE, Jin L, Dykes Hoberg M, Vidensky S, Chung DS, Toan SV, Bruijn LI, Su ZZ, Gupta P, Fisher PB (2005) Beta-lactam antibiotics offer neuroprotection by increasing glutamate transporter expression. Nature 433:73–77
196. Li Y, Sattler R, Yang EJ, Nunes A, Ayukawa Y, Akhtar S, Ji G, Zhang PW, Rothstein JD (2011) Harmine, a natural beta-carboline alkaloid, upregulates astroglial glutamate transporter expression. Neuropharmacology 60:1168–1175
197. Colton CK, Kong Q, Lai L, Zhu MX, Seyb KI, Cuny GD, Xian J, Glicksman MA, Lin CL (2010) Identification of translational activators of glial glutamate transporter EAAT2 through cell-based high-throughput screening: an approach to prevent excitotoxicity. J Biomol Screen 15:653–662
198. Guo H, Lai L, Butchbach ME, Stockinger MP, Shan X, Bishop GA, Lin CL (2003) Increased expression of the glial glutamate transporter EAAT2 modulates excitotoxicity and delays the onset but not the outcome of ALS in mice. Hum Mol Genet 12:2519–2532
199. Saito Y, Yokota T, Mitani T, Ito K, Anzai M, Miyagishi M, Taira K, Mizusawa H (2005) Transgenic small interfering RNA halts amyotrophic lateral sclerosis in a mouse model. J Biol Chem 280:42826–42830
200. Joseph D'Ercole A, Ye P (2008) Expanding the mind: insulin-like growth factor I and brain development. Endocrinology 149:5958–5962
201. Lunn JS, Sakowski SA, Kim B, Rosenberg AA, Feldman EL (2009) Vascular endothelial growth factor prevents G93A-SOD1-induced motor neuron degeneration. Dev Neurobiol 69:871–884
202. Murphy M, Dutton R, Koblar S, Cheema S, Bartlett P (1997) Cytokines which signal through the LIF receptor and their actions in the nervous system. Prog Neurobiol 52:355–378
203. Pierce RC, Bari AA (2001) The role of neurotrophic factors in psychostimulant-induced behavioral and neuronal plasticity. Rev Neurosci 12:95–110
204. Saarma M, Sariola H (1999) Other neurotrophic factors: glial cell line-derived neurotrophic factor (GDNF). Microsc Res Tech 45:292–302
205. Schulte-Herbruggen O, Braun A, Rochlitzer S, Jockers-Scherubl MC, Hellweg R (2007) Neurotrophic factors–a tool for therapeutic strategies in neurological, neuropsychiatric and neuroimmunological diseases? Curr Med Chem 14:2318–2329
206. Sendtner M, Carroll P, Holtmann B, Hughes RA, Thoenen H (1994) Ciliary neurotrophic factor. J Neurobiol 25:1436–1453

Part VI

Exercise in Multiple Sclerosis – An Integral Component of Disease Management

Andrea Döring, Caspar F. Pfueller, Friedemann Paul, and Jan Dörr

Abstract Multiple sclerosis (MS) is the most common chronic inflammatory disorder of the central nervous system (CNS) in young adults. Depending on the underlying pathology, its localization and further characteristics the disease causes a wide spectrum of symptoms and disabilities. Up to date management of MS comprises both drug-based and non-drug approaches complementarily to reduce existing symptoms and to prevent sequelae, such as infections or orthopedic diseases. To improve the individual patient's outcome substantially, it is necessary to adapt physical therapy like exercise and physiotherapy to the individual needs. However, high quality systematic data on physical therapy in MS, that support this process of customization, are rare.

This chapter summarizes the current knowledge on the influence of physical activity and exercise on disease-related symptoms and physical restrictions in MS patients. Drug-based treatment strategies, cognitive training or other treatment strategies are not in the focus of interest in this chapter.

Keywords Multiple sclerosis • Physical therapy • Exercise • Prevention of sequelae • Personalized treatment

1 Background of MS

MS is a chronic inflammatory disease of the CNS that is characterized by multifocal CNS demyelination along with astrocytic gliosis and neuroaxonal degeneration in the brain and spinal cord. MS is one of the most common causes of non-traumatic

A. Döring (✉) • C.F. Pfueller • F. Paul • J. Dörr
NeuroCure Clinical Research Center and Clinical and Experimental Research Center for Multiple Sclerosis, Charité – Universitätsmedizin Berlin, Charitéplatz 1, 10117 Berlin, Germany
e-mail: andrea.doering@charite.de

S. Mandel (ed.), *Neurodegenerative Diseases: Integrative PPPM Approach as the Medicine of the Future*, Advances in Predictive, Preventive and Personalised Medicine 2, DOI 10.1007/978-94-007-5866-7_12,

disability in young adults, and approximately 1–2.5 million people world-wide are estimated to be affected [1, 2]. Women are more often affected than men, reflected in a male:female ratio of 1:2 to 1:3. The first manifestation of MS usually takes place in young adults between the age of 20–40 years, only rarely during childhood or in elder people. The most common initial disease course is the relapsing-remitting form with a possible conversion into a secondary progressive form after a varying period of time. Alternatively, patients can present with a primary progressive disease course right from the start. So far, the precise etiology and specific trigger factors of MS still remain unclear. Environmental and genetic factors that initiate autoimmune reactions against CNS structures leading to CNS tissue damage and associated neurological impairment are considered to be the underlying pathomechanism [2].

Differences in localization of inflammation and neuroaxonal degeneration and tissue damage characteristics in both white and gray brain matter finally lead to a variety of symptoms and signs, such as impaired vision, dysarthria and dysphagia, paresis, spasticity, disturbed coordination and balance, ataxia, pain, sensory impairment, bladder, bowel and sexual dysfunction (reviewed in [3]). Fatigue, emotional and cognitive changes are also frequently present in MS [4, 5]. Combination of these symptoms, often accompanied by a lack of self-confidence, and inability to manage the symptoms, lead to impaired functional capacity, and subsequently to a reduced physical and sporting activity and a reduced quality of life [6–9]. This lack of physical activity can lead to secondary sequelae such as obesity, osteoporosis and/or cardiovascular damage putting patients at risk of tertiary complications such as thrombosis, pulmonary embolisms, upper respiratory or urinary tract infections, or decubital ulcers [7, 8, 10].

To date, a disease-modifying therapy using immunomodulatory drugs such as interferon-β or glatiramer acetate is the first line of drug-based therapeutic interventions in MS. If these drugs are not sufficiently effective in a specific patient, escalation therapy with immunosuppressive substances (mitoxantrone), monoclonal antibodies (natalizumab) or the recently approved sphingosine phosphate receptor modulator fingolimod may be required (Fig. 1) [11, 12].

2 Definitions

In this chapter the terms movement, physical activity, exercise, physical function, physical therapy, physiotherapy and sport will be defined as follows (see also Tables 1 and 2): Referring to the motor system, the term "movement" represents an actively or passively induced change in the body position. Regular exercise and physical activity are decisive factors in a person's quality of life and contribute in preserving and improving health and wellbeing while preventing diseases at all stages of life. Physical activity encompasses any type of physical movements, which consume energy, regardless of the underlying motivation, whereas sport refers to activities focusing on physical achievement, competition and fun. Both leisure time activities (e.g. sport) and everyday activities (e.g. climbing stairs) are summarized

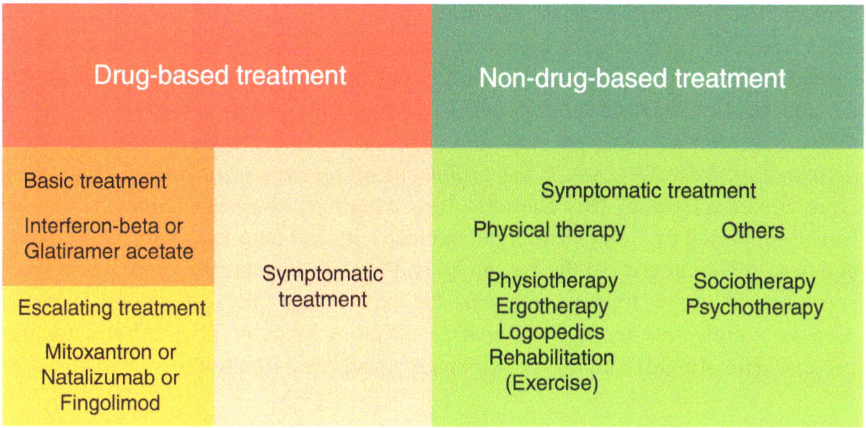

Fig. 1 Drug-based and non-drug-based symptomatic treatment approaches for MS complement each other at almost every stage of disease. Drug-based strategies encompass basic treatments (interferon-β or glatiramer acetate) and – if these drugs are not sufficiently effective – escalation therapy with immunosuppressive substances (mitoxantrone), monoclonal antibodies (natalizumab) or the sphingosine phosphate receptor modulator fingolimod. Non-drug strategies like physical therapy (physiotherapy, ergotherapy, logopedics, rehabilitation) and occupational therapy (sociotherapy and psychotherapy) are used complementarily at all stages of the disease

Table 1 Definitions of different types of physical movements

Definitions	
Movement	Active or passive change in the position of the body
Physical activity	Any type of physical movement that consumes the subject's energy
Physical function	A series of increasing steps, with the highest level consisting of the most advanced activities of daily life, the fulfillment of social roles and the pursuit of recreational activities
Exercise	Planned performance of systematically repeated movements to accomplish skills, maintain and strengthen physical condition, and improve performance
Sport	Exercise with a focus on physical achievement, competition, and fun

Table 2 Types of human movements, sorted according to intensity

Type of human movements
Movement → Physical activity → Physical function → Exercise → Sport

in the term "health-enhancing physical activity". In contrast to general physical activity, exercise encompasses the planned performance of systematically repeated movements to accomplish skills, maintain and strengthen physical condition, and improve performance, whereas athletics, more specifically, aim to improve general flexibility and include endurance training to maintain performance over longer periods of time at a high level and strength training to increase muscle strength. Often, the terms aerobic training and endurance as well as resistance and strength training are used synonymously. Physical function encompasses "*a series of increasingly integrated steps, with the highest level consisting of the most advanced activities of daily life (ADL), the fulfillment of societal roles and the pursuit of recreational activities*" [8]. The term "physiotherapy" includes manual skills that can be supported by water, heat, light, or electricity and aims to restore functionality and perceptive integrity of the human body. Physiotherapeutic methods can contain active and/or passive training programs. On the contrary "physical therapy" is rather used as a collective term to summarize different kinds of physical activity like exercise, (functional) training, physiotherapy, and rehabilitation.

3 Symptomatic Treatment of MS – Aiming at a Personalized Modification of Symptoms and Outcome

Drug-based and non-drug-based symptomatic treatment approaches for MS complement each other. Apart from counseling and nursing care, non-drug strategies encompass physical therapy like physiotherapy, logopedics, occupational therapy including living and mobility aids, sociotherapy and psychotherapy (Fig. 1). These measures should be applied in parallel so that different approaches are combined in a patient's treatment strategy and complement drug therapy [11, 13, 14]. Depending on the patient's individual symptoms physical therapies are adapted and have the potential to positively affect several factors multimodally. These strategies do not only reduce symptoms, enhance mobility, improve quality of life, and confer as much independence as possible, for example by functional training of ADLs, such as washing, eating, drinking, dressing, and performing household chores. Symptomatic therapies may rather prevent potentially life-threatening sequelae [7, 15]. Physical therapies can be applied throughout all stages of disease – from the symptom onset to states of pronounced functional impairment and in palliative conditions. Exercise is, in contrast to physiotherapy, not a regular component of therapeutic approaches offered to MS patients, although it might be a promising and cost-effective tool to improve various functional deficits in MS patients.

4 Exercise in MS Patients – Effects on Clinical Parameters (Table 3)

Impairment of MS patients like limited mobility, spasticity or paresis is primarily a consequence of the disease itself, but it can be aggravated by reduced physical activity [6, 16]. Exercise has been shown to improve various clinical symptoms in

Table 3 Overview of selected studies on exercise in multiple sclerosis

Form of training	Author	Method	(Major) Endpoints	Sample size	EDSS	Main results	Comments
Aerobic training	Newman et al. [29]	Treadmill walking	Gait parameters, Fatigue	16	<7	Improvement in some gait parameters, fatigue unchanged	Repeated measures design and blinded assessments
	Pilutti et al. [19]	Bodyweight supported treadmill training (BWSTT)	Functional ability, quality of life, Fatigue	6	5,5–8	Improvement in some functional abilities and some parameters of quality of life, fatigue non-significantly reduced	Patients with progressive MS of high disability
	Rampello et al. [32]	Aerobic training program compared with neurological rehabilitation	Walking parameters, maximal exercise tolerance, quality of life, fatigue	19	<6	Improvement of some walking parameters after aerobic training, fatigue after aerobic training and neurological rehabilitation comparable	Only 11 patients completed
	Schulz et al. [28]	Aerobic training	Immune-endocrine parameters, neurotrophic factors, quality of life, coordinative function	67	<5	Lactate levels lowered, quality of life increased and coordinative increased	

(continued)

Table 3 (continued)

Form of training	Author	Method	(Major) Endpoints	Sample size	EDSS	Main results	Comments
	van den Berg et al. [86]	Aerobic treadmill training	Walking parameters, fatigue	19	Walk 10 m in 60 s, using aid if necessary	Improvement of some walking parameters after aerobic treadmill training, fatigue not significantly reduced	Prospective, randomized controlled trial with blinded assessments, 16 patients completed
	Mostert and Kessel-ring [31]	Aerobic training	Aerobic fitness, fatigue, health perception, activity level	26	<6.5	Improvement of health perception, activity level	Randomized trial
Resistance Training	Dodd et al. [20]	Progressive resistance strength-ening	Physical, psychological, social factors	8	No information	Fatigue reduced significantly	Semi-structured interviews
	Dalgas et al. [18]	Progressive resistance training	Muscle strength, functional capacity	38	3–5.5	Muscle strength and functional capacity increased	Randomized controlled trial including follow-up
	Harvey et al. [25]	General phys-iotherapy exercises, strength-ening training	Muscle strength, functional activities	19	Ambulant with or without the use of walking aids	Muscle strength and functional activities improved	
	White et al. [15]	Progressive resistance training	Lower extremity strength, ambulatory function, fatigue, disability	8	1–5	Lower extremity strength increased	

	Taylor et al. [26]	Progressive resistance training	Maximal muscle force, muscle endurance, functional activity, psychological function	9	Able to walk at least 200 m without aid or rest	Improvement in muscle strength, muscle endurance, and functional activity	Pre/post single group research design
	Gutierrez et al. [33]	Resistance training	Kinematic gait parameters, isometric strength, stepping, fatigue, disability	8	2.5–5.5	Increases in some kinematic gait parameters, fatigue decreased	
	Broekmans et al. [68]	Resistance training	Muscle strength, functional mobility	36	2–6.5	Improvement in muscle strength and some functional parameters	Randomized controlled trial, long-term investigation (20 weeks)
Combined training	Romberg et al. [22]	Strength, aerobic training	Walking speed, lower extremity strength, upper extremity endurance, dexterity, static balance	95	1–5.5	Walking speed improved	Six-month exercise program, randomized study, 91 patients completed
	Cakit et al. [23]	Cycling progressive resistance training, balance exercises	Walking parameters, balance, fatigue, fear of falling, depression, quality of life	45	≤6.0, ability to stand independently in upright position for > 3 s	Improvement in walking parameters, fatigue, fear of falling, depression	Randomized (two exercise training and one control group), only 33 patients completed

(continued)

Table 3 (continued)

Form of training	Author	Method	(Major) Endpoints	Sample size	EDSS	Main results	Comments
	Smith et al. [87]	Strengthening, stretches, fitness exercises	Function, fatigue, sensory symptoms	34	0–6	>40% temporary increased sensory symptoms (number or intensity)	Single exercise session with follow up, all measures self-rated
	Surakka et al. [34]	Aerobic and strength exercise	Motor fatigue	95	1–5.5	Motor fatigue reduced in women (not in men)	Randomized controlled trial
Others	Motl et al. [9]	Wearing an accelerometer	Physical activity, quality of life, disability, fatigue, mood, pain, self-efficacy, social support	292	no information	Improvement in disability, fatigue, depression, pain, self-efficacy, social support	No specific training-protocol, completed self-report measures
	Rasova et al. [27]	Neurophysio-logically based phys-iotherapy, aerobic training, combined therapy	Impairment, disability, handicap, quality of life, fatigue, depression, respiratory function, physical fitness	112	0–6.5	Improvement in training groups with different impact on parameters, fatigue reduced	Randomized (three exercise training and one control group), only 95 patients completed
	Wiles et al. [35]	Physiotherapy	Mobility, mood	42	4.0–5.5	Improvement in mobility, subjective wellbeing, and mood	Randomized controlled crossover trial

MS patients, particularly inactivity-related impairment [16]. Still, recommendations on exercise for patients with MS have to face several limitations: Although a large number of published studies provide suggestions on therapeutic exercise, many of these studies have methodological limitations, including small sample sizes, lack of an appropriate control group, unblinded study design, and failure to account for different courses and stages of the disease. Randomized controlled and blinded study designs are only seldom applied. Applied training regimes are often not standardized or not described in detail. Different treatment intensities, frequencies and durations ranging from weeks to months in distinct studies further limit comparability. Long term outcomes of therapeutic interventions are rarely reported [6, 17, 18]. Furthermore, studies on the effect of exercise focus almost exclusively on MS patients with slight or moderate impairment (score on the expanded disability status scale (EDSS) less than 7) [6]. To date, to our knowledge only one recently published study examined highly impaired MS patients with an EDSS of 5–8 [19].

Apart from the methodological quality most of these studies including resistance training (e.g. progressive resistance exercise, walking mechanics), endurance training (e.g. bicycle ergometry, arm or arm-leg ergometry, aquatic exercise, treadmill walking) as well as combined resistance and endurance training provided evidence for a benefit of exercise in MS patients [6, 7, 17, 20, 21]. These training programs are referred to in more detail below. All training programs have shown to be well tolerated by the patients. Almost all inpatient participants and 59–96% participants of home-based trials completed without occurrence of adverse events [22–24].

4.1 Endurance Training

Moderate endurance training increased muscle strength and functional outcome parameters of both lower and upper extremities such as walking speed, fatigue, and overall quality of life [7, 17, 18, 22]. Beneficial effects on chair transfer, improved gait, stair climbing, and timed up and go test (standing up from a chair, walking 3 m, turning around and seat again) were reported by several authors [6, 23, 25, 26]. However, in some occasions varying and contradictory results were found. For example, some authors reported marked improvements in aerobic capacity, measured by maximal oxygen uptake (VO_2-max), whereas others did not observe significant improvements [6, 27, 28].

This also applies to fatigue as some authors provided evidence for an improvement of fatigue by endurance training, whereas other studies could not show a statistically significant improvement or could not reveal any differences at all [6, 15, 20, 29].

Conflicting data have been reported on various items of health related quality of life like vitality, social functioning, mood, energy, anger, sexual function, bladder and bowel function, and depression [6, 28, 30–32].

One group analyzed the effect of a 6 months outpatient aerobic training program in MS patients with mild to moderate disability (EDSS ranging from 1.0 to 6.0) and observed a trend for larger benefits in more severely disabled than in less affected patients. Again, this study was limited by the small sample size of 19 patients of which only 11 patients finally completed the study [32]. Therefore, these results have to be handled with care and further studies are required.

4.2 Resistance Training

Resistance training enhances muscle strength in healthy people and also in MS patients and showed beneficial effects on walking-speed, stepping endurance, stair climbing, timed up and go test, self-reported disability, and self-reported fatigue have been described in MS patients [23, 26]. As well, significant improvements in gait disturbances, measured by Dynamic Gait index, could be shown [23, 33].

Different forms of resistance training can be distinguished: one form, for example, constitutes progressive resistance exercise (PRE), which according to Taylor et al. comprises the following three principles: "*1. perform a small number of repetitions with relatively high loads until muscle fatigue is reached, 2. allow sufficient rest between exercise for recovery, and 3. increase the load as the ability to generate muscle force development*" [26].

Cakit et al. examined the effect of PRE by means of cycling progressive resistance training and lower-limb strengthening, both combined with balance exercise in a prospective randomized controlled trial of 45 MS patients [23]. After 8 weeks, patients in the two training groups performed better with respect to 10 m walking test, duration of exercise, and timed up and go test compared with patients in the control group without intervention. Moreover, the training groups showed improved balance, fatigue, depression, and fear of falling.

Taylor and colleagues investigated the effect of a 10 week PRE program on maximal muscle force, muscle endurance, functional activity, and overall psychological function in MS patients [26]. They reported significant improvements of arm strength, leg endurance, and fast walking speed, and a trend towards improvement in the 2-min walk-test and day-to-day life function.

Other training forms besides PRE like strategies to promote proper gait mechanics, focusing on weight bearing, weight shifting, and body positioning, or weightlifting have been used [33]. For example, Pilutti et al. examined the effect of resistance exercise in six severely disabled patients (EDSS 5.0–8.0) with progressive MS (five patients with primary progressive, one patient with secondary progressive disease course) by means of a 12 week course of bodyweight supported treadmill training performed three times weekly for 30 min [19]. The patients improved in terms of training intensity, treadmill walking speed and required bodyweight support as well as in physical and mental subscales of a quality of life questionnaire. No reduction of fatigue was reported.

4.3 Combined Endurance and Resistance Training

Only few studies reported on the effect of combined resistance and endurance training in MS. Small improvements both in muscle strength and gait velocity have been described [6, 22, 34]. Interestingly, in a comparatively large study on 95 MS patients, Surakka et al. observed significant training effects after 6 months of combined resistance and endurance training only in women, but not in men, which might be explained by a 25% higher exercise activity in women [34]. Furthermore, Romberg et al. reported significant improvements in walking speed and upper extremity endurance following 6 months combined exercise training, whereas lower extremity strength, VO_2-max, static balance, and manual dexterity did not improve [22].

In 2005 the Cochrane Collaboration published a first systematical review on the effects of exercise on ADL and health-related quality of life (HRQoL) and the effects of physical therapy on various symptoms in MS patients [21]. Only controlled, randomized clinical studies on adult MS patients not experiencing an exacerbation at the time were included. Six studies, of which four have so far only been published as an abstract, analyzed the effects of physical therapy (rehabilitation, physiotherapy, exercise, functional training, independent home-based training, aquatic exercise) on several disease-related variables compared to a control group without any physical therapy [24, 25, 30, 35]. Three other studies compared the results of two different physical therapy programs, that could show improvement in muscle strength, movement parameters (changing and maintaining posture, walking, moving around, timed transfer, walking cadence), and exercise tolerance tests (modified graded exercise test, VO_2-max, and physiological cost index), whereas mood-related parameters (fear, depression) were only moderately improved, and EDSS, fatigue, cognitive parameters, and ADL remained unchanged [31, 36, 37].

In a post-hoc analysis of the methodological quality of selected randomized controlled trials (RCT) on exercise interventions in MS carried out from 1950 to 2007 Asano et al. found evidence for positive effects of exercise on physical and psychosocial functioning and quality of life, but highlighted a great need for high quality RCTs in this field [17].

5 Exercise in MS Patients – The Impact of Body Temperature on Disability

In 1890 the German ophthalmologist Wilhelm Uhthoff (1853–1927) first described visual impairment and paresis occurring after physical activity. Because the patients' body temperature was not recorded, Uhthoff assumed that the described symptoms were caused by the physical activity itself and not by the resulting increased body temperature. Consequently, MS patients were advised for many decades not to engage in exercise [6, 38, 39]. In fact, 60–80% of MS patients experience a reversible (re)occurrence or aggravation of neurological symptoms in situations with increased body temperature, for example during vigorous physical activity, fever, or a hot bath [6, 38, 39]. As a reference to the first description, the eponym "Uhthoff's phenomenon" has been coined. The underlying cause is thought to be a temperature dysregulation due to dysautonomia with subsequent temperature–dependent impairment of the conduction velocity of partially demyelinated axons [7, 38, 40]. Another argument for MS patients to avoid exercise was the assumption that a "waste" of energy might aggravate fatigue and reduce ADLs which, however, has never been confirmed [6]. In fact, a detrimental effect of physical activity itself on CNS structures or an activity-mediated increase of the relapse rate has never been demonstrated [7, 41].

6 Exercise in MS Patients – Effects on the Immune System

It is a well-known phenomenon that exercise may influence the susceptibility to common infectious diseases like upper respiratory tract infections in different directions [42]. Whereas vigorous physical activity such as competitive sport can lead to an increased occurrence of infections, moderate exercise may in turn contribute to their prevention [7, 10, 42]. On the immune cell level in healthy subjects, physical strain has been demonstrated to initially increase the peripheral lymphocyte count which subsequently falls to below the initial level after cessation of the physical activity [10, 43, 44]. This transient lymphocyte reduction was short-lasting with a maximum duration of 3–24 h and was shown to be more pronounced in Th1 than in Th2 cells [44–46]. As Th1 cells primarily secrete pro-inflammatory cytokines like IFN-γ, IL-2 and TNF-α, whereas Th2 cells rather secrete anti-inflammatory cytokines such as IL-4, IL-5 and IL-10, exercise can promote a shift from a Th1-mediated pro-inflammatory to a rather anti-inflammatory Th2-mediated cytokine milieu which is of particular interest because an imbalance of Th1- and Th2-cells is considered relevant in the MS pathogenesis and a possible target to modulate the autoimmune reaction towards a more anti-inflammatory status [42, 44, 45, 47]. Since established immunomodulatory drugs such as IFN-β or glatiramer acetate exert similar effects on the immune system, drug treatment and physical activity may complement each other in terms of modulating the immune system. The fact that effects of exercise on the immune cell compartment are short lasting argues in favor of regular and frequent training intervals [10]. An evolving concept is the important role of another pro-inflammatory T helper cell subset in neuroinflammation – Th17 cells, that are characterized by their IL-17 production [47]. However, so far no data is available on a possible impact of exercise on IL-17 secretion or Th17 cells.

The effects of exercise on measurable cytokine production and response upon trigger events are less clear and often contradictory which can in part be explained by different populations studied, different training protocols and/or different readout parameters and paradigms [28, 43, 45, 48]. Therefore, these data have to be interpreted with care. For example, Heesen et al. found comparable IFN-γ, TNF-α and IL-10 serum concentrations in trained and untrained MS patients, whereas White et al. reported reduced resting plasma concentrations of IL-4, IL-10, C-reactive protein (CRP) and IFN-γ and a tendency for decreased TNF-α in MS patients upon 8 weeks of PRE [45, 49]. Muscle contractions are thought to stimulate secretion of IL-6 [28, 50]. Likewise, contradictory data have been published on the effect of exercise on immunoregulatory IL-6 in MS patients [28, 48].

Given the evident neurodegenerative component of MS, the effect of physical activity, particularly of exercise on nerve growth factors is of particular importance. In rodents, exercise has been shown to stimulate the release of brain-derived neurotrophic factor (BDNF), insulin-like growth factor 1 (IGF-1) and vascular endothelial growth factor (VEGF), all of which support cell proliferation, synaptic plasticity, neuroprotection, and neurogenesis in both physiological and neuroinflammatory conditions [49, 51–54]. Exercise seems to modify the secretion of neuroactive proteins also in humans [6, 49]. In both healthy participants and MS patients 30 min of moderate ergometry-based exercise increased the serum concentrations of BDNF and nerve growth factor (NGF) [49, 55, 56]. Since the hippocampus is crucially involved in learning and memory tasks as well as mood regulation, these findings might connect exercise with slowing of cognitive impairment and stabilization of affect in MS patients [49]. An increased secretion of IGF-1 has so far been demonstrated in healthy people after exercise [49, 57, 58]. IGF-1 as an important factor in development supports cell survival, brain growth and CNS myelination. During later phases of life IGF-1 might play a role in neuroprotection and synaptic and cognitive plasticity. Exercise also increased the activity of antioxidant enzymes, which might further support the role of exercise in neuroprotection [49].

7 Exercise in MS Patients – Effects on Morphology and Imaging Findings

Repetitions of stereotypic motor programs strengthen the cortical engrams and cause neuroplastic and adaptive processes like improved motor unit activation and synchronization of firing rates, in contrast, periods of inactivity are associated with opposite effects [23, 33, 59].

Although data on the effect of physical activity on brain structural features are sparse, some evidence indicates that physiotherapy and regular fitness training counteract the degeneration of brain tissue in patients with relapsing-remitting MS and might have a neuroprotective impact. Both grey and white matter atrophy occur already in early stages of relapsing-remitting MS [60]. However, patients with a higher level of aerobic fitness were shown to have a comparatively larger local volume of grey matter in the right post-central gyrus and midline cortical structures including the frontal medial and the anterior cinguli gyrus and the precuneus somatosensory cortex than "unfit" patients. Furthermore, higher fitness levels were associated with greater recruitment of cortical regions whereas lower fitness levels were associated with enhanced anterior cingulated cortex activity [61]. These data should however be treated with caution as they are based on a small sample of 24 female MS patients with a wide range in disability (EDSS 0–6.0) and disease duration (1–18 years).

Compared to healthy controls MS patients have been shown to activate more widespread, often even bilateral brain areas when performing motor and cognitive tasks, possibly as an expression of neuroplasticity [5, 62–64]. The degree of ipsilateral activation appears to correlate with the disease course and severity and is considered to reflect cortical adaptive or compensatory reorganization processes [62, 64, 65]. For example, in MS patients with primary progressive disease course movement-associated cortical activation involved "non-motor" areas like the insula and several multimodal cortical regions in the temporal, parietal, and occipital lobes in addition to the "classic" areas of motor planning and execution regions (including the supplementary motor area and the cingulate motor area) [65]. Morgen et al. reported that thumb movements of untrained MS patients elicited a more prominent activation of the contralateral dorsal premotor cortex in fMRI than in healthy controls which in contrast to healthy controls was not attenuated upon repetitive thumb movements [64].

In MS patients the corpus callosum is typically affected. Besides callosal lesions detected by standard MRI sequences, diffusion tensor imaging (DTI) sequences show ultrastructural damage, reflected by a reduced fractional anisotropy and increased mean diffusivity [59, 66, 67]. Interestingly, in a small study comprising 11 MS patients and healthy controls, Ibrahim et al. described a significant increase of fractional anisotropy and mean diffusivity in the corpus callosum after a 2 months physiotherapy program of 2 h per week, suggesting that physiotherapy may influence the brain microstructure in MS [59]. In summary, some data suggest that effects of exercise in MS patients may be reflected by morphological changes in the CNS which may be detectable by advanced imaging techniques. However, existing data are not yet sufficient to unequivocally prove an impact of exercise on brain structure in MS.

8 Personalized Exercise in MS Patients – General and Specific Recommendations

At the start of the 1990s the German Federal Health Monitoring System's general recommendation of performing a specific health-related training program at least three times a week was replaced by a more global perspective, namely the integration of everyday physical activities. In the situation of MS patients with their often reduced everyday activity, regular exercise is particularly important.

Apart from improving muscle strength, exercise is intended to improve endurance, muscle tone, posture stability, the degree of flexibility, and endurance and should involve both the agonists and antagonists [7, 23]. A physical training program needs to be customized to the individual needs and symptoms of a patient. Factors to be considered include the course and stage of disease, the degree and pattern of disability, age as well as concomitant diseases and sequelae. Importantly, it has to be ensured that the patient is not overstrained [6, 7, 8].

Compared to healthy people MS patients have a reduced aerobic capacity, decreased muscle strength, retarded rate of muscle tension development, reduced muscle endurance, and impaired balance [6, 7, 68]. A relationship between gait speed and strength parameters has been postulated [69]. Petajan and White have illustrated the level of muscular fitness and physical activity of MS patients in two "pyramids": passive range of motion (ROM) forms the basis of the muscular fitness pyramid and can minimize the risk of contractures when practiced regularly. The next step in the pyramid comprises active flexibility and resistance exercise against or without gravity to maintain muscle integrity, for example to enable the patient carrying out essential daily functions. A well-rounded program of muscle strengthening exercise represents the top of the muscular fitness pyramid. ADLs form the basis of the physical activity pyramid, followed by built-in inefficiencies, active recreation, and structured aerobic training programs. Again, design, frequency, and intensity of training programs have to be adapted to the individual patient. Weight-supported exercises like ergometry and water exercise are particularly recommended for patients with motor deficit or balance disturbances [8].

There are no universally applicable specific recommendations for exercise treatment. However, general therapeutic recommendations can be defined. Since exercise programs have not sufficiently been investigated in more severely disabled patients, these recommendations are restricted to MS patients with a maximum EDSS score of 7.0 [6, 7, 22]. Any new exercise program should be initialized by a physiotherapist or an exercise physiologist familiar with the disease [6]. A brief anamnesis including impairments in particular within daily activities should be made [8]. Regardless of the type of exercise, training programs should be simple enough and comprehensible to the patients. If necessary, it might be advisable to explain training programs in an illustrated or written form [7]. Patients should be supervised until they can perform the program adequately and independently [6, 7, 16]. Exercise programs should specifically target weaker muscles and encompass preferably multi-segmental complex movements [7, 23]. The intensity should be increased only slowly and should not reach the point of pain. Special care should be paid to peripheral nerves, particularly overstretching should be avoided [7]. Training sessions are recommended to start at a low level, include a light warm-up, progress according to the patients' clinical state and specific problems, and finally reach light to moderate intensity [6, 7, 16]. 10–15 min of daily stretching to maintain and improve flexibility of muscles and tendons and recovery time between training sessions of 24–48 h are recommended [7]. Immobilized patients or those with severe clinical symptoms should be individually assisted. Some authors advise that cardiopulmonary function and VO_2-max should be assessed prior to treatment start since MS patients may have reduced heart rate responses in graded exercise testing, possibly as an expression of cardiovascular dysautonomia, although this probably can hardly be implemented in the daily routine [7, 8]. Regarding endurance training and according to the American College of Sports Medicine, White and Dressendorfer recommend using the actual heart rate response to graded exercise testing for finding the ideal target heart range for training [7]. No symptoms should appear and "moderate intensities" ought to be strived, for example by means of the Borg scale of perceived exertion, which ranges from 6 to 20 (6 means "no exertion at all", 20 means "maximal exertion"). For a moderate intensity a range of 11–14 is aspired [7, 70]. Depending on the symptoms and the training program, exercises should be performed at home, either individually, with a training partner, or with a training group, and may include training equipment such as elastic bands, additional weights and pulley systems. Due to its social support a training group seems to be favorable in terms compliance and motivation. To achieve similar effects in homebased training programs, patients should be closely supervised, for example by visits or telephone calls [8, 20]. Most importantly, the training sessions have to be performed regularly [6–8].

Some special recommendations for MS patients regarding exercise training have been published. It has to be emphasized that these recommendations mostly represent the personal experience of the authors and are not always justified or supported by high standard clinical trials. Dalgas et al., for example, recommended endurance training of approximately 10–40 min duration, with an initial training intensity of 50–70% of VO_2-max corresponding to 60–80% of maximum heart rate [6]. According to these authors, resistance training is recommended to initially comprise 8–15 repetitions which can then be increased over several months. The training should start with 1–3 sets, later 3–4 sets with a 2–4 min break between sets and should be performed two or three times per week. For heat-sensitive patients and those who regularly develop Uhthoff's phenomenon exercise training in the morning or in water at temperatures of 27–28°C could be preferable since body temperature is physiologically lower early in the day and heat generated by physical activity is quickly dissipated in water. Alternatively, cooling before exercise and/or during physical activity for example by cold packs may help to prevent Uhthoff's phenomenon [7, 8, 39]. Also, resistance instead of endurance training could be preferable for heat-sensitive patients [6].

9 Physical Therapy Approaches to Prevent or Alleviate Individual Target Symptoms and Signs in MS

9.1 Fatigue

Fatigue, defined as an extreme physical and mental tiredness inadequate to the preceding demand, is a frequent, often very debilitating symptom in MS, which is generally difficult to treat [71, 72]. Approximately 75–90% of all MS patients experience fatigue during disease progression and some MS patients end up in a vicious circle: out of a wish to reduce fatigue they decrease physical activity and thereby reduce endurance, muscle strength, and quality of life which may enhance fatigue over time [8, 73, 74]. This in turn further limits physical activity and social life [32, 33, 75]. Apart from cooling, moderate exercise, particularly aerobic training, seems to have a positive effect on fatigue [15, 23, 76]. Because fatigue often increases over the day, training sessions should be performed in the morning and must not overexert the patient [77]. Special support like participation in a training group or attending psychological support to increase motivation for continuation of training over time could be advantageous [8]. Energy saving strategies are also applied, in which the patient learns to prioritize and to perform everyday tasks with a minimum of exertion [8, 13, 29, 72]. Although a beneficial effect of moderate exercise on fatigue has been described by some authors, effects are usually not pronounced enough to be reflected as significant improvements in current fatigue scales [6, 9, 23, 76]. Other studies completely failed to detect any improvements (e.g. [17]). One explanation for contradicting results could be the use of different fatigue scales, which focus on physical symptoms, or in attendant sleep disturbances such as insomnia, sleep related breathing disorders, restless legs syndrome, or periodic limb movement disorder [71, 72, 77]. In conclusion, there is some however not unequivocal evidence for low to moderate beneficial effects of moderate exercise on fatigue.

9.2 Spasticity

With a lifetime prevalence of about 90% spasticity is a frequent symptom in MS and has a potential to significantly reduce quality of life [77]. It leads to limitations in the range and normal pursuit of movements, results in malpositioning of the joints and is often accompanied by pain [14]. Controlled studies on exercise and physiotherapy for MS-related spasticity are rare; however some evidence for improvements has been reported [77].

Physical therapy measures include active and passive exercise (e.g. targeted positioning of the patient, passive exercise using motorized cycles, active treadmill exercise) which can be assisted by a training partner or training equipment such as elastic bands. Physiotherapeutic techniques according to Bobath or Vojta and proprioceptive neuromuscular facilitation (PNF) are among the treatments applied. None of these measures has been proven to be superior [77, 78]. It is most important to carry them out regularly and with a sufficient intensity [13, 77]. Light stretching of the affected muscle groups with duration of approximately 20–60 s should be performed prior to and after exercise [7].

9.3 Pareses

Pareses lead to various physical disabilities, such as difficulty in walking and fine-motor dysfunction. A relationship between gait speed and muscle strength in MS patients has been shown [6]. As no drug treatment for pareses exists and antispastic drugs such as baclofen may also lead to a worsening of existing pareses, physical and occupational therapy techniques are the sole treatment option. Just recently, 4-aminopyridine became available for symptomatic improvement of walking [79, 80]. However, only 35% of patients showed a clinical response, clinical effects can wear off in some patients or use of 4-aminopyridine might be limited by adverse reactions [80]. Because of reduced impact of gravity aquatic training allows patients with even severe pareses of the lower extremities to perform standing and moving exercises. A standing frame can help patients who are unable to stand, to train torso, limb, and respiratory muscles and protects against cardiovascular dysregulation. For immobilized patients, passive range of motion exercises proximal to the paralyzed region is recommended [7, 8]. Various studies have shown a significant improvement of muscle strength due to exercise [17, 23, 26, 68]. Furthermore, some authors reported beneficial effects in walking speed, stepping endurance, stair climbing, and timed up and go test [23, 26, 33]. In summary, evidence suggests that exercise is beneficial in the treatment of MS-related pareses, however again, only few, partially inconsistent data are available. Moreover, effects of exercise have been studied almost exclusively in MS patients with mild or moderate impairment.

9.4 Coordination and Balance Dysfunction

Abnormalities in balance control are frequent symptoms in MS patients, restrict their daily living activities and increase their risk of falls. Balance skills like standing and walking, as well as the patients' perception of their own balance are important to assess [81]. The sitting position of cycling training is advantageous for unsteady patients [7, 8]. Only a few studies investigated the influence of exercise programs on balance and coordination in MS and very few have chosen these variables as primary outcome parameter. Catteneo et al., for example, investigated the effect of balance training in 44 MS patients in a randomized controlled trial. Two treatment groups received particular balance rehabilitation for 3 weeks, a third (control) group participated an unspecific training program. In both treatment groups a reduction of the number of falls and an improvement in clinical tests of static (Berg Balance Scale) and dynamic balance (Dynamic Gait Index) could be detected. However, in self-assessment scales patients did not report significant improvements [81]. Data from another controlled study do not support a beneficial effect of exercise training on static balance [22].

9.5 Cognitive and Mood Disturbances

Depending on the disease course and stage 45–70% of MS patients are affected by cognitive impairments like reduced information processing speed, attentional deficits and episodic memory deficits [5, 14, 77, 82]. Up to 60–70% of patients suffer from mood disturbances [5, 83, 84]. Some evidence for a positive correlation between aerobic exercise and cognition or brain function in healthy people has been described [85]. In MS patients, beneficial effects of regular physical activity and exercise on both mood and quality of life have been repeatedly reported [7, 22, 31]. Valid data on the effect on cognitive function are however hardly available.

10 Conclusion and Outlook

Several lines of evidence suggest that MS patients benefit from regular physical activity and exercise with respect to clinical, imaging and physiological parameters. However, the quality of so far realized clinical trials on exercise training in MS do not always satisfy the requirements of a high standard study. Moreover, because of different treatment paradigms and endpoints, data are often difficult to compare. Thus, many questions remain still unanswered. In consequence, there is a great need for standardized high quality and well described studies that address both short and long term effects of exercise on clinical and paraclinical parameters in MS patients with different disease courses and different grades of disability.

Acknowledgments The authors report no conflicts of interest. This work was supported by the DFG (Exc 257).

References

1. Kantarci O, Wingerchuk D (2006) Epidemiology and natural history of multiple sclerosis: new insights. Curr Opin Neurol 19:248–254
2. Compston A, Coles A (2002) Multiple sclerosis. Lancet 359:1221–1231
3. Compston A, Coles A (2008) Multiple sclerosis. Lancet 372:1502–1517
4. Ziemssen T (2009) Multiple sclerosis beyond EDSS: depression and fatigue. J Neurol Sci 277(Suppl 1):S37–S41
5. Chiaravalloti N, Hillary F, Ricker J, Christodoulou C, Kalnin A, Liu W-C, Steffener J, DeLuca J (2005) Cerebral activation patterns during working memory performance in multiple sclerosis using FMRI. J Clin Exp Neuropsychol 27:33–54
6. Dalgas U, Stenager E, Ingemann-Hansen T (2008) Multiple sclerosis and physical exercise: recommendations for the application of resistance-, endurance- and combined training. Mult Scler 14:35–53
7. White LJ, Dressendorfer RH (2004) Exercise and multiple sclerosis. Sports Med 34:1077–1100
8. Petajan JH, White AT (1999) Recommendations for physical activity in patients with multiple sclerosis. Sports Med 27:179–191

9. Motl RW, McAuley E, Snook EM, Gliottoni RC (2009) Physical activity and quality of life in multiple sclerosis: intermediary roles of disability, fatigue, mood, pain, self-efficacy and social support. Psychol Health Med 14:111–124
10. Waschbisch A, Tallner A, Pfeifer K, Mäurer M (2009) Multiple sclerosis and exercise: effects of physical activity on the immune system. Nervenarzt 80:688–692
11. Thompson AJ (2005) Neurorehabilitation in multiple sclerosis: foundations, facts and fiction. Curr Opin Neurol 18:267–271
12. Multiple Sclerosis Therapy Consensus Group (MSTCG), Wiendl H, Toyka KV, Rieckmann P, Gold R, Hartung H-P, Hohlfeld R (2008) Basic and escalating immunomodulatory treatments in multiple sclerosis: current therapeutic recommendations. J Neurol 255:1449–1463
13. Henze T, Rieckmann P, Toyka KV (2006) Symptomatic treatment of multiple sclerosis. Multiple Sclerosis Therapy Consensus Group (MSTCG) of the German Multiple Sclerosis Society. Eur Neurol 56:78–105
14. Kesselring J, Beer S (2005) Symptomatic therapy and neurorehabilitation in multiple sclerosis. Lancet Neurol 4:643–652
15. White LJ, McCoy SC, Castellano V, Gutierrez G, Stevens JE, Walter GA, Vandenborne K (2004) Resistance training improves strength and functional capacity in persons with multiple sclerosis. Mult Scler 10:668–674
16. Gallien P, Nicolas B, Robineau S, Pétrilli S, Houedakor J, Durufle A (2007) Physical training and multiple sclerosis. Ann Readapt Med Phys 50(373–376):369–372
17. Asano M, Dawes DJ, Arafah A, Moriello C, Mayo NE (2009) What does a structured review of the effectiveness of exercise interventions for persons with multiple sclerosis tell us about the challenges of designing trials? Mult Scler 15:412–421
18. Dalgas U, Stenager E, Jakobsen J, Petersen T, Hansen HJ, Knudsen C, Overgaard K, Ingemann-Hansen T (2009) Resistance training improves muscle strength and functional capacity in multiple sclerosis. Neurology 73:1478–1484
19. Pilutti LA, Lelli DA, Paulseth JE, Crome M, Jiang S, Rathbone MP, Hicks AL (2011) Effects of 12 weeks of supported treadmill training on functional ability and quality of life in progressive multiple sclerosis: a pilot study. Arch Phys Med Rehabil 92:31–36
20. Dodd KJ, Taylor NF, Denisenko S, Prasad D (2006) A qualitative analysis of a progressive resistance exercise programme for people with multiple sclerosis. Disabil Rehabil 28: 1127–1134
21. Rietberg MB, Brooks D, Uitdehaag BMJ, Kwakkel G (2005) Exercise therapy for multiple sclerosis. Cochrane Database Syst Rev 1:CD003980
22. Romberg A, Virtanen A, Ruutiainen J, Aunola S, Karppi S-L, Vaara M, Surakka J, Pohjolainen T, Seppänen A (2004) Effects of a 6-month exercise program on patients with multiple sclerosis: a randomized study. Neurology 63:2034–2038
23. Cakit BD, Nacir B, Genç H, Saraçoğlu M, Karagöz A, Erdem HR, Ergün U (2010) Cycling progressive resistance training for people with multiple sclerosis: a randomized controlled study. Am J Phys Med Rehabil 89:446–457
24. DeBolt LS, McCubbin JA (2004) The effects of home-based resistance exercise on balance, power, and mobility in adults with multiple sclerosis. Arch Phys Med Rehabil 85:290–297
25. Harvey L, Smith AD, Jones R (1999) The effect of weighted leg raises on quadriceps strength, EMG parameters and functional activities in people with multiple sclerosis. Physiotherapy 85:154–161
26. Taylor NF, Dodd KJ, Prasad D, Denisenko S (2006) Progressive resistance exercise for people with multiple sclerosis. Disabil Rehabil 28:1119–1126
27. Rasova K, Havrdova E, Brandejsky P, Zálisová M, Foubikova B, Martinkova P (2006) Comparison of the influence of different rehabilitation programmes on clinical, spirometric and spiroergometric parameters in patients with multiple sclerosis. Mult Scler 12:227–234
28. Schulz K-H, Gold SM, Witte J, Bartsch K, Lang UE, Hellweg R, Reer R, Braumann K-M, Heesen C (2004) Impact of aerobic training on immune-endocrine parameters, neurotrophic factors, quality of life and coordinative function in multiple sclerosis. J Neurol Sci 225:11–18

29. Newman MA, Dawes H, van den Berg M, Wade DT, Burridge J, Izadi H (2007) Can aerobic treadmill training reduce the effort of walking and fatigue in people with multiple sclerosis: a pilot study. Mult Scler 13:113–119
30. Petajan JH, Gappmaier E, White AT, Spencer MK, Mino L, Hicks RW (1996) Impact of aerobic training on fitness and quality of life in multiple sclerosis. Ann Neurol 39:432–441
31. Mostert S, Kesselring J (2002) Effects of a short-term exercise training program on aerobic fitness, fatigue, health perception and activity level of subjects with multiple sclerosis. Mult Scler 8:161–168
32. Rampello A, Franceschini M, Piepoli M, Antenucci R, Lenti G, Olivieri D, Chetta A (2007) Effect of aerobic training on walking capacity and maximal exercise tolerance in patients with multiple sclerosis: a randomized crossover controlled study. Phys Ther 87:545–555
33. Gutierrez GM, Chow JW, Tillman MD, McCoy SC, Castellano V, White LJ (2005) Resistance training improves gait kinematics in persons with multiple sclerosis. Arch Phys Med Rehabil 86:1824–1829
34. Surakka J, Romberg A, Ruutiainen J, Aunola S, Virtanen A, Karppi S-L, Mäentaka K (2004) Effects of aerobic and strength exercise on motor fatigue in men and women with multiple sclerosis: a randomized controlled trial. Clin Rehabil 18:737–746
35. Wiles CM, Newcombe RG, Fuller KJ, Shaw S, Furnival-Doran J, Pickersgill TP, Morgan A (2001) Controlled randomised crossover trial of the effects of physiotherapy on mobility in chronic multiple sclerosis. J Neurol Neurosurg Psychiatr 70:174–179
36. Lord SE, Wade DT, Halligan PW (1998) A comparison of two physiotherapy treatment approaches to improve walking in multiple sclerosis: a pilot randomized controlled study. Clin Rehabil 12:477–486
37. Solari A, Filippini G, Gasco P, Colla L, Salmaggi A, La Mantia L, Farinotti M, Eoli M, Mendozzi L (1999) Physical rehabilitation has a positive effect on disability in multiple sclerosis patients. Neurology 52:57–62
38. Guthrie TC, Nelson DA (1995) Influence of temperature changes on multiple sclerosis: critical review of mechanisms and research potential. J Neurol Sci 129:1–8
39. White AT, Wilson TE, Davis SL, Petajan JH (2000) Effect of precooling on physical performance in multiple sclerosis. Mult Scler 6:176–180
40. Rasminsky M (1973) The effects of temperature on conduction in demyelinated single nerve fibers. Arch Neurol 28:287–292
41. Schulz K-H, Heesen C (2005) Effects of exercise in chronically ill patients. Examples from oncology and neurology. Bundesgesundheitsblatt Gesundheitsforschung Gesundheitsschutz 48:906–913
42. Gleeson M (2007) Immune function in sport and exercise. J Appl Physiol 103:693–699
43. White LJ, Castellano V (2008) Exercise and brain health–implications for multiple sclerosis: part 1–neuronal growth factors. Sports Med 38:91–100
44. Steensberg A, Toft AD, Bruunsgaard H, Sandmand M, Halkjaer-Kristensen J, Pedersen BK (2001) Strenuous exercise decreases the percentage of type 1 T cells in the circulation. J Appl Physiol 91:1708–1712
45. Heesen C, Gold SM, Hartmann S, Mladek M, Reer R, Braumann K-M, Wiedemann K, Schulz K-H (2003) Endocrine and cytokine responses to standardized physical stress in multiple sclerosis. Brain Behav Immun 17:473–481
46. Lancaster GI, Halson SL, Khan Q, Drysdale P, Wallace F, Jeukendrup AE, Drayson MT, Gleeson M (2004) Effects of acute exhaustive exercise and chronic exercise training on type 1 and type 2 T lymphocytes. Exerc Immunol Rev 10:91–106
47. Bennett JL, Stüve O (2009) Update on inflammation, neurodegeneration, and immunoregulation in multiple sclerosis: therapeutic implications. Clin Neuropharmacol 32:121–132
48. Castellano V, Patel DI, White LJ (2008) Cytokine responses to acute and chronic exercise in multiple sclerosis. J Appl Physiol 104:1697–1702
49. White LJ, Castellano V (2008) Exercise and brain health–implications for multiple sclerosis: part II–immune factors and stress hormones. Sports Med 38:179–186

50. Ozenci V, Kouwenhoven M, Link H (2002) Cytokines in multiple sclerosis: methodological aspects and pathogenic implications. Mult Scler 8:396–404
51. Vaynman S, Ying Z, Gomez-Pinilla F (2004) Hippocampal BDNF mediates the efficacy of exercise on synaptic plasticity and cognition. Eur J Neurosci 20:2580–2590
52. Fabel K, Fabel K, Tam B, Kaufer D, Baiker A, Simmons N, Kuo CJ, Palmer TD (2003) VEGF is necessary for exercise-induced adult hippocampal neurogenesis. Eur J Neurosci 18: 2803–2812
53. Swain RA, Harris AB, Wiener EC, Dutka MV, Morris HD, Theien BE, Konda S, Engberg K, Lauterbur PC, Greenough WT (2003) Prolonged exercise induces angiogenesis and increases cerebral blood volume in primary motor cortex of the rat. Neuroscience 117:1037–1046
54. Christie BR, Eadie BD, Kannangara TS, Robillard JM, Shin J, Titterness AK (2008) Exercising our brains: how physical activity impacts synaptic plasticity in the dentate gyrus. Neuromolecular Med 10:47–58
55. Gold SM, Schulz K-H, Hartmann S, Mladek M, Lang UE, Hellweg R, Reer R, Braumann K-M, Heesen C (2003) Basal serum levels and reactivity of nerve growth factor and brain-derived neurotrophic factor to standardized acute exercise in multiple sclerosis and controls. J Neuroimmunol 138:99–105
56. Griffin ÉW, Mullally S, Foley C, Warmington SA, O'Mara SM, Kelly AM (2011) Aerobic exercise improves hippocampal function and increases BDNF in the serum of young adult males. Physiol Behav 104:934–941
57. Frystyk J (2010) Exercise and the growth hormone-insulin-like growth factor axis. Med Sci Sports Exerc 42:58–66
58. Nishida Y, Matsubara T, Tobina T, Shindo M, Tokuyama K, Tanaka K, Tanaka H (2010) Effect of low-intensity aerobic exercise on insulin-like growth factor-I and insulin-like growth factor-binding proteins in healthy men. Int J Endocrinol. doi:10.1155/2010/452820
59. Ibrahim I, Tintera J, Skoch A, Jirů F, Hlustik P, Martinkova P, Zvara K, Rasova K (2011) Fractional anisotropy and mean diffusivity in the corpus callosum of patients with multiple sclerosis: the effect of physiotherapy. Neuroradiology 53:917–926
60. Chard DT, Griffin CM, Parker GJM, Kapoor R, Thompson AJ, Miller DH (2002) Brain atrophy in clinically early relapsing-remitting multiple sclerosis. Brain 125:327–337
61. Prakash RS, Snook EM, Motl RW, Kramer AF (2010) Aerobic fitness is associated with gray matter volume and white matter integrity in multiple sclerosis. Brain Res 1341:41–51
62. Cifelli A, Matthews PM (2002) Cerebral plasticity in multiple sclerosis: insights from fMRI. Mult Scler 8:193–199
63. Rocca MA, Matthews PM, Caputo D, Ghezzi A, Falini A, Scotti G, Comi G, Filippi M (2002) Evidence for widespread movement-associated functional MRI changes in patients with PPMS. Neurology 58:866–872
64. Morgen K, Kadom N, Sawaki L, Tessitore A, Ohayon J, McFarland H, Frank J, Martin R, Cohen LG (2004) Training-dependent plasticity in patients with multiple sclerosis. Brain 127:2506–2517
65. Rocca MA, Falini A, Colombo B, Scotti G, Comi G, Filippi M (2002) Adaptive functional changes in the cerebral cortex of patients with nondisabling multiple sclerosis correlate with the extent of brain structural damage. Ann Neurol 51:330–339
66. Roosendaal SD, Geurts JJG, Vrenken H, Hulst HE, Cover KS, Castelijns JA, Pouwels PJW, Barkhof F (2009) Regional DTI differences in multiple sclerosis patients. NeuroImage 44:1397–1403
67. Coombs BD, Best A, Brown MS, Miller DE, Corboy J, Baier M, Simon JH (2004) Multiple sclerosis pathology in the normal and abnormal appearing white matter of the corpus callosum by diffusion tensor imaging. Mult Scler 10:392–397
68. Broekmans T, Roelants M, Feys P, Alders G, Gijbels D, Hanssen I, Stinissen P, Eijnde BO (2011) Effects of long-term resistance training and simultaneous electro-stimulation on muscle strength and functional mobility in multiple sclerosis. Mult Scler 17:468–477
69. Mevellec E, Lamotte D, Cantalloube S, Amarenco G, Thoumie P (2003) Relationship between gait speed and strength parameters in multiple sclerosis. Ann Readapt Med Phys 46:85–90

70. Meyer T, Broocks A (2000) Therapeutic impact of exercise on psychiatric diseases: guidelines for exercise testing and prescription. Sports Med 30:269–279
71. Veauthier C, Radbruch H, Gaede G, Pfueller CF, Dörr J, Bellmann-Strobl J, Wernecke K-D, Zipp F, Paul F, Sieb JP (2011) Fatigue in multiple sclerosis is closely related to sleep disorders: a polysomnographic cross-sectional study. Mult Scler 17:613–622
72. Veauthier C, Paul F (2012) Fatigue in multiple sclerosis: which patient should be referred to a sleep specialist? Mult Scler 18:248–249
73. Fisk JD, Pontefract A, Ritvo PG, Archibald CJ, Murray TJ (1994) The impact of fatigue on patients with multiple sclerosis. Can J Neurol Sci 21:9–14
74. Krupp LB, Alvarez LA, LaRocca NG, Scheinberg LC (1988) Fatigue in multiple sclerosis. Arch Neurol 45:435–437
75. MacAllister WS, Krupp LB (2005) Multiple sclerosis-related fatigue. Phys Med Rehabil Clin N Am 16:483–502
76. Stroud NM, Minahan CL (2009) The impact of regular physical activity on fatigue, depression and quality of life in persons with multiple sclerosis. Health Qual Life Outcome 7:68
77. de Sa JCC, Airas L, Bartholome E, Grigoriadis N, Mattle H, Oreja-Guevara C, O'Riordan J, Sellebjerg F, Stankoff B, Vass K, Walczak A, Wiendl H, Kieseier BC (2011) Symptomatic therapy in multiple sclerosis: a review for a multimodal approach in clinical practice. Ther Adv Neurol Disord 4:139–168
78. Paci M (2003) Physiotherapy based on the Bobath concept for adults with post-stroke hemiplegia: a review of effectiveness studies. J Rehabil Med 35:2–7
79. Goodman AD, Brown TR, Krupp LB, Schapiro RT, Schwid SR, Cohen R, Marinucci LN, Blight AR (2009) Sustained-release oral fampridine in multiple sclerosis: a randomised, double-blind, controlled trial. Lancet 373:732–738
80. Goodman AD, Brown TR, Edwards KR, Krupp LB, Schapiro RT, Cohen R, Marinucci LN, Blight AR (2010) A phase 3 trial of extended release oral dalfampridine in multiple sclerosis. Ann Neurol 68:494–502
81. Cattaneo D, Jonsdottir J, Zocchi M, Regola A (2007) Effects of balance exercises on people with multiple sclerosis: a pilot study. Clin Rehabil 21:771–781
82. Weinges-Evers N, Brandt AU, Bock M, Pfueller CF, Dörr J, Bellmann-Strobl J, Scherer P, Urbanek C, Boers C, Ohlraun S, Zipp F, Paul F (2010) Correlation of self-assessed fatigue and alertness in multiple sclerosis. Mult Scler 16:1134–1140
83. Lincoln NB, Yuill F, Holmes J, Drummond AER, Constantinescu CS, Armstrong S, Phillips C (2011) Evaluation of an adjustment group for people with multiple sclerosis and low mood: a randomized controlled trial. Mult Scler 17:1250–1257
84. Feinstein A (2006) Mood disorders in multiple sclerosis and the effects on cognition. J Neurol Sci 245:63–66
85. Prakash RS, Snook EM, Erickson KI, Colcombe SJ, Voss MW, Motl RW, Kramer AF (2007) Cardiorespiratory fitness: a predictor of cortical plasticity in multiple sclerosis. NeuroImage 34:1238–1244
86. van den Berg M, Dawes H, Wade DT, Newman M, Burridge J, Izadi H, Sackley CM (2006) Treadmill training for individuals with multiple sclerosis: a pilot randomised trial. J Neurol Neurosurg Psychiatr 77(4):531–533
87. Smith RM, Adeney-Steel M, Fulcher G, Longley WA (2006) Symptom change with exercise is a temporary phenomenon for people with multiple sclerosis. Arch Phys Med Rehabil 87(5):723–727

Bridging the Informatics Gap Between Bench and Bedside: Implications to Neurodegenerative Diseases

Amnon Shabo (Shvo) and Maurizio Scarpa

Abstract Realizing personalized medicine is dependent on effective use of *omics* data and knowledge along with the patient's medial history. Such history includes clinical data sets that need to be as rich as possible and more importantly, its semantics should be made explicit and machine-processable. Often, semantics of research & clinical data is merely implied and needs to be resolved from separate descriptions of the data which are typically unstructured (e.g., clinical database schema textual definition in healthcare or study protocol description in research). Furthermore, the nature of patient's data is changing: a constantly growing stream of raw data with preliminary analysis is available today in both research and clinical environments, e.g., DNA sequences along with rare variants or homecare devices data along with personal alerts. The representation of all data types should adhere as much as possible to agreed-upon international standards in order to assure interoperability across the translational health domain. Dispersed and disparate medical records of a patient are often inconsistent and incoherent. A patient-centric longitudinal electronic health record could provide coherent and explicit representation of the data semantics, including *omics* information resulting from personalized analysis of the patient raw *omics* data and its association to phenotypic data.

Keywords Bioinformatics • Integrative medicine • Patient modeling • Neuroegenerative disease • Omics • Clinical implication • Healthcare • Personalized medicine • Standards • Translational medicine • Medical records

---

A. Shabo (Shvo) (✉)
IBM Haifa Research Lab, Haifa, Israel
e-mail: shabo@il.ibm.com

M. Scarpa
Women's and Children's Health Department, University of Padova, Padova, Italy

S. Mandel (ed.), *Neurodegenerative Diseases: Integrative PPPM Approach as the Medicine of the Future*, Advances in Predictive, Preventive and Personalised Medicine 2, DOI 10.1007/978-94-007-5866-7_13,

1 Introduction

The personalized medicine approach [1–3] suggests that the care of an individual patient can be improved if it is adjusted according to the patient's unique genetic makeup, along with the common information used at the point of care, e.g., clinical, social, psychological, and demographic data. Ideally, the patient's unique genetic makeup should be associated with the downstream *omics* consequences (e.g., gene and protein expression or affected biological pathways), and more importantly, should be associated with phenotypic data, preferably validated clinical observations, along with preferences set by the patient and health organizations. Associating different types of information and making them meaningful to the patient poses a great informatics challenge which is the focus of this chapter.

Translational medicine cycles create an abundance of information represented in a variety of formats. Clinical research studies produce results that get peer-reviewed and published, typically in scientific papers which contain mostly free text and charts. Some studies get reproduced by new studies that confirm the results of previous studies and are published in the same venues. Eventfully, an evidence is generated that might be used in clinical practice. Feedback from actual use of new evidences in clinical environments should be conveyed to researchers in order to refine the evidence and make it more helpful to clinicians and patients. Such feedback is typically conveyed via clinical data formats used by healthcare providers who are using the evidence in their practice. Furthermore, some evidences scale out to the community when population-based studies or comparative-effectiveness research are conducted and produce successful results. These studies are published with epidemiological orientation which yet has its own typical notation of information representation. Feedback from these studies needs to be conveyed back to the researchers to further refine the evidence. Finally, some evidences get widely adopted in a variety of clinical environments and consequently are being incorporated to the clinical guidelines published by the medical professional societies. These guidelines have yet their typical venues of publication and representation formats.

Only a few of the abovementioned information representation formats are semantically interoperable across the various information systems used by healthcare stakeholders, e.g., research organizations, healthcare providers, public health agencies and medical professional societies. Semantic interoperability means that an information system and its inner algorithms can parse incoming information as if it were originally created in that system. For example, to reproduce successful results of a study, it could be beneficial for the reproducing researchers to have the study setup described in a way that their information systems can receive (electronically) and operate accordingly. Semantic interoperability can also enable the transmission of clinical data from healthcare provider information systems back to the research organization, providing feedback on the actual use of new evidences in clinical environments. Such clinical data should be represented in a way that does not require human intervention in its utilization within the study environment.

The realization of personalized medicine relies on the ability to bring together three distinct types of information: patient clinical data, patient genetic data and relevant knowledge. If representation of core concepts (e.g., observation) is similar among these three types of information and there are sufficient touch points among them, clinical decisions support systems (CDSS) might be able to reason about the consolidated information and provide clinicians at the point of care with advice that is tailored to the individual patient.

The more granular the patient genetic data is (e.g., full DNA sequences), the more personalized the advice provided by CDSS could be, if in-depth analyses of the data are conducted against relevant knowledge sources and *omics* ontologies. First-line analyses of the patient genetic data (e.g., specific genes sequencing) could identify different types of variants such as known mutations (inherited and somatic) and variants that are not described in the scientific literature or have unknown significance. The *omics* downstream consequences of these genetic variants could be observed (e.g., gene and protein expression levels) or deduced by means of functional analysis along with the biological pathways stemming from the inspected genes and genetic biomarkers. Finally, clinical implications of the analyses could be suggested.

Recent advancements in DNA sequencing technologies now enable the whole-exome sequencing for about 1,000 dollars. There are already clinical use cases where such mass sequencing of the patient genetic makeup is used to help shed light on difficult clinical cases where common genetic testing (e.g., known genetic mutations) did not explain the clinical status of the patient.

Therefore, high-throughput screening used thus far in research settings is now moving to clinical environments and can significantly facilitate personalized medicine beyond the classification of a patient to a genetic profile like ‘rapid metabolizer’ in pharmacokinetics. The goal is to start with the unique genetic makeup of the patient and go all the way to the phenotypes through biological reasoning. This is a long journey since at this stage most of the generated throughput cannot be fully understood, however, computational biology method and data-driven approaches (e.g., statistical and machine learning analysis) can yield useful insights both as cues to scientists to guide their research as well as to clinicians when coping with fuzzy decision points.

2 Encapsulate and Bubble-Up

New clinical-genomics information standards are being developed that enable the encapsulation of key genomic data sets in for interoperability use cases. These key data sets are represented in common bioinformatics formats and (1) point to the mass genomic data from which the key data was extracted and (2) are referenced by genomic data objects that are associated with phenotypic information. This design principle creates standards that can be used by the various workflows of both research and clinical environments.

Genomic data sets vary in its granularity, starting from raw and mass genomic data (e.g., results of 1M SNP genotyping, whole-exome sequencing or expression arrays), and ending in summarized data (e.g., human-readable reports targeted at the 'end user'). In-between the two extremes there are intermediate data sets, typically 'key data' resulting from analysis of the mass data and selection of the most significant data items out of the mass data, based on relevancy to the subject in a research setting or to the patient at the point of care. Furthermore, these key data sets are possibly associated with phenotypic data (either interpretive or observed) and thus make up Clinical Genomic Statements (CGS) that can be better 'understood' by decision support applications for meaningful use of the data, because semantics is explicitly represented within the CGS.

The Health Level Seven (HL7) standards developing organization develops Clinical Genomics specifications [4] that are capable of 'encapsulating' such intermediate data sets (e.g., sequences of genes in a genetic testing panel) in their common representations like bioinformatics formats within the HL7 instance, where certain items of these data sets (e.g., certain DNA variations) were placed in other HL7 objects that were associated with phenotypic information (e.g., pharmacogenomics interpretation, or eventual observation of drug responsiveness). Links are available from the HL7 objects holding the 'bubbled up' data' back to the encapsulated data, and the latter could have back references to the mass data.

Matching of patient data with the most up-to-date knowledge underpins personalized medicine. Consider the following description on the use of Azathioprine (AZA) for the treatment of various neuroimmunological disease of neuromuscular transmission (e.g., myasthenia gravis, Lambert-Eaton myasthenic syndrome), muscles (e.g., polymyositis, dermatomyositis), peripheral nerves (e.g., chronic inflammatory demyelinating polyneuropathy), and CNS (e.g., sarcoid, vasculitis, multiple sclerosis [5]). Hematological toxicity that can be fatal in about 0.3% of cases is the most severe adverse effect [6]. Methylation by thiopurine-methyl transferase (TPMT) is the rate limiting step in the conversion of thiopurines to inactive metabolites. Due to the strong genotype-phenotype correlation between TPMT-deficiency and thiopurine toxicity, AZA therapy is an excellent example of the potential impact of pharmacogenetics in clinical practice with a recommendation for TPMT genotyping and/or phenotyping by the FDA (http://www.fda.gov/Drugs/ScienceResearch/ResearchAreas/Pharmacogenetics/ucm083378.htm; Table of Pharmacogenomic Biomarkers in Drug Labels). AZA treatment is contraindicated in patients with low/absent TPMT activity (<0.3%) due to potentially life threatening myelotoxicity. In these "poor TPMT metabolizers" two inactive alleles lead to less than 5 U/ml blood TPMT activity. More common is partial TPMT deficiency (<10%) with intermediate TPMT activity (5–10 U/ml) reflecting only one active allele; here, slow dose escalation is recommended, starting with 50% of the normal dosage. TPMT deficiency can be detected by genotyping of the three common deficiency alleles with high sensitivity [7].

Matching patient's data with this knowledge could be done by clinical decision support if both data and knowledge are represented in a similar fashion, and more importantly, in a machine-processable way that has all the information components

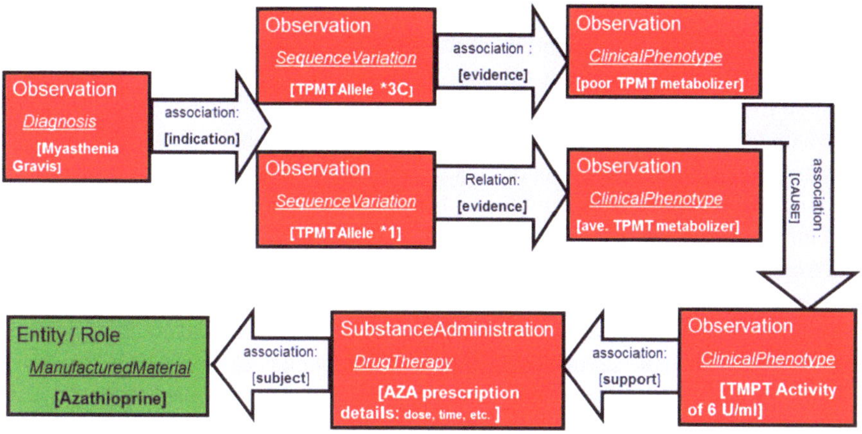

Fig. 1 A 'Clinical Genomic Statement' example where observations and substance administration classes are associated with each other along with a relevant entity, in order to explicitly represent the type of metabolizer the patient is and consequently the prescribed drug with dosing that is adjusted to the clinical phenotype of TMPT activity level

represented explicitly in a coherent 'clinical genomic statement'. Figure 1 shows a conceptual diagram of such clinical genomic statement that describes a patient with just one TPMT active allele and consequently a TMPT activity of 6 U/ml. The class stubs in this diagram are instantiation of the HL7 Clinical Genomics specifications. These classes were created as a result of constraining core classes of the Reference Information Model (RIM) of HL7 [8]. For example, Observation is a core class of the RIM. In the Clinical Genomics specifications, there is a SequenceVariation class that constrains the Observation core class and adjusts it to the requirements of the genomic domain. Other standards are derived from the RIM using the same methodology and thus specialized observations in all standards share the same Observation base structure, for example, the way an observation instance is uniquely identified (the 'id' attribute), the way an observation is typed (e.g., this is a SNP type of sequence variation), the way the observation is valued, the way the observation method is described, the way the timing aspects of the observation are represented, and so forth. This leads to semantic consistency across all healthcare domains that could lay the ground to better integration of disparate patient data sets and its matching to relevant knowledge.

3 The Labyrinth of Data Workflows

When dealing with mass amount of heterogeneous data, it could be useful that all parties exchanging data adhere to the same workflows. Such data workflows need to allow flexibility in terms of how much of the data is shared, who could have access to the data and what is the authorization mechanism agreed by all parties. The IHE XDS [9] integration profile provides a mechanism to register metadata and have links to the full blown data. That is, the XDS Registry (shared by a number of enterprises) registers clinical document headers and has links to the full documents held in some document repository (typically within an enterprise). In this framework, the HL7 Genetic Testing Report (GTR) standard could play an important role in being the root through which one can backtrack the data in any resolution needed, whether for evidence or re-interpretation or even re-analysis of the mass data [10]. The GTR can be handled by the IHE XDS integration profile as follows: GTR headers could be registered in an IHE XDS Registry which can be queried by an XDS Consumer. Retrieving the full GTR document could be done by accessing the document repository. The GTR document body could have structured entries pointing to HL7 instances encapsulating key genomic data (as previously described). Finally, the encapsulated data could have the links to the mass genomic data (when available).

4 The Legacy of EHR Systems

In the context of IHE XDS, an EHR system (or an EHR ancillary system) can play the role of an XDS Repository and thus participates in a sharing environment. However, the XDS profile doesn't mandate the internal structure of the document body so it is quite limited in this sense, and also it is typically implemented in regional sharing efforts. An EHR system usually serves a single enterprise and has its own set of functions and proprietary information models. These systems are powerful in performing the basic functions but quite often are limited in specialized areas such as genomics. In addition, mass data is typically set aside and handled in dedicated systems. A genetic data repository that serves EHR systems is somewhat analogous to a medical imaging archive. So, an EHR should mainly be a single point of access to the patient health data, along with communication components to seamlessly work with specialized systems, services and repositories. For interoperability with disparate systems, an EHR system should be able to exchange the entire health record of the patient, including the mass data. The latter is definitely a challenge in field of genomics; however the key genomic data encapsulated within the clinical data provide the essential data as well as pointers to the full data set. In the field of medical imaging, this is already done through the HL7 Diagnostic Imaging Report (DIR) standard [11] that includes standard access specification to key images using WADO (web Access to DICOM Objects [12]). This is analogous to the GTR, that is, a human readable report points to key data sets out of the mass data. However, since raw genomic data formats are much more diverse than medical images and the field of genomic repositories has no equivalent to WADO, the encapsulation of key genomic data within the health record is still needed.

5 Health Data Integration

The completeness of patient's data available for analysis is another challenge in realizing personalized medicine. When complete data is not available, the power of analysis is reduced, especially because the knowledge is not available for understanding many of the data items, and having incomplete data makes it more difficult as if the number of variables is increased is the equation you are trying to solve. In particular, the clinical data is often incomplete and sparse since the world of healthcare information is extremely fragmented [13, 14]. This situation has resulted in incomplete patient data at the point of care [15]. For example, episodic data such as discharge summaries created in one hospital are typically not available in another hospital. Medical records are handled by enterprise healthcare information systems, which are centered on the healthcare enterprise requirements such as those related to billing and administration. The needs of the patient as well as the secondary uses of medical records, such as clinical research/trials and drug development, are not properly addressed by the current constellation of information in healthcare [16]. Adding personal genetic data into this reality of dispersed and dissimilar patients' medical records makes the integration challenge much harder. This is because genetic data typically lacks standard representation and, in particular, standard description of the testing methods on which the data quality is highly dependent.

The vision of independent health record banks [17, 18] could provide the constellation by which complete health records will be sustained over the lifetime of individuals and be available for both healthcare and research. In addition, these records could be represented following the EHR standard developed by the European standards body [19] and thus better serve semantic interoperability.

6 Conclusions

This chapter explored the informatics challenges in providing proper infrastructure to personalized medicine. It focused on several issues, beginning from the base representations of genotype-phenotype associations suggesting an approach that allows the encapsulation of raw *omics* data in its native format in patient's medical records, while surfacing up certain data items that are associated with phenotypic data and analysis results. Using these base representations enables semantic interoperability among healthcare organizations as well as among all health stakeholders including research and public health entities. The chapter suggests interoperability standards that can make use of the base representations, eventually resulting in a more coherent and longitudinal health record for the patient that can be sustained over lifetime.

References

1. Ruano G (2004) Quo vadis personalized medicine? Personalized Med 1(1):1–7
2. Davis RL, Khoury MJ (2005) The journey to personalized medicine. Personalized Med 2(1): 1–4
3. Vazquez ES (2004) Personalized therapy: an interdisciplinary challenge. Personalized Med 1(1):127–130
4. HL7 Clinical Genomics Standards. http://www.hl7.org/Special/committees/clingenomics/index.cfm. Accessed 10 Dec 2012
5. Chan A, Pirmohamed M, Comabella M (2011) Pharmacogenomics in neurology: current state and future steps. Ann Neurol 70:684–697
6. Connell WR, Kamm MA, Ritchie JK et al (1993) Bone marrow toxicity caused by azathioprine in inflammatory bowel disease: 27 years of experience. Gut 34:1081–1085
7. Wusk B, Kullak-Ublick GA, Rammert C et al (2004) Thiopurine S-methyltransferase polymorphisms: efficient screening method for patients considering taking thiopurine drugs. Eur J Clin Pharmacol 60:5–10
8. Core Principles and Properties of HL7 Version 3 Models. http://wiki.hl7.org/images/d/dc/CorePrinciples_v005h.pdf. Accessed 10 Dec 2012
9. IHE (Integrating the Healthcare Enterprise) XDS (Cross-enterprise Document Sharing) Integration Profile. http://www.ihe.net/Technical_Framework/index.cfm#IT. Accessed 10 Dec 2012
10. Shabo A (2010) Meaningful use of pharmacogenomics in health records: semantics should be made explicit. Pharmacogenomics 11(1):81–87
11. HL7 CDA (Clinical Document Architecture) Implementation guide for diagnostic imaging reports. www.hl7.org. Accessed 10 Dec 2012
12. Digital Imaging and Communications in Medicine (DICOM) Part 18: Web Access to DICOM Persistent Objects (WADO). http://medical.nema.org/dicom/2004/04_18PU.PDF. Accessed 10 Dec 2012
13. Dietrich L, Rothschild A, Ball MJ (2005) Patient-centered clinical documentation -problem-driven electronic health records can help defragment healthcare. Healthcare Informatics Online.http://www.healthcareinformatics.com/newsclips/newsclips03_3_05.htm. Accessed 23 Apr 2005
14. Lawrence D (2002) From chaos to care: the promise of team-based medicine. Perseus Publishing, Cambridge, MA
15. Shabo A (2005) The implications of electronic health records for personalized medicine. Personalized Med 2(3):251–258
16. Knaup P, Ammenwerth E, Brandner R, Brigl B, Fischer G, Garde S, Lang E, Pilgram R, Ruderich F, Singer R, Wolff AC, Haux R, Kulikowski C (2004) Towards clinical bioinformatics: advancing genomic medicine with informatics methods and tools. Methods Inform Med 43:302–307
17. Shabo A (2006) A global socio-economic-medico-legal model for the sustainability of longitudinal electronic health records. Part 1. Methods Inf Med 45(3):240–245
18. Shabo A (2006) A global socio-economic-medico-legal model for the sustainability of longitudinal electronic health records. Part 2. Methods Inf Med 45(5):498–505
19. Health Informatics – Electronic healthcare record communication: EN13606, Committee European Normalisation, CEN/TC 251 Health Informatics Technical Committee, 2000. http://www.centc251.org/

Part VII

Personalized Medicine in Rare Pediatric Neurometabolic Diseases

Rosella Tomanin, Cinzia Maria Bellettato, Francesca D'Avanzo, Alessandra Zanetti, Adriana Ceci, David Begley, and Maurizio Scarpa

Abstract Enormous advances in life sciences and molecular biology mean that research today generates massive amounts of data that can potentially serve to advance the development of individualized therapies. Such treatments are based on the principles of disease prevention, prediction, early diagnosis and reliable monitoring of therapeutic efficacy with the goal of providing personalized treatments that improve the health of individuals. The development of personalized medicine is of considerable importance for pediatric patient populations, and represents a move away from the use of treatment dosages based on experience with the same compounds in adults. Currently, however, we know little about developmental pharmacogenomics and, although many biomarkers are available for use in clinical research, there have been few applications in the management of pediatric diseases. This chapter reviews the requirements for pediatric personalized medicine and describes a group of diseases that require such an approach. Personalized medicine is particularly relevant for the treatment of rare childhood diseases, and the group of life-threatening neurological diseases known as lysosomal storage diseases (LSDs) represents a potential study population. The genetic bases of these disorders are generally well defined, there is the potential for diagnosis at birth or prenatally, and there are several therapeutic options available or under development.

R. Tomanin • C.M. Bellettato • F. D'Avanzo • A. Zanetti • M. Scarpa (✉)
Women's and Children's Health Department, University of Padova, Via Giustiniani 3, 35128 Padova, Italy
e-mail: maurizio.scarpa@unipd.it

A. Ceci
Gianni Benzi Pharmacological Research Foundation Onlus, Valenzano, Bari, Italy

D. Begley
King's College London, London, UK

S. Mandel (ed.), *Neurodegenerative Diseases: Integrative PPPM Approach as the Medicine of the Future*, Advances in Predictive, Preventive and Personalised Medicine 2, DOI 10.1007/978-94-007-5866-7_14,

Keywords Personalized medicine • Rare disease • Pediatric population • Neurological impairment • Biomarker • 'Omics' technologies • Lysosomal storage diseases

Abbreviations

ADRs	Adverse Drug Reactions
ALL	Acute Lymphoblastic Leukaemia
CNS	Central Nervous System
CPNDS	Canadian Pharmacogenomics Network for Drug Safety
CYP	Cytochrome P450
ERT	Enzyme Replacement Therapy
EU	European Union
HLA	Human Leukocyte Antigen
LSDs	Lysosomal Storage Diseases
SRT	Substrate Reduction Therapy
TEDDY	Task force in Europe for Drug Development for the Young
TPMT	Thiopurine methyltransferase
UGT	UDP-glucuronosyltransferase

1 Introduction

Recent advances in human genomics and the availability of powerful molecular biology tools, in particular 'omics' technologies such as transcriptomics, proteomics and metabolomics, are giving rise to new possibilities for personalized medicine [1]. It is anticipated that the integrated and translational use of these new methodologies and technologies in personalized medicine will help to elucidate disease predisposition and enable us to achieve the best medical outcomes in the management of disease. According to its original description, 'personalized medicine' consists of preventing, diagnosing, treating, and monitoring diseases in ways that achieve optimal individual healthcare decisions [2]. It goes in the direction of improving patients health, in particular child health, in a flow going from the basic research, aiming at understanding mechanisms determining or determined by a specific pathologic condition, up to the evaluation of the patient compliance (Fig. 1). The main concept at the foundation of personalized medicine is individual variability and the potential that this affords to treat each person according to his or her unique genetic variation. In simpler words, the term personalized medicine can be defined as the right treatment for the right person at the right time (Fig. 1). As a result, the current methodological approach to diagnosis and treatment requires updating to take individual variability into account. In this context, biomarkers play a pivotal role as they can potentially help in identifying those patients who are most likely to respond favorably to a given therapy.

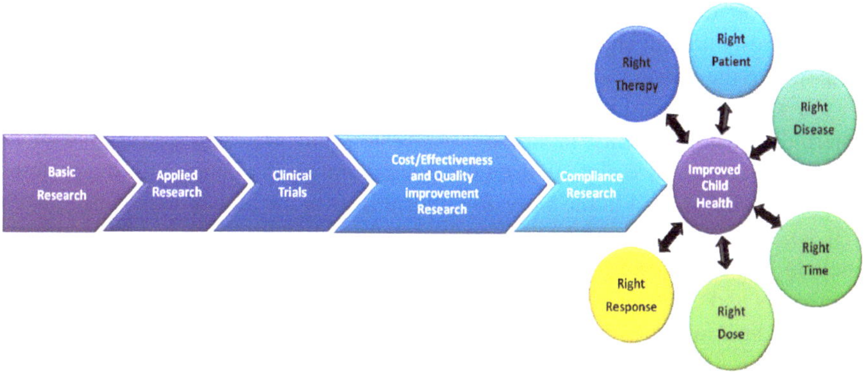

Fig. 1 Research flow towards improving child health and a personalized medicine approach

Personalized medicine, and in particular the integrated use of biomarkers as predictors of treatment response, is of particular interest for children. In fact, the effects of genetic polymorphisms that impact on drug actions are generally enhanced in the pediatric population because of the natural physiological processes of growth and development. Thus, during childhood there is a greater degree of variability in metabolism and, as a consequence, in drug response.

2 Biomarkers

The term biomarker is defined as an 'objectively measurable biological characteristic that can either serve as an indicator of normal or pathogenic processes in the body, or as a tool to track pharmacological responses to therapeutic drugs' [3, 4]. It can refer to a specific cell, molecule, gene, gene product, enzyme, hormone, strand of mRNA or other signaling molecule used by cells. A wider view of the term, released by the Parliamentary Office of Science and Technology in London (UK), includes complex organ functions or general characteristic changes in biological structures [5].

Today, there are several new sciences and technologies that are helpful in identifying and measuring biomarkers. Among the first to be developed were pharmacogenetics, which involves studying the relationship between variant forms of human genes and related interindividual variability in drug response, and pharmacogenomics, which is directed at tailoring pharmaceutical treatment to a particular genomic profile. While pharmacogenetics is focused on single genes and aims to find 'the right drug for the right patient', pharmacogenomics focuses on defining 'the right drug for the right disease' by studying how gene interaction determines individual drug responses, in order to maximize efficacy and to reduce toxicity and adverse drug reactions (ADRs) [6]. Unfortunately, 90% of the drugs available on the market to date work in only 40% of treated individuals [7], thus, representing an enormous waste of money and above all a highly inadequate response to the still unmet treatment needs of the diseased population.

For many adult and some childhood conditions, biomarkers that reliably reflect interindividual variability in disease expression and that also correlate with or predict outcomes have already been discovered among the genes, mRNA, proteins and metabolites found in the body. Polymorphisms in the human leukocyte antigen (HLA), for example, have been identified as risk factors for severe ADRs: these include the association between HLA-B*1502 and carbamazepine-induced skin necrolysis and Stevens–Johnson syndrome [8], as well as the association between HLA-B*5701 and hypersensitivity reactions to the HIV drug, abacavir [9, 10]. Screening for both of these alleles is recommended before starting the respective treatment and, in the case of HLA-B*5701, screening prior to abacavir initiation has been suggested to produce cost savings [10, 11]. In future, it is expected that such biomarkers will be detected more readily thanks to the new high-throughput technologies and powerful analytical approaches employed within the fields of pharmacogenomics, proteomics, metabolomics and other so-called 'omics'. This type of genetic knowledge may ultimately be included within the drug development process where biomarkers will serve as identifiers of treatment response. As a result, it will become important to include population screening early in the clinical trial development phase in order to establish which subpopulations receiving an experimental drug are most/least likely to benefit from, or to experience adverse reactions associated with that treatment.

Refinement of this approach will lead to more targeted patient recruitment and should reduce the likelihood of new therapies failing at the clinical trial stage. Most importantly, it should reduce the risk of unsuccessful treatments or undesired secondary adverse effects in clinical practice. Whether pharmacodiagnostic (theranostic) applications can simplify the selection of drugs for clinical testing, and help to develop more focussed clinical trials that provide enhanced knowledge about the effects of a drug in humans within shorter time frames, are currently being explored by various organizations such as the National Institutes of Health (NIH) [12].

3 The Application of Biomarkers to the Pediatric Population

The pediatric population constitutes a wide spectrum of different physiologies including a range of individuals from the very small preterm newborn infant to the adolescent [13]. Furthermore, this age group is subject to substantial changes in body proportions and composition accompanying the dynamic process of growth and development. These changes influence the efficacy, toxicity and dosing of medicines used in children [14].

The use of biomarkers as indicators of genetic factors affecting drug response is of particular interest for children. Pediatric populations represent approximately 20% of the entire population [15]. Historically, despite the need for specific treatments for conditions affecting children, ethical, economic and methodological reasons have limited clinical trials in young populations [16, 17]. In fact, only about 30% of marketed drugs in Europe include information on pediatric use in their documentation (SmPC or Package Leaflet) [17]. The number of pediatric medicines available varies considerably for each therapeutic area, with the lack of medicines tailored for children being particularly serious for neonates [17]. As a result, off-label drug use in pediatric patients has been relatively common [16, 18, 19]. An example of this is the fact that metabolic capacity is different in children and adults. Hepatic glucoronidation activity of the UDP-glucuronosyltransferase (UGT) family of enzymes that are needed for the elimination of certain xenobiotic and endogenous substances, for instance, is lower in children than adults, which may have implications for drug detoxification [20]. UGT activity towards bilirubin at birth, for example, is approximately 1% of the level observed in adults [21, 22], although activity increases rapidly and reaches levels equivalent to those in adults 14 weeks after birth [23, 24]. Total UGT activity has been reported to develop to adult levels by 20 months of age; however, the age at which enzymes reach adult levels of activity may not be universally indicative of *in vivo* drug clearance [25]. This study showed that intrinsic hepatic clearance of the antipsychotic drug trifluoperazine did not appear to reach adult levels until 18.9 years of age despite UGT being maximally active well before this time. These types of infancy/childhood-specific developmental changes are clearly complex and likely underlie some of the variability in drug response seen in neonates and children. Failing to take the potential impact of these differences on drug action into account places children at risk of ADRs.

In recent years there has been a rise in the number of pediatric drug safety and efficacy studies, as well as changes in drug labeling for young patients showing that unique pediatric doses are often necessary [26, 27], although most of the studies so far have been focussed on pediatric cancer [27, 28]. Furthermore, some diseases that occur in children either do not develop or have a different effect/prevalence in adults, making it necessary to develop drugs specifically for children. Despite the clear need for individualized medicine for pediatric patient populations, we still know very little about developmental pharmacogenomics and there have been few applications of biomarkers in the management of pediatric disease [28]. Results of a literature search conducted by the Task force in Europe for Drug Development for the Young (TEDDY) European Network of Excellence demonstrated that throughout the world more than 50% of pharmacogenomic/pharmacogenetic research activities in children during the last 10 years were related to predisposition, which means the investigation of the correlation between genetic traits and the probability of or susceptibility for a given pathology or disease [29]. These types of exploratory studies provide little or no insight into mechanisms of disease or drug action, and their results cannot be used to improve medical practice or to support therapeutic solutions. Only a small number of polymorphisms that are linked to therapeutic response have been studied in children and, even when studies demonstrate an impact on treatment response, their use in clinical practice remains limited. This is highlighted by the fact that very few FDA drug label amendments for pharmacogenomic testing refer to pediatric populations [16].

A major consideration that has arisen from analyses of publications on pharmacogenetics and pharmacogenomics is that studies often show contradictory or inconsistent results. This is particularly the case for studies in pediatric populations which, by definition, generally have limited patient numbers. A systematic review, conducted in 2009 on pharmacogenetic studies in general, suggested that the limited translation of pharmacogenetic research into practice may be explained by several factors including limited primary research, small sample sizes, a mainly candidate gene approach, the use of surrogate markers and paucity of meta-analyses [30].

Clear analysis of the situation is further complicated by the use of different research methods across studies, the lack of both standardized outcome measures and study replication, and the low level of consideration given to potential covariates such as comorbidities. Other investigators have similarly raised concerns about the need to improve the quality of studies and prevent false-positive associations [28]. In conclusion, while the new 'omic' technologies offer great promise for the identification of biomarkers of drug response, these studies are subject to many of the same pitfalls that apply in general to pediatric and other clinical trials.

Translational research aimed at understanding the complexities of disease etiology, with the goal of improving healthcare practice, has some important limitations in pediatric populations for which the study of the relative interactions between genetic background (genotype), lifestyle (behavior) and environmental exposure represent a difficult target to achieve [31]. In pediatrics, such research has been hindered by factors such as the small size of the pediatric patient population, as well as the many practical and ethical considerations related to tissue procurement from children. This has served to delay the establishment of biobanks that will allow the shared use of human tissues for translating research, which makes the process of obtaining useful biomarkers more difficult and complex [31].

Notwithstanding the limited experience and considerations above, some exciting examples of the usefulness of biomarkers in clinical practice can be derived from recent published literature. Childhood disorders in which pharmacogenetic studies have most commonly been conducted include acute lymphoblastic leukemia (ALL), attention–deficit hyperactivity disorder, growth hormone deficiency, juvenile idiopathic arthritis, atopic dermatitis and asthma (see [28, 32, 33] for reviews). A good example of a childhood condition where there will be benefits from the use of pharmacogenetics is ALL. Current treatment for patients with ALL results in event-free survival greater than 75% for most patients; however, approximately 25% fail to respond to treatment [34]. One of the reasons for treatment failure is the development of toxicity due to the absence or presence of only low levels of enzymes that metabolize chemotherapies. Thiopurines (azathioprine, mercaptopurine and thioguanine), for example, are often used to treat ALL and their safety has been shown to be related to thiopurine methyltransferase (TPMT) genotype. Genotypes resulting in reduced TPMT enzyme activity have been shown to be associated with dose-related side effects/toxic events to thiopurines in children with

ALL [35–37]. Thus, TPMT genotype may be used to help choose the appropriate dose of thiopurine for an individual child. In the US, the drug labels for azathioprine and 6-mercaptopurine now give the FDA recommendation for TPMT testing [9]. Although this biomarker is not yet reliably assessed in children within Europe, because TPMT deficiency can now be easily detected, this knowledge should lead to improvements in the safety of treatment in children receiving thiopurine chemotherapy. The recent publication of Clinical Pharmacogenetics Implementation Consortium Guidelines for TPMT genotyping and thiopurine dosing may also facilitate the routine incorporation of TPMT genotyping in clinical practice for pediatric patients receiving these compounds [38].

4 Toward the Development of Pediatric Personalized Medicine

There are several pharmacogenomic/pharmacogenetic networks that aim to help collect, coordinate or communicate information about genetic determinants of drug response. In Europe, for example, there is the European Network for Pharmacogenetics (http://www.epr-network.org/), in the US there is the NIH Pharmacogenomics Research Network (http://www.nigms.nih.gov/Initiatives/PGRN), in Canada the Canadian Pharmacogenomics Network for Drug Safety (CPNDS), and there is also the International HapMap project (http://hapmap.ncbi.nlm.nih.gov/) that represents a partnership of researchers and funding agencies from several countries. Each of these networks is distinct; however, the activities of the CPNDS illustrate how the establishment of networks may play an important role in the development of personalized medicine. The CPNDS is an active surveillance network that collects information on ADRs, and operates predominantly within Canadian pediatric teaching hospitals [16, 39]. The aim of the network is to improve drug safety by identifying genetic markers that can predict which patients are likely to experience specific serious ADRs, as well as those who may experience decreased therapeutic response to a drug. To help determine the role of genetic variations in the development of ADRs, the network has created a database of clinical ADRs and a biobank of DNA and other samples. Among other examples, this approach has proven to be useful by contributing to the development of a research protocol to validate reports of a potential association between genetic variations in enzymes in the CYP and UGT families in breastfeeding women taking codeine and the development of codeine-induced central nervous system (CNS) depression in their babies [40].

The development of databases containing information about genomic positions and biobanks containing tissue samples from large numbers of individuals also makes it possible to conduct genome-wide association studies [28, 41]. Rather than selecting candidate genes, this type of investigation can look at thousands of single nucleotide polymorphisms across the genome in one investigation, which may represent an important approach for studying complex, common diseases where multiple genetic variations may contribute to drug response.

As mentioned previously, despite the fact that many biomarkers are currently available for use within the clinical research setting, few have been integrated into pediatric clinical practice. In addition, the validation process of such markers in children remains an issue. As a consequence, it is important to encourage the establishment of pediatric biobanks, as well as the connection of existing biobanks through intelligent computing platforms in order to share data, integrate results and promote the use of biomarkers in drug development and clinical trials. Furthermore, the uptake of biomarkers within the drug development process requires investment and research on the part of the pharmaceutical industry. As of 2007, very few pharmaceutical companies indicated ongoing pharmacogenomic- or pharmacogenetic-related research in children [29]. However, more recently an investigation performed by Tufts suggests a growing interest and investment in this area. Their report shows that 12–50% of current clinical pipelines of the companies interviewed involve personalized medicines [42]. Thus, although at present pediatric drug development and utilization cannot benefit fully from pharmacogenetic/pharmacogenomic findings there is hope for the future. Further efforts are needed in order to reduce the gap between promise and application.

Particular actions that have to be undertaken in order to reduce the existing gaps include:

- the identification of the most promising research approaches for use in pediatric populations from among those based on 'omics' sciences;
- the development of model-based methodologies for clinical trials in children based on the use of 'omics'-derived biomarkers;
- the initiation and completion of pilot studies designed with the goal of developing pediatric personalized medicines in a specific sector such as oncology, respiratory disease, neurodegenerative disease or infective disease.

The development of personalized medicine is of relevance for all pediatric patient groups, and in particular for those populations that are generally deprived of drugs when the traditional clinical trial approach is applied. Examples of populations for whom the personalized medicine approach will be particularly important are neonates and children with rare diseases.

5 Personalized Medicine for Children with Rare Neurological Diseases

In the European Union (EU), a rare disease is defined as one that affects fewer than 5 people per 10,000 inhabitants [43]. Most of these conditions are genetically inherited and, importantly, many develop during childhood. Although individually uncommon, collectively there are thousands of rare diseases, making them an

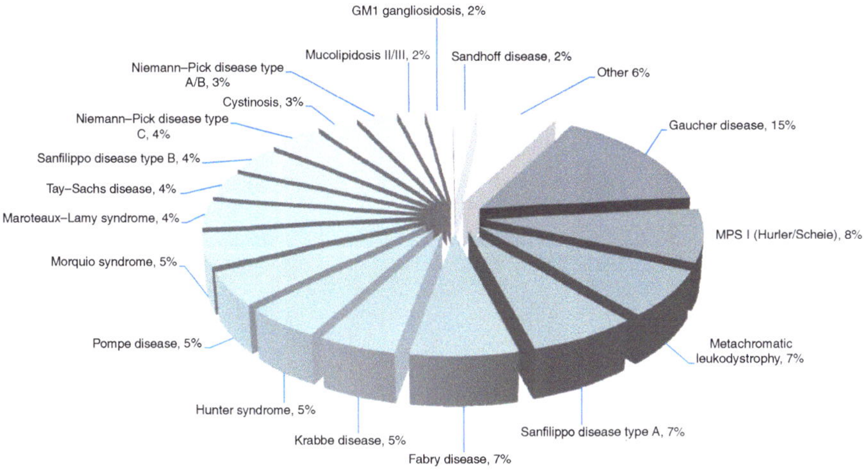

Fig. 2 Proportion of patients diagnosed with different lysosomal storage diseases in Australia between 1980 and 1996. Based on data from Meikle et al. [45] (modified from Scarpa et al. [47] with permission from EPMA Journal)

important group of disorders that affect a large number of patients throughout Europe. In the context that we have so far identified somewhere in the region of 5,000–8,000 distinct rare diseases, between 27 and 36 million people in the EU could be affected (6–8% of the population), making the management of these conditions an important issue [43]. For these reasons, the EU takes the position that rare diseases are a serious public health concern and should be a priority in EU health and research programs.

Although neurodegenerative diseases are most prevalent in the elderly, in rare cases they also affect individuals early in life. In children, neurodegeneration leads to severe mental retardation and premature death with devastating consequences on their immediate environment. These conditions also have relatively high costs for society. In total, there are thought to be about 600 different genetically inherited metabolic diseases, approximately 400 of which involve the CNS with different degrees of severity and rates of degeneration. Lysosomal Storage Diseases (LSDs) are a subgroup of approximately 50 of these metabolic disorders that affect about 1 in 7,500 newborns [44, 45], although more recently a combined incidence as high as 1 in 5,000 newborns has been suggested [46]. The proportion of patients diagnosed with different LSDs in Australia between 1980 and 1996 is shown in Fig. 2 [47].

These conditions are caused by the lack of certain lysosomal enzymes or lysosome components, thus preventing the complete degradation of macromolecules and the recycling of their constituents. As a result of the failure to degrade these molecules, intermediate degradation products begin to accumulate within the cells, affecting the function of lysosomes and possibly other cellular organelles. Each disorder is classified on the basis of the type of molecules that accumulate. The accumulation of these substrates starts immediately after birth and continues

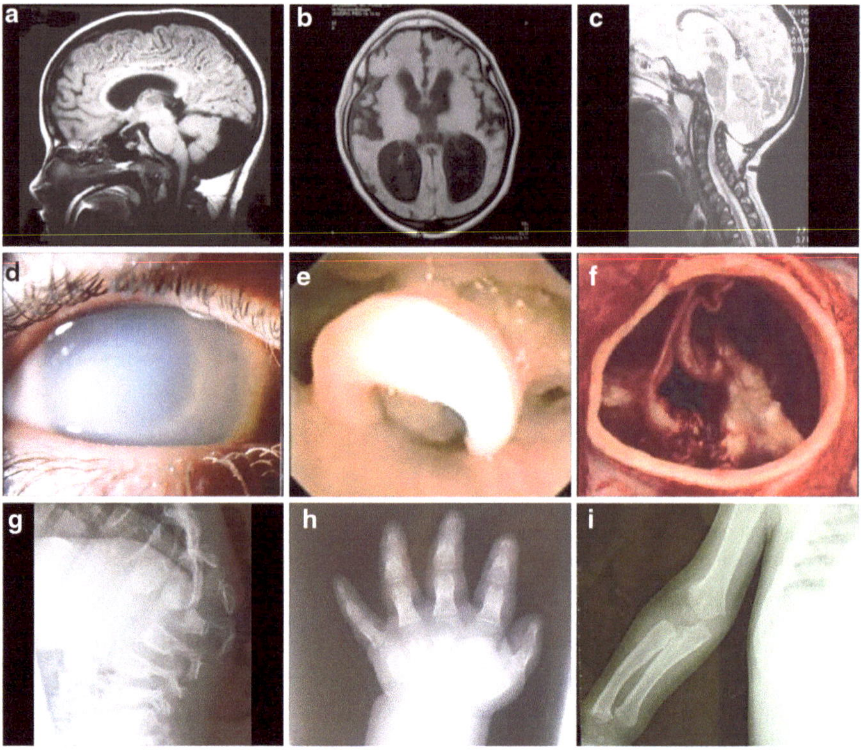

Fig. 3 Impact of the accumulation of undegraded substrates in lysosomal storage diseases. Brain scans show: (**a**) atrophy with enlarged Virchow–Robin spaces and ventricles, (**b**) communicating hydrocephalus, and (**c**) spinal cord compression at the cervical junctions. Other images show: (**d**) corneal clouding, (**e**) larynx thickening, (**f**) valvular alterations due to the accumulation of undegraded substrates, and (**g–i**) dysostosis multiplex of back bone, hand and long bones (reproduced from Scarpa et al. [47] with permission from EPMA Journal)

throughout life, often altering the structure and impairing the function of several organs, including the CNS (Fig. 3). CNS pathology causes mental retardation and progressive neurodegeneration that ultimately ends in early death of these young patients [48–52].

Our understanding of the pathogenesis of the LSDs has grown dramatically over the last 10–20 years and there is already a strong community supporting research and scientific developments in this area, which includes physicians, researchers, patient advocates and industry. Thanks largely to advances in cellular and molecular biology, and the incentives for drug development provided by the introduction of the Orphan Drug Regulation in Europe, we are now moving closer to being able to provide therapies for neurological signs and symptoms in patients with these conditions. An advanced state of knowledge on the mechanisms involved in LSDs pathogenesis, in particular, will allow the elucidation of features of neurodegenerative diseases that will most likely be useful in the study of more prevalent neurodegenerative diseases that predominantly affect the elderly (e.g. Alzheimer's and Parkinson's diseases). Furthermore, secondary events that lead to neurodegeneration (e.g. brain inflammation, alteration of intracellular trafficking, impairment of autophagy, or oxidative stress) are common to both pediatric and adult neurodegenerative diseases (for a review see [51]). Therefore, at least some of the observations relative to loss and recovery of brain plasticity in pediatric neurodegeneration can be extrapolated to adult disorders, and findings will provide unique insights into the pathophysiology and restorative capacities of neurodegenerative diseases in general. In fact, suppressing the primary cause of neurodegeneration (e.g. by supplementing the missing enzyme) in a young brain has the potential to maximize benefit as the brain retains considerable plasticity at this stage of development.

There are several factors that make LSDs good candidates for pharmacogenetic studies and personalized medicine. The first is that the genetic basis of most of these conditions is well defined; they are monogenic disorders and many disease-causing mutations have already been characterized. Although these mutations are often private, they all result in reduced enzyme levels and storage to varying degrees. Despite a current general lack of awareness of these conditions among physicians, which can lead to delayed diagnosis, there is the potential for presymptomatic and, in many cases, prenatal diagnosis [53–55]. This means that once biomarkers that predict therapeutic response have been identified, it will be possible to initiate the appropriate treatment at an early stage of disease. Furthermore, for many LSDs there is some overlap in the pathways involved in the degradation of the substrates that accumulate, making it likely that any biomarkers discovered may apply in more than one condition. Another factor is that patients show considerable phenotypic variability (even those with the same underlying genetic mutation) and there is similarly variation between individuals in terms of response to therapy. At present, several different therapeutic approaches have been proposed for different disorders, including replacement of the missing enzyme (enzyme replacement therapy, ERT), reduction of levels of the enzyme substrate (substrate reduction therapy, SRT), the use of chemical chaperones to stabilize existing but unstable enzyme (chaperone therapy), bone marrow transplantation, and gene therapies [56–58] (Fig. 4).

Based on current knowledge, predicting which patients will respond to which treatment remains a challenge [59]. In order to make personalized medicine for children with rare neurological disorders a reality it is necessary to refine the techniques used to ensure that therapy gains access to the CNS. Although it is known which enzyme deficiency causes each of the LSDs, supplementation of the functional lysosomal enzyme intravenously only allows for modification of the natural history of these diseases in the *peripheral* organs (e.g. liver, spleen, joints, etc.). The reason for this is that the therapeutic enzyme molecules used in ERT are too large to cross the blood–brain barrier which protects the brain by allowing only those molecules required naturally by the CNS to pass into the brain from the peripheral circulation.

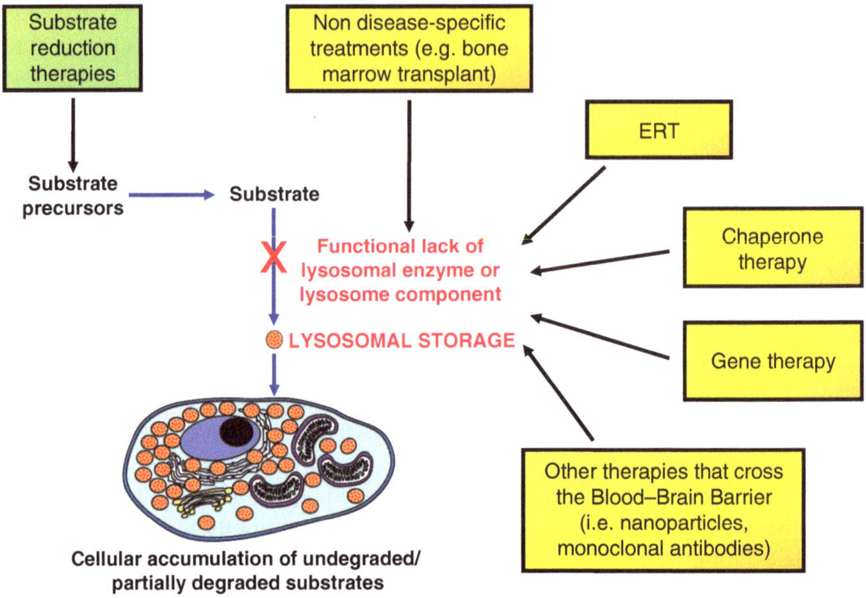

Fig. 4 Routes of intervention for the treatment of patients with lysosomal storage diseases (modified from Scarpa et al. [47] with permission from EPMA Journal)

Proposed methods to circumvent the blood–brain barrier in patients with LSDs include the direct injection into the brain of either the recombinant enzyme itself or of gene transfer vehicles able to induce the synthesis of the enzyme in genetically deficient brain cells [60]. Another approach is to inhibit enzymes early in the biosynthetic pathway of the accumulating protein(s) to reduce the number of molecules that require degradation in the lysosome. Small pharmaceutical molecules that are capable of crossing the blood–brain barrier are used to achieve this. In cases where the affected lysosomal enzyme is present but does not function because it is unstable, small pharmaceutical molecules known as chemical chaperones that can cross the blood–brain barrier may be used to stabilize the enzyme, thus at least partially restoring its function. As the process of neurodegeneration observed in LSDs is characterized by secondary events such as neuroinflammation, glutamatergic activation and oxidative stress, the use of small molecules with anti-inflammatory and/or neuroprotective properties may also be a useful adjunctive approach to treat LSDs. To date, the utility of most of these approaches has already been demonstrated in animal models.

Looking to the future, there is also a need to establish and validate biomarkers that will allow us to understand both disease pathophysiology and which patients are most likely to respond to the different therapies available. Although we are currently able to quantify the stored protein material and to assess changes in storage in response to treatment, in general, these variables have not proved to be reliable biomarkers for the LSDs. Other candidate biomarkers are already under investigation. Hopefully in near future, the availability of biomarkers other than the stored material, will allow the choice of the most appropriate therapy (ERT, chaperone therapy or SRT) when taking into account patient phenotype and expected disease severity and course of progression. Once a therapy is developed that can slow or prevent the progression of neurodegeneration, it will become extremely important to facilitate early detection of the genetic defect by newborn screening to enable early intervention.

6 The Need for Data Sharing

The development of new tools/technologies to diagnose patients and to predict disease severity and progression, as well as to assess new therapies, can be achieved only through collaboration between existing stakeholders. Beyond the existence of biobanks for sample collection that were started several years ago, in the last few years the generation of huge amounts of data obtained by the application of the different 'omics' technologies has increased the need for data biobanks. In fact, an increasing consciousness of the wide gap which is developing between obtained data and the appropriate analysis and interpretation needed to generate useful information from these data is coming to light. For rare diseases in general, and even more specifically for those involving pediatric populations, the organization of accessible biobanks and subsequent data sharing are of utmost importance. Only the sharing of data obtained from as many different sources as possible, through properly protected systems that respect patient privacy, will give us sufficient information on rare diseases to allow the development of more reliable diagnostic procedures and a greater understanding of therapeutic applications. It is anticipated that these advances will improve patient compliance. In addition, data sharing will afford the opportunity to consider the same condition from several different points of view, thus enriching our horizon on individual pathogenic conditions and on specific aspects expressed from different sources. All of this will be enormously important when tailoring a personalized medicine approach (Fig. 5).

The first steps towards establishing a community to focus efforts on developing therapies for rare neurological diseases of childhood were taken in December 2010 at a meeting held at the European Parliament in Brussels. This meeting was organized by the Brains for Brain Foundation (www.brains4brain.eu), the European Brain Council, the Lysosomal Storage Disease Patient Collaborative and the Veneto Region, and sponsored by Members of the European Parliament. It brought together stakeholders from 13 EU countries, as well as the US and Brazil, and included political representatives, members of scientific organizations, several European family organizations and biotechnology companies, and over 40 European Universities. This project will unite all interested parties, who are working to increase the visibility, recognition and awareness of rare neurological disorders in order to facilitate early diagnosis of these conditions; to promote and facilitate

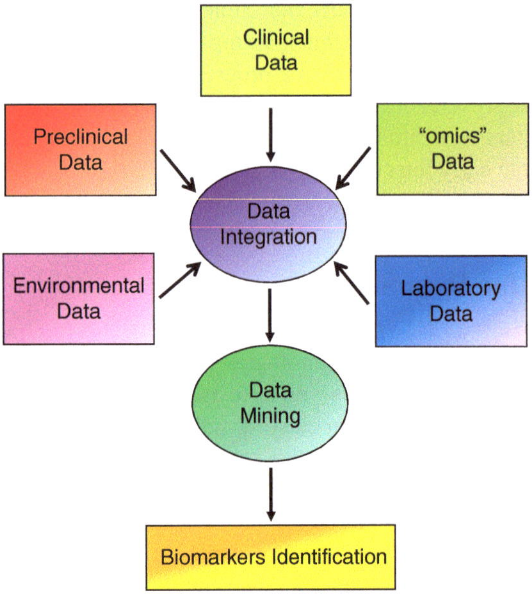

Fig. 5 Data sharing requirements within a personalized medicine approach

partnership and collaboration between physicians, researchers, patient advocates, carers, policy- and decision-makers and industry. This first step will hopefully be followed by several others with the aim to encourage and support research and to translate scientific breakthroughs into clinical practice; to contribute to the establishment of a standard of care for patients with rare neurological disorders which is agreed across Europe; and to ensure equity of access to diagnosis, treatment and care.

Acknowledgements The authors would like to thank all of the members of the Brains for Brain Foundation and in particular the patients and families affected by lysosomal storage disorders; Harriet Crofts (Oxford PharmaGenesis Ltd, Tubney, Oxfordshire, UK) for kindly revising the manuscript.

Medical writing support to the authors for the original manuscript on which this book chapter is based (Scarpa et al. [47]) was sponsored by Shire HGT. The sponsor had no influence on the contents or direction of that manuscript. The authors take full responsibility for the content of the paper.

Conflict of interest Maurizio Scarpa has received research grants, honoraria and travel support for speaking engagements from Actelion, Shire HGT, Genzyme Corporation and BioMarin. David Begley has received research grants, honoraria and travel support for speaking engagements from Actelion, Shire HGT, Genzyme Corporation, BioMarin and GlaxoSmithKline. Adriana Ceci, Rosella Tomanin, Cinzia Maria Bellettato, Francesca D'Avanzo and Alessandra Zanetti have no conflicts of interest.

References

1. Tanaka H (2010) Omics-based medicine and systems pathology. A new perspective for personalized and predictive medicine. Methods Inf Med 49(2):173–185
2. Lesko LJ (2007) Personalized medicine: elusive dream or imminent reality? Clin Pharmacol Ther 81:807–816
3. Biomarkers Definitions Working Group (2001) Biomarkers and surrogate endpoints: preferred definitions and conceptual framework. Clin Pharmacol Ther 69(3):89–95
4. Chakravarty A (2003) Surrogate markers: their role in regulatory decision process. Food and Drug Administration. Washington DC
5. Parliamentary Copyright (2009) The Parliamentary Office of Science and Technology. http://www.parliament.uk/post. Accessed 10 Dec 2012
6. Klein TE, Chang JT, Cho MK, Easton KL, Fergerson R, Hewett M, Lin Z, Liu Y, Liu S, Oliver DE, Rubin DL, Shafa F, Stuart JM, Altman RB (2001) Integrating genotype and phenotype information: an overview of the PharmGKB project. Pharmacogenetics Research Network and Knowledge Base. Pharmacogenomics J 1(3):167–170
7. Editorial (2012) What happened to personalized medicine? Nat Biotechnol 30(1):1
8. Hung SI, Chung WH, Jee SH, Chen WC, Chang YT, Lee WR, Hu SL, Wu MT, Chen GS, Wong TW, Hsiao PF, Chen WH, Shih HY, Fang WH, Wei CY, Lou YH, Huang YL, Lin JJ, Chen YT (2006) Genetic susceptibility to carbamazepine-induced cutaneous adverse drug reactions. Pharmacogenet Genomics 16:297–306
9. Hetherington S, Hughes AR, Mosteller M, Shortino D, Baker KL, Spreen W, Lai E, Davies K, Handley A, Dow DJ, Fling ME, Stocum M, Bowman C, Thurmond LM, Roses AD (2002) Genetic variations in HLA-B region and hypersensitivity reactions to abacavir. Lancet 359:1121–1122
10. Kauf TL, Farkouh RA, Earnshaw SR, Watson ME, Maroudas P, Chambers MG (2010) Economic efficiency of genetic screening to inform the use of abacavir sulfate in the treatment of HIV. Pharmacoeconomics 28:1025–1039
11. Chung WH, Hung SI, Chen YT (2010) Genetic predisposition of life-threatening antiepileptic-induced skin reactions. Expert Opin Drug Saf 9:15–21
12. Vogenberg FR, Isaacson Barash C, Pursel M (2010) Personalized medicine: part 1: evolution and development into theranostics. P T 35:560–576
13. Leeder JS, Lantos J, Spielberg SP (2010) Conference scene: pediatric pharmacogenomics and personalized medicine. Pharmacogenomics 11(12):1691–1702
14. ICH E 11 (2000) Clinical investigation of medicinal products in the paediatric population. Published in the UK. In: EU: Note for guidance on clinical investigation of medicinal products in the paediatric population (CPMP/ICH/2711/99). London; and in the US. In: US: International conference on harmonisation. Guidance on E 11 clinical investigation of medicinal products in pediatric population; Notice. Federal Register, 65:78493–78494
15. Boat TF (2007 Nov) The future of pediatric research. J Pediatr 151(5 Suppl):S21–S27
16. Loo TT, Ross CJ, Sistonen J, Visscher H, Madadi P, Koren G, Hayden MR, Carleton BC (2010) Pharmacogenomics and active surveillance for serious adverse drug reactions in children. Pharmacogenomics 11:1269–1285
17. Ceci A, Felisi M, Baiardi P, Bonifazi F, Catapano M, Giaquinto C, Nicolosi A, Sturkenboom M, Neubert A, Wong I (2006) Medicines for children licensed by the European Medicines Agency (EMEA): the balance after 10 years. Eur J Clin Pharmacol 62:947–952
18. Leeder JS, Kearns GL, Spielberg SP, van den Anker J (2010) Understanding the relative roles of pharmacogenetics and ontogeny in pediatric drug development and regulatory science. J Clin Pharmacol 50:1377–1387
19. Kearns GL, Abdel-Rahman SM, Alander SW, Blowey DL, Leeder JS, Kauffman RE (2003) Developmental pharmacology – drug disposition, action, and therapy in infants and children. N Engl J Med 349:1157–1167

20. Strassburg CP, Strassburg A, Kneip S, Barut A, Tukey RH, Rodeck B, Manns MP (2002) Developmental aspects of human hepatic drug glucuronidation in young children and adults. Gut 50:259–265
21. Yokoi T (2009) Essentials for starting a pediatric clinical study (1): pharmacokinetics in children. J Toxicol Sci 34(2):SP307–SP312
22. Odell GB (1967) "Physiologic" hyperbilirubinemia in the neonatal period. N Engl J Med 277:193–195
23. Kawade N, Onishi S (1981) The prenatal and postnatal development of UDP-glucuronyltransferase activity towards bilirubin and the effect of premature birth on this activity in the human liver. Biochem J 196:257–260
24. Miyagi SJ, Collier AC (2011 May) The development of UDP-glucuronosyltransferases 1A1 and 1A6 in the pediatric liver. Drug Metab Dispos 39(5):912–919
25. Miyagi SJ, Collier AC (2007) Pediatric development of glucuronidation: the ontogeny of hepatic UGT1A4. Drug Metab Dispos 35:1587–1592
26. Rodriguez W, Selen A, Avant D, Chaurasia C, Crescenzi T, Gieser G, Di Giacinto J, Huang SM, Lee P, Mathis L, Murphy D, Murphy S, Roberts R, Sachs HC, Suarez S, Tandon V, Uppoor RS (2008) Improving pediatric dosing through pediatric initiatives: what we have learned. Pediatrics 121:530–539
27. Paolucci P, Jones KP, del Carmen Cano Garcinuno M, Catapano M, Iolascon A, Ceci A (2008) Challenges in prescribing drugs for children with cancer. Lancet Oncol 9:176–183
28. Russo R, Capasso M, Paolucci P, Iolascon A (2011) Pediatric pharmacogenetic and pharmacogenomic studies: the current state and future perspectives. Eur J Clin Pharmacol 67(Suppl 1):17–27
29. Krekels E, Ceci A, Iolascon A, Girotto S, Della Pasqua O (2009) The role of paediatric pharmacogenetic studies in Europe. Pharm Policy Law 11:23–39
30. Holmes MV, Shah T, Vickery C, Smeeth L, Hingorani AD, Casas JP (2009) Fulfilling the promise of personalized medicine? Systematic review and field synopsis of pharmacogenetic studies. PLoS One 4:e7960
31. Brisson AR, Matsui D, Rieder MJ, Fraser DD (2012 Jan) Translational research in pediatrics: tissue sampling and biobanking. Pediatrics 129(1):153–162
32. Becker ML, Leeder JS (2010) Developmental pharmacogenetics in pediatric rheumatology: utilizing a new paradigm to effectively treat patients with juvenile idiopathic arthritis with methotrexate. Hum Genomic Proteomic 2010:257120
33. Arnold D, Jones BL (2009) Personalized medicine: a pediatric perspective. Curr Allergy Asthma Rep 9:426–432
34. Carroll WL, Bhojwani D, Min DJ, Raetz E, Relling M, Davies S, Downing JR, Willman CL, Reed JC (2003) Pediatric acute lymphoblastic leukemia. Hematol Am Soc Hematol Educ Progr 2003:102–131
35. Yates CR, Krynetski EY, Loennechen T, Fessing MY, Tai HL, Pui CH, Relling MV, Evans WE (1997) Molecular diagnosis of thiopurine S-methyltransferase deficiency: genetic basis for azathioprine and mercaptopurine intolerance. Ann Intern Med 126:608–614
36. Lennard L, Lilleyman JS, Van Loon J, Weinshilboum RM (1990) Genetic variation in response to 6-mercaptopurine for childhood acute lymphoblastic leukaemia. Lancet 336:225–229
37. Relling MV, Hancock ML, Rivera GK, Sandlund JT, Ribeiro RC, Krynetski EY, Pui CH, Evans WE (1999) Mercaptopurine therapy intolerance and heterozygosity at the thiopurine S-methyltransferase gene locus. J Natl Cancer Inst 91:2001–2008
38. Relling MV, Gardner EE, Sandborn WJ, Schmiegelow K, Pui CH, Yee SW, Stein CM, Carrillo M, Evans WE, Klein TE (2011) Clinical pharmacogenetics implementation consortium guidelines for thiopurine methyltransferase genotype and thiopurine dosing. Clin Pharmacol Ther 89:387–391
39. Carleton B, Poole R, Smith M, Leeder J, Ghannadan R, Ross C, Phillips M, Hayden M (2009) Adverse drug reaction active surveillance: developing a national network in Canada's children's hospitals. Pharmacoepidemiol Drug Saf 18:713–721

40. Madadi P, Ross CJ, Hayden MR, Carleton BC, Gaedigk A, Leeder JS, Koren G (2009) Pharmacogenetics of neonatal opioid toxicity following maternal use of codeine during breastfeeding: a case–control study. Clin Pharmacol Ther 85:31–35
41. Crowley JJ, Sullivan PF, McLeod HL (2009) Pharmacogenomic genome-wide association studies: lessons learned thus far. Pharmacogenomics 10:161–163
42. (2010) November/December Tufts CSDD impact report
43. (2009) Council recommendation on action in the field of rare diseases. http://www.consilium.europa.eu/uedocs/cms_data/docs/pressdata/en/lsa/108383.pdf. Accessed 10 Dec 2012
44. Poorthuis BJ, Wevers RA, Kleijer WJ, Groener JE, de Jong JG, van Weely S, Niezen-Koning KE, van Diggelen OP (1999) The frequency of lysosomal storage diseases in The Netherlands. Hum Genet 105:151–156
45. Meikle PJ, Hopwood JJ, Clague AE, Carey WF (1999) Prevalence of lysosomal storage disorders. JAMA 281:249–254
46. Fuller M, Meikle PJ, Hopwood JJ (2006) Epidemiology of lysosomal storage diseases: an overview. In: Mehta A, Beck M, Sunder-Plassmann G (eds) Fabry disease: perspectives from 5 years of FOS. Oxford PharmaGenesis Ltd, Oxford, pp 9–20
47. Scarpa M, Ceci A, Tomanin R, Mincarone P, Begley D (2011) Personalised medicine in paediatrics: individualising treatment in children with rare neurological diseases. EPMA J 2(2):231–239
48. Neufeld EF, Muenzer J (2001) The mucopolysaccharidoses. In: Scriver CR, Beaudet AL, Sly WS, Valle D (eds) The metabolic and molecular bases of inherited disease. McGraw-Hill, New York, pp 3421–3452
49. Gieselmann V, Krageloh-Mann I (2010) Metachromatic leukodystrophy – an update. Neuropediatrics 41:1–6
50. Parkinson-Lawrence EJ, Shandala T, Prodoehl M, Plew R, Borlace GN, Brooks DA (2010) Lysosomal storage disease: revealing lysosomal function and physiology. Physiology (Bethesda) 25:102–115
51. Bellettato CM, Scarpa M (2010) Pathophysiology of neuropathic lysosomal storage disorders. J Inherit Metab Dis 33:347–362
52. Vitner EB, Platt FM, Futerman AH (2010) Common and uncommon pathogenic cascades in lysosomal storage diseases. J Biol Chem 285:20423–20427
53. Duffner PK, Caggana M, Orsini JJ, Wenger DA, Patterson MC, Crosley CJ, Kurtzberg J, Arnold GL, Escolar ML, Adams DJ, Andriola MR, Aron AM, Ciafaloni E, Djukic A, Erbe RW, Galvin-Parton P, Helton LE, Kolodny EH, Kosofsky BE, Kronn DF, Kwon JM, Levy PA, Miller-Horn J, Naidich TP, Pellegrino JE, Provenzale JM, Rothman SJ, Wasserstein MP (2009) Newborn screening for Krabbe disease: the New York State model. Pediatr Neurol 40:245–252
54. Schroder W, Petruschka L, Wehnert M, Zschiesche M, Seidlitz G, Hopwood JJ, Herrmann FH (1993) Carrier detection of hunter syndrome (MPS II) by biochemical and DNA techniques in families at risk. J Med Genet 30:210–213
55. Meikle PJ, Grasby DJ, Dean CJ, Lang DL, Bockmann M, Whittle AM, Fietz MJ, Simonsen H, Fuller M, Brooks DA, Hopwood JJ (2006) Newborn screening for lysosomal storage disorders. Mol Genet Metab 88:307–314
56. Kaye EM (2001) Lysosomal storage diseases. Curr Treat Options Neurol 3:249–256
57. Anson DS, McIntyre C, Byers S (2011) Therapies for neurological disease in the mucopolysaccharidoses. Curr Gene Ther 11:132–143
58. Neufeld EF (2006) Enzyme replacement therapy – a brief history. In: Mehta A, Beck M, Sunder-Plassmann G (eds) Fabry disease: perspectives from 5 years of FOS. Oxford PharmaGenesis Ltd, Oxford, pp 93–99
59. Escolar ML, Poe MD, Martin HR, Kurtzberg J (2006) A staging system for infantile Krabbe disease to predict outcome after unrelated umbilical cord blood transplantation. Pediatrics 118:e879–e889
60. Begley DJ, Pontikis CC, Scarpa M (2008) Lysosomal storage diseases and the bloodbrain barrier. Curr Pharm Des 14:1566–1580

Author Index

S. Mandel (ed.), *Neurodegenerative Diseases: Integrative PPPM Approach as the Medicine of the Future*, Advances in Predictive, Preventive and Personalised Medicine 2, DOI 10.1007/978-94-007-5866-7,

D

E

H

K

Q

R

U

V

W

X

Y

Z

Subject Index

S. Mandel (ed.), *Neurodegenerative Diseases: Integrative PPPM Approach as the Medicine of the Future*, Advances in Predictive, Preventive and Personalised Medicine 2, DOI 10.1007/978-94-007-5866-7,

B

Q

R

MIX
Papier aus verantwortungsvollen Quellen
Paper from responsible sources
FSC® C105338

If you have any concerns about our products,
you can contact us on
ProductSafety@springernature.com

In case Publisher is established outside the EU,
the EU authorized representative is:
Springer Nature Customer Service Center GmbH
Europaplatz 3, 69115 Heidelberg, Germany

Printed by Libri Plureos GmbH
in Hamburg, Germany